Courtesy Edward E. Sjogren. Photograph by Luella A. Greenleaf.

Dendritic Growth in Magnesium.

ELEMENTARY

Metallurgy and Metallography

ARTHUR M. SHRAGER, A.B., B.S.

Member, American Society for Metals

Third Revised Edition

DOVER PUBLICATIONS, INC.

NEW YORK

To

Ida, Paulette, Laura

Published in Canada by General Publishing Company, Ltd., 30 Lesmill Road, Don Mills, Toronto, Ontario.

Published in the United Kingdom by Constable and Company, Ltd., 10 Orange Street, London WC 2.

This work, originally published in 1949 by The Macmillan Company, was revised and expanded by the author for the second edition (Dover, 1961). For this third edition, first published in 1969, the author has again revised the text, and has added a new preface, several new tables, and three new chapters (Appendices 1-3).

Standard Book Number: 486-60138-2
Library of Congress Catalog Card Number: 73-101599

Manufactured in the United States of America
Dover Publications, Inc.
180 Varick Street
New York, N.Y. 10014

Preface to the Third Edition

For this new edition I have updated the discussion of atomic theory, added material on the wave theory of electric conductivity and on vacancies and dislocations, and provided a table of "standard classification for copper and copper alloys." Three new appendices introduce the reader to maraging steels, plasma arcs, and lasers. I would like to express my appreciation to the International Nickel Co., Inc., the Union Carbide Corporation, the American Society for Metals, and the Copper & Brass Research Association for permission to reproduce material in this edition.

A.M.S.

Brooklyn, New York
January, 1969

Preface to the Second Edition

This Dover edition is a thoroughly revised version of the 1949 edition. Almost every page has been altered, and many topics, such as a discussion of the properties of metals, have been added. I have been particularly interested in discussing the important developments in the field since 1949 and feel that the book now offers a good coverage of the most modern methods of the day.

I have been fortunate in obtaining many excellent new illustrations for this edition and am grateful to the many individuals, technical journals and industrial firms for their splendid cooperation in making these pictures available to me. I am especially grateful to Mr. W. J. Lincoln who has drawn most of the highly instructive charts and diagrams in this book.

A.M.S.

Brooklyn, New York
May, 1961

Preface to the First Edition

This book is intended primarily to serve as an introduction to the metallurgy and metallography of the common metals and alloys. The information has been presented in such form that the student who has a background of one year of physics and one year of chemistry may learn the basic facts required for a general knowledge of metallurgy, and thus be prepared to pursue the subject further in more advanced texts.

It is hoped that the book may be of value also to men engaged in branches of metallurgical work, but who have had no technical training. Workers in heat-treating departments, for example, should find the discussion of basic principles of heat treatment of assistance in their daily work.

Finally, the book may be of interest to persons connected with the metal industries in nontechnical positions. These may include executives, purchasing agents, salesmen, and perhaps clerical workers, to whom a general knowledge of the technical phase of their business would be of assistance.

A.M.S.

Brooklyn, New York
1949

Table of Contents

1 / The Structure of Metals

Atomic Structure

Metals are widely used because, under most circumstances, a) they are rigid below, but quite plastic above, a certain stress; b) they become liquid above their melting point and can then be shaped by casting. In addition metals are superior to nonmetals in the ease with which they can be worked into a large variety of forms and shapes.

For an adequate understanding of the constitution of metals, and of their uses, fabrication, and treatment, it is necessary to know something of the atom, which constitutes the fundamental unit structure of the metal and which displays all its chemical characteristics. Even at the magnifications produced by the most powerful microscopes, great numbers of atoms are present in the field of view, and no single atom can be distinguished and studied separately.

The atom is not a solid body. It consists of a dense *nucleus* composed of positively charged *protons* and uncharged neutrons, surrounded by a cloud of negatively charged *electrons*. The nucleus is very small, about 10^{-13} inch in diameter; the electron cloud is 20,000 to 40,000 times as large, consisting mainly of empty space. It is fairly accurate to think of the atom as a miniature solar system, with the nucleus as the sun and the electrons as planets. The positive charge of the nucleus exactly balances the negative charge of the electron cloud, and the atom is held together by the attractive forces between these positive and negative charges. The electrons of an atom revolve around the nucleus in well defined circular orbits, each orbit associated with a particular energy-level. An electron in the innermost orbit has the smallest amount of energy. Energy must be supplied to this electron to raise it to a higher energy-level in an orbit further from the nucleus, and when an electron drops from an outer to an inner orbit it loses energy.

1

When atoms are grouped closely together, as in the crystalline state, electrons are attracted not only by the nucleus of the atoms to which they belong, but by the nuclei of adjacent atoms as well. This interaction is the basis of the strength and rigidity of solid materials. The loosely held outer electrons of metals easily free themselves from their atoms, and can pass readily from atom to atom; they are so completely shared that they are no longer associated with individual atoms. It is these shared electrons which give metals their properties, such as conductivity of heat and electricity, which distinguish them from nonmetals.

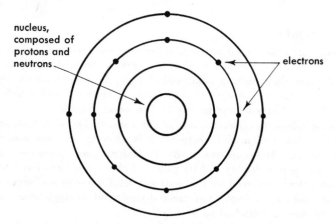

nucleus, composed of protons and neutrons

electrons

Figure 1. Theoretical concept of the atom.

Crystallization

Atoms of matter in the liquid state move freely, but as the temperature is reduced they lose energy and their motion becomes increasingly sluggish until the freezing point is reached. At that point the energy of motion begins to be exceeded by a force of attraction between atoms, which tends to hold them in a definite orderly crystalline pattern. As the liquid starts to solidify, groups of atoms form nuclei (minute crystals) simultaneously at many starting points in the liquid, and these crystal nuclei increase in size by progressive addition of atoms.

The Space Lattice

As solidification proceeds, the crystal nuclei build out in different

directions and the atoms tend to become disposed in a definite geometrical pattern which is known as the *space lattice*. The space lattice is
of such small dimensions that it cannot be resolved by ordinary methods
of microscopy. The principal cubic space lattices are the body-centered
cubic (B.C.C.), the face-centered cubic (F.C.C.), and the close-packed
hexagonal (C.P.H.) lattices. At room temperatures, the body-centered

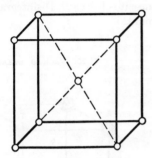 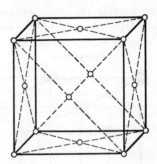

Figure 2. Body-centered cubic unit cell.   Figure 3. Face-centered cubic unit cell.

structure includes some of the stronger common metals: chromium, iron,
molybdenum, tungsten, vanadium, and zirconium. Such ductile metals
as aluminum, copper, gold, lead, nickel, and platinum crystallize in the
face-centered structure. Metals that crystallize in the close-packed hexagonal structure include antimony, beryllium, cadmium, cobalt, magnesium, titanium, and zinc. Metals in the latter group generally lack
plasticity or lose plasticity rapidly during cold forming. A single body-
centered cube may be considered representative of the smallest possible
unit cell, because it is the smallest grouping of atoms characteristic of
crystalline arrangement within the lattice. The formation of the space
lattice, and therefore of the crystal, results from multiplication of these
unit cells in all three directions, along their axes and at right angles to
each other. The growth of a space lattice may be compared to that of a
tree, with trunk, branches, and twigs.

The atoms that constitute a space lattice are separated by definite
intervals in a repeating pattern, and are held in position by mutual
attractive and repulsive forces acting between them. For example, each
atom in the interior of a body-centered cubic lattice is surrounded by
eight equidistant atoms, each separated from it by a distance of one-half

the diagonal of the cube, and is therefore kept in equilibrium by these eight equal forces. The repeating atomic bonds (represented by the connecting lines in Figures 2 and 3) may be considered analogous to springs capable of resisting both tensile and compressive stresses. When the stress is moderate the bonds stretch or contract, but if the stress is increased by pulling, hammering, or rolling, it is possible to rupture some of the bonds and thus produce permanent deformation of the metal. During heating, the lattice expands because it is loosened. Disintegration of the lattice takes place at the melting point.

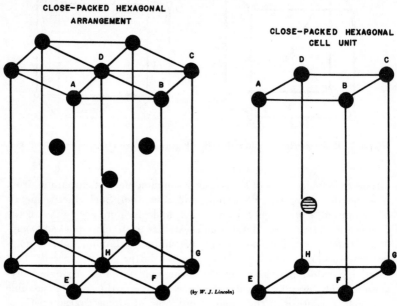

CLOSE-PACKED HEXAGONAL ARRANGEMENT

CLOSE-PACKED HEXAGONAL CELL UNIT

(by W. J. Lincoln)

Figure 4. Formation of a six-sided crystal common to zinc and magnesium.

Figure 5. Single cell unit of lattice. Shaded central atom not on a lattice point. Compare this with lettered parallelepiped in six-sided crystal.

Characteristic Properties of Metals

Metals do not act like individual molecules or ions. Their characteristic properties are determined by the *metallic bond*, a bond caused by the fact that all the atoms in a metal share all the outer—or valence—electrons. Thus the properties which are most useful in metals belong to the whole metal rather than to its individual atoms or simple molecules.

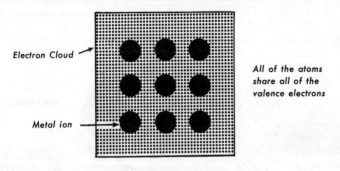

Figure 5a. Metallic bond.

Mechanical Properties

Mechanical or working properties are most important to engineers and metallurgists because they determine the ability of a metallic part to withstand breaking or tearing when a mechanical force is applied.

Compressive strength is a measure of the ability of a material to withstand squeezing or pressing (as in pillars and supports) without failure.

Tensile strength is a measure of the maximum normal load per unit area that a material can withstand without rupturing. Tensile strength is known also as *maximum strength* or *ultimate strength*. It is lowest in a metal in the annealed state.

Toughness is the ability to withstand load without breaking. It is the reverse of brittleness, and is commonly interpreted as ability to withstand sudden shock.

Brittleness is the tendency to fracture suddenly without appreciable deformation and under low stress.

Malleability is the property of being permanently deformable by rolling, forging, extrusion, etc., without rupture and without pronounced increase in resistance to deformation (as in the case of ductility). Malleability generally increases at elevated temperatures. Examples of malleable metals are aluminum, copper, gold, magnesium, and silver, all of which can be rolled into thin sheets or foil.

Ductility is the property that permits permanent deformation by stress in tension without rupture. Ductility determines the degree to which a metal can be cold worked without rupturing. Examples are found in wire drawing and the making of curved fenders of automobiles. Metals are ductile because of the free electrons between the close-packed rows of metal atoms; since the atoms are not bonded together directly, but merely held together by the free electrons, the rows of atoms can slide past each other very easily without separating.

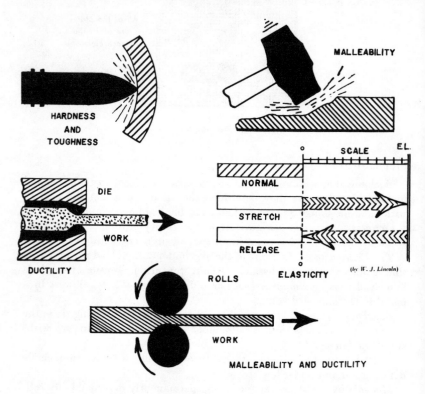

Figure 6. Mechanical properties of metals.

Impact strength measures the ability of a metal to withstand shock without breaking.

Fatigue is the phenomenon of progressive fracture of metal by a crack that spreads under repeated cycles of stress.

Elastic limit is the limit of an applied stress which, if exceeded, will cause permanent deformation. For commercial purposes, elastic limit is considered *yield strength*.

Creep is the slow permanent deformation in a metallic specimen produced by a relatively small steady force, below the elastic limit, acting for a long period of time.

Figure 7. Mechanical properties of metals in specific applications.

Physical Properties

Electric conductivity measures the ability of a metal to conduct electric current; it is the reciprocal of resistance. Pure metals have higher electric conductivities than do alloys. Metals conduct electricity because the outer electrons are free to move in an electric field. Electrical energy is conducted in the form of *waves*. Waves are able to pass through a regular array of obstacles, but not through an irregular array. For this reason electrical conductivity is much higher in pure, cold, and crystalline metals than in warm metals or alloys, in which atoms are no longer in regular arrays.

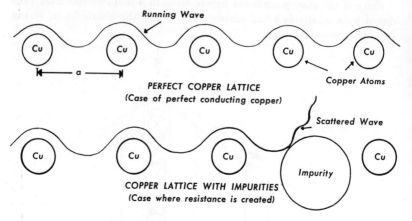

Figure 7a. Wave theory of conductivity.

Thermal conductivity is a measure of the ability of a metal to conduct heat. The thermal conductivities of metals are high because the free outer electrons can transfer thermal energy.

Melting point is the temperature at which the solid phase of a substance changes into the liquid phase. Pure metals, eutectics, and some intermediate constituents melt at constant temperatures. Most alloys melt over a range of temperatures.

Density is the weight per unit volume of a substance; it is generally expressed as pounds per cubic foot.

Color is the sensation produced upon the human eye by light of a specified wavelength. In metallurgical practice, color is used to describe the surface appearance of a metal, which is determined by the ability of the metal to absorb or reflect light of various wavelengths.

Magnetic properties are limited to the so-called ferromagnetic metals. These include only iron, cobalt, nickel, and certain alloys of these metals or of manganese and chromium.

Chemical Properties

Oxidation, in the restricted sense in which the term is used in a discussion of the smelting processes, is the combination of an element with oxygen to form an oxide, or the combination of an oxide with more oxygen to form a higher oxide.

Reduction, in a metallurgical discussion, is the partial or complete removal of oxygen from an oxide.

Corrosion is an undesired chemical attack on a metal, and a consequent wasting away of the metal. It may be atmospheric, submarine, subterranean, or electrolytic.

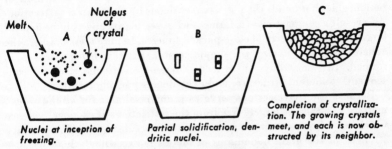

Figure 8. Stages of crystallization.

Reproduced by permission from G. E. Doan and E. M. Mahla, *The Principles of Physical Metallurgy*, copyright 1941 by the McGraw-Hill Book Co., Inc., New York.

The Process of Solidification of a Metal

The transition of a metal from the liquid to the solid state takes place in two stages. In the first stage a few atoms arrange themselves in the geometrical pattern characteristic of the metal, and as they solidify give up to their surroundings a part of the energy which made it possible for them to move freely through the liquid. These atoms therefore become relatively stable and serve as the crystal nuclei to which other atoms become attached in the same geometrical pattern to form crystals. During the second stage of crystal growth the crystals feed upon the molten metal by attracting other atoms to their space lattices, and throw out arms into the surrounding liquid. Other arms grow from these first arms and form branches, which in turn throw out spines of their own. This method of growth results in the formation of crystal skeletons which resemble pine trees, and which therefore are called *dendrites*. The dendritic system occurs in space within the liquid as trees grow in a forest, but during the formation of the crystal skeletons, trunks extend in all directions, not only upward as in the case of trees. The breadth of the normal crystalline grain in a metal is between 0.004 and 0.008 inch. This is equivalent to approximately 520,000 atoms in a row. Dendrites frequently can be seen with the naked eye on the surface of slowly cooled steel.

Eventually adjacent crystals meet and interfere with each other. As a result of this mutual interference on every side the external form of the crystals after completion of solidification is geometrically irregular.

Crystal Structure Formation

The negatively charged gas of mobile electrons serves as a kind of mobile glue that bonds the positive metal ions by electrostatic attraction. As a result a compact mass is formed, whose structure and volume depend largely on the geometry of close-packed spheres. In pure metals, where the spheres are equal in size, the simple crystal structure being discussed is formed. In some alloys there is a difference in atomic size and consequently different structures and even denser packing result.

Because the free electrons serve as a universal glue for all the atoms, metallic crystals are largely free from the restrictions of chemical valence which are so important in most nonmetallic substances. The cohesive strength of the grain boundaries in metals is exceedingly high because different metallic crystals can be bonded readily by their free electrons. Hence it is extremely difficult to break a cold metal along its grain boundaries unless the boundaries have been contaminated by impurities.

In a perfect crystal all atoms would be identical and in their correct positions. Actually, this condition is impossible, because

1. atoms vibrate at all temperatures; this prevents them from remaining stationary and in their correct positions in the lattice structure.

2. *Vacancies*, sites where atoms are missing, exist in crystal structures; the number of vacancies increases with an increase in temperature, reaching 1 in 400 and 1 in 1000 atoms at the melting point. Vacancies can be considered to circulate very slowly in a crystal; what really happens is that an adjacent atom moves over to fill the vacancy, leaving its own original site vacant. Thus if an atom moves toward the left into a vacancy, the vacancy seems to move toward the right. In this way the presence of vacancies permits the circulation of atoms in a crystal.

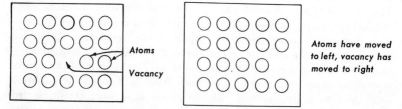

Figure 8a. Circulation of atoms in a crystal with vacancies.

Crystals and Grains

Although the practicing metallurgist has little occasion to distinguish between crystals and grains, there is technically a difference. A *crystal* is a homogeneous solid of regular geometrical structure peculiar to the element, compound, or isomorphous mixture of which it is composed; within each crystal the atoms are spaced in characteristic pattern. A *grain* is a larger structure built both by accretions from the molten metal and by the uniting of crystals. Although each of the particles which compose a grain has a definite geometrical form, the external shape of the grain is usually not symmetrical because irregular surfaces are formed by the mutual interference of adjacent crystals during crystallization.

This distinction is illustrated by the solidification of bismuth. If molten bismuth is cooled rapidly, each individual solid particle which appears has the regular cubic form which is characteristic of bismuth; each such particle is a crystal. As the crystals combine with or attach themselves to other crystals the larger structures known as grains are formed.

Steel and most other industrial metals do not occur as regular crystals, but as aggregations of irregularly shaped grains. In each grain there is a single and continuous orientation of the array of atoms. The orientation in adjacent grains obviously must be different, however, because otherwise the grains would merge. Steel differs from most other metals in that it can exist in any of several types of lattice structures, and there are wide variations in the size, shape, and constitution of the grains of different varieties of steel. The characteristic properties of steel and of other metals and alloys are determined by their grain structures.

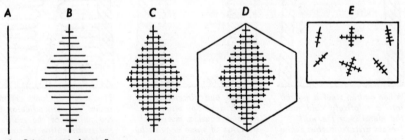

A   Primary axis forms first.
B   Primary and secondary axes follow quickly.
C   Primary, secondary, and ternary axes follow quickly.
D   Completed dendrite and boundary after interstices are filled up.
E   Dendritic segregation in steel.

Figure 9. Formation of a dendrite.

Amorphous substances are those that exhibit no crystalline structure because the atoms of which they are composed are not arranged in the geometrical structure of a space lattice. An amorphous substance has no definite melting point or range as has a metal or other crystalline substance, but softens gradually as the temperature is raised until its viscosity is lowered to a point at which it flows. Glass and pitch are typical examples of amorphous materials; they may be considered solid solutions that can be supercooled to any degree without the occurrence of crystallization.

Grain Boundaries

The *grain boundary* is a layer only a few atoms thick which constitutes the zone of contact between adjacent grains having differently oriented space lattices.

As a result of the interference between differently oriented grains during the process of solidification, the atoms of the grain boundary are not arranged in the geometrical pattern that is the space lattice characteristic of the metal. The metal in the grain boundary therefore has characteristics different from those of the metal in the grain.

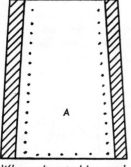 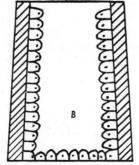 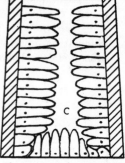

When molten steel is poured into a relatively cold mold the metal near the wall freezes first. Numerous nuclei form and begin to grow in all directions, at right angles to the faces of the mold chamber walls.

Vertical and circumferential growth develops about each nucleus, causing mutual obstruction of these crystals by their neighbors. However, they are at liberty to grow inward toward the center of the mold as the metal cools.

The result is long crystals, radial or perpendicular to the surface of the mold. In the solidification of castings of large cross section the central portion of the metal freezes independently of any chilling action, and equiaxed grains form.

Figure 10. Crystal formation in a chill mold.

Courtesy Gathmann Engineering Company.

At moderate temperatures the material of the grain boundary is harder than that of the grain. At high temperatures, however, the effect of this difference in constitution becomes important because the atomic bonds which hold the disarranged atoms are the first to loosen. The disarranged atoms acquire freedom of motion as the bonds loosen, and the metal tends to creep or flow when placed under stress. Under these conditions the grain boundary becomes the weakest part of the metal.

Effect of Cooling Rate on Grain Size

The rate of cooling is of great importance because it governs the size, shape, and arrangement of the grains of a metal casting. In sections that are cooled slowly through the solidification range, grains that start on one side increase in size until they meet other grains that have started from the other side, and the resulting structure is coarse grained. If cooling is rapid, as in the case of thin sections, or if the metal is poured into chill molds, solidification starts at a greater number of different points and a larger number of grains develops, producing a finer grain structure. If further control of grain size is required, a foreign material such as alumina may be added to the melt to provide additional nuclei for crystallization.

Availability of Metals

Of the 102 chemical elements that have been isolated, approximately 70 are metallic, and about 40 of the metals are used commercially. Only 8 elements are used in common engineering alloys: aluminum, copper, iron, lead, magnesium, nickel, tin, and zinc. The other 32 are not considered basic engineering metals, either because the supplies are limited or because extraction from their ores is expensive. These metals, however, are important in metallurgy because they are responsible for the important properties of many alloys. Among the most important are antimony, beryllium, calcium, cobalt, chromium, manganese, mercury, molybdenum, titanium, tungsten, vanadium, and zirconium. Some of these metals are not available in the United States, and must be imported.

Of the basic engineering metals only aluminum, iron, and magnesium are present in the earth's crust to the extent of or more than 2 per cent. They are found in physical or chemical combination with other elements, and must be extracted from their ores and purified for use. Many of our most useful metals, including copper, lead, and zinc, are present in large amounts, although in much less quantities than are aluminum, iron, and magnesium. Such metals as gold, platinum, and silver occur in only relatively very small amounts.

TABLE 1.   COMPARISON OF PLASTICS AND METALS IN ENGINEERING APPLICA-
TIONS

Advantages of Plastics	Advantages of Metals
Lower density	Higher mechanical strength
Higher dielectric strength	Higher modulus of elasticity
Lower thermal conductivity	Less thermal expansion
Higher damping properties	Lower cold flow
Greater ease of fabrication	Higher heat resistance
Unlimited color range*	Higher hardness
Transparency*	Lower water absorption
Higher resistance to chemical attack*	Ductility
	Treatable for localized changes in properties
	Suitable structural material because of strength, toughness, low cost

*Does not apply to all plastics.

Questions

1 Define the following terms:
   a. Electron   e. Space lattice
   b. Atom   f. Crystal
   c. Crystal nucleus   g. Grain
   d. Dendrite   h. Amorphous

2 Describe the structure of the atom.
3 What is the basis of the strength and rigidity of solid materials?
4 What is the chief characteristic that distinguishes a metal from a nonmetal?
5 Why are the mechanical or working properties of a metal so valuable to an engineer?
6 Distinguish between malleability and ductility.
7 Define corrosion, and explain how it affects metals.
8 Describe the transition of a metal from the liquid to the solid state.
9 Explain the difference between a body-centered cubic and a face-centered cubic space lattice, and illustrate with sketches.
10 Describe the process involved in the formation of a dendrite.
11 State the difference between a crystal and a grain, and describe an experiment designed to illustrate this difference.
12 How does the grain boundary form, and why is it important?
13 In what respect does an amorphous body differ from a crystalline substance?

14 Name the 8 elements used in common engineering alloys.
15 Compare the use of plastics and metals in engineering applications.
16 Why are metals widely used?
17 What is meant by the metallic bond?
18 Why is electrical conductivity highest in pure, cold, and crystalline metals?
19 Why are there no perfect crystals?
20 What are vacancies?

2 / Slip, Plastic Deformation, and Recrystallization

The strength of a metal is determined partly by the amount of deformation necessary to initiate *plastic flow*, i.e., to make the atoms slip. Below this necessary amount deformation is *elastic*: the metal returns to its original shape and condition when the external force is removed. Beyond this amount the force causes planes of the metallic crystal to shift with respect to one another. This shifting is known as *slip*.

Slip and Plastic Deformation

Slip depends upon the perfectly repetitive structure of the metallic crystal, which permits the atoms in one face of a slip plane to shear away from their original neighbors in an adjacent face; the atoms slide along this face and join a new set of neighbor atoms. Since the crystal is a repetitive structure the new bond is identical to the old, and the original properties and internal structure of the crystal are retained.

The arrangement of atoms in the space lattice of a metal is a determining factor in slip. Various groups of parallel planes may be considered as passing through the grain, and every atom of the grain must necessarily lie in one plane of each group. Depending upon the type of lattice, certain groups of parallel planes are most favorable to the occurrence of slip; they are usually those groups in which the parallel planes are spaced the greatest distance apart, and in which the planes contain the greatest number of atoms per unit of surface. This may be explained by the fact that the interplanar atomic bonds are longest and the cohesive force therefore weakest between planes separated by the greatest distances, whereas the planes of greatest atomic density are the strongest because the bonds between the atoms included in them are short and therefore strong. The closely packed atoms in the strong planes tend to resist

16

displacement, whereas the relatively weak bonds between two parallel planes give way under the stress, and slip results. Metals which crystallize in the face-centered cubic form, for example, have twelve planes of easy slip.

Slip starts in that grain or in those grains in which a group of planes of easy slip is most favorably oriented with respect to the applied force (about 45 degrees), and the entire crystal lattice turns in such a direction that the planes tend to become parallel to the direction of flow. As slip progresses in the plane in which it started, resistance to further slip develops in that plane. When resistance has increased sufficiently in this first plane to stop slip in it, the stress is increased in adjacent parallel planes, and they progressively start to slip. This progressive slipping of

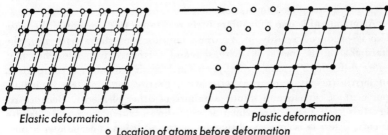

Elastic deformation   Plastic deformation

o Location of atoms before deformation
• Location of atoms after deformation

Figure 11. Deformation of a simple space lattice.

Reproduced by permission from R. S. Williams and V. O. Homerberg, *Principles of Metallography*, 4th ed. copyright 1939 by the McGraw-Hill Book Co., Inc., New York.

parallel planes is the phenomenon which determines the ductility of a metal; ductility is greatest in metals which have the greatest number of planes of easy slip. When one grain is deformed in this manner, it necessarily exerts pressure upon adjacent grains and thus initiates slip in them. Eventually slip is produced in many grains, and blocks of unit cells slide past each other. The effect is similar to the displacement produced in a vertical pile of cards by pushing horizontally in such a way that the cards slide over each other.

The resistance initially offered to slip, and consequently to deformation, is the result of changes produced in the relative positions of the atoms and the consequent changes in the interatomic forces. After slip has started in the most favorably oriented grains, other grains, the slip planes of which are differently oriented, exert a retarding effect on the initial slip, because greater force is required to initiate slip; thus the resistance to slip is augmented. Further resistance is offered by the grain boundaries, in which the atoms are not in the regular array of the space lattice, and which are stronger than the grain. Eventually slipping is not restricted to the planes of easy slip of the variously oriented grains, but extends to other planes. As slipping progresses, the shear strength on the slip planes gradually increases until slip ceases.

This oversimplified explanation of plastic deformation and slip is designed to give a general idea of the phenomenon. Actually the subject is far more complicated, and none of the hypotheses which have been offered so far have been accepted generally.

The self-stopping power of slip and the tendency of adjacent fragments to cling together instead of separating as a fracture are the chief characteristics of metallic deformation.

A soft metal is one that offers little resistance to deformation because slip is produced readily in it; when anything interferes with slip, the hardness of the metal is increased and further deformation becomes more difficult. The effect of slip on hardness is shown in a comparison of ferrite (nearly pure iron) with steel. Ferrite contains less than 0.05 per cent of carbon and minor amounts of other elements. Because its crystals have a structure which permits slip to take place readily, ferrite is soft. If ferrite is converted to steel by the addition of carbon with suitable heating, and if the resultant steel is cooled slowly, a harder product results. The addition of carbon results in the formation of particles of pearlite, an intimate mixture of ferrite and iron carbide; these particles, which are harder than ferrite, interfere with slip and make it self-stopping. It is because of this self-stopping property that steel is harder than ferrite. The extent to which hardness is increased depends upon the fineness of the internal structure of the pearlite grains. The finer the grain structure, the greater is the number of grain boundaries, and therefore the greater the resistance to slip. Any strain or distortion within a lattice increases the resistance of the lattice to plastic deformation, and thereby increases its strength and hardness.

Dislocations

Dislocations are irregularities of crystal structure which permit slip to take place more easily. Metals with wide dislocations have easy slip, and are therefore soft. Copper, aluminum, and gold are extremely soft because of wide dislocations; their close-packed structure and free electron bonding permit atoms to slide over one another with relative ease. Hardness can be increased by the blocking of dislocations. Alloying, for example, introduces foreign atoms, which form clumps that offer resistance to dislocations. Duralumin is made hard by such alloying. Since dislocations accumulate at grain boundaries, a reduction of grain size of dislocations will also increase the hardness of a metal. Work hardening (see below), by which a metal acquires resistance to further slip, is caused by the creation of new dislocations. When there is a sufficient number of these they move along intersecting slip planes, thus obstructing one another's movements and preventing further slip.

Cold Working

Hardening takes place in a metal when work of any sort, such as bending, rolling, hammering, drawing, punching, and the like, is done at a temperature below that at which recrystallization takes place. Any work which is done at such a temperature is called *cold working*, and the result is known as *work hardening*. It is well illustrated by the example of a piece of copper wire; when the wire is bent for the first time it is still soft and bends easily, but if bent repeatedly at the same spot, bending becomes more difficult, and the copper becomes harder and more brittle until it finally breaks. The amount of cold work that can be done on a metal is a measure of the metal's ductility. High ductility means that the metal will work harden locally rather than break when it is subjected to unusually high local loading. Some materials possess no ductility; for example, if one attempts to bend a cold glass rod it will break. Such materials are said to be brittle.

Cold working sometimes is carried out deliberately to harden or strengthen a piece of metal; common examples are the drawing of copper wire and the cold rolling of steel. In other operations the hardening produced by necessary shaping operations is detrimental, or impedes further work, so that the metal must be softened repeatedly during the shaping process.

Recrystallization and Grain Growth

When a metal has been subjected to a great degree of work hardening its grains are deformed and broken up, and the metal as a result is in a condition of internal strain. This strain can be relieved by *annealing* (heating) because the added energy imparted to the atoms by the heat gives them some power of motion, and they tend to return to their normal stable positions. As the temperature is raised, strain is partially removed

DIAGRAM A

Ferrite grains are interspersed by a few pearlite grains (darkened areas).

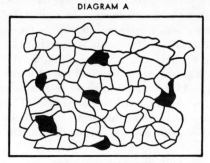

ORIGINAL STRUCTURE

DIAGRAM B

Grain is elongated after drawing.

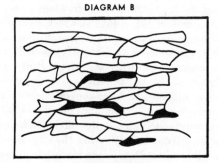

AFTER DRAWING

DIAGRAM C

Reheating gives the ferrite grains an opportunity to resume some of their original shape. The pearlite grains and impurities were not affected, as shown by dark areas.

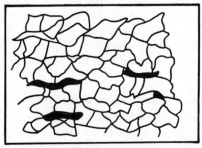

AFTER REHEATING

Figure 12. Result of work hardening followed by annealing.

Reproduced by permission from W. T. Frier, *Elementary Metallurgy*, 1st ed., copyright 1942 by the McGraw-Hill Book Co., Inc., New York.

with little or no change at first in the strength or hardness of the metal. When the temperature is raised still further, a point is reached at which new nuclei are formed among the deformed grains, and if the metal is kept for a sufficient time at just above that temperature, *recrystallization* takes place, and new grains crystallize from the old deformed grains in solid metal in substantially the same way in which crystallization takes place from molten metal. The new grains begin to form in regions where plastic deformation has introduced the greatest strains in the metal, generally at the grain boundary.

When all the old grains have been replaced completely by new ones, recrystallization is considered complete. The temperature at which recrystallization occurs varies with the amount of work hardening which took place and with the size of the original grain before deformation; the greater the amount of deformation and the smaller the grain size before deformation, the lower is the temperature required for recrystallization. The new grain structure is largely dependent upon the amount of deformation which took place, the fineness of grain increasing with the amount of deformation that existed before recrystallization.

If metal is raised to a temperature above that necessary for recrystallization and is kept at that temperature for a sufficient length of time, the grains formed by recrystallization begin to grow rapidly; this is called *grain growth*, and is actually a continuation of the process of recrystallization. The size that the crystals eventually attain depends upon the temperature to which the metal is heated; if the metal is kept at this temperature for a longer time than is required for this maximum growth, no further growth takes place. If the temperature is allowed to rise too high, grain growth may proceed too far so that the crystals grow to an excessive size, and the metal consequently may be weakened. An interesting example is the tungsten filament of an incandescent lamp bulb. Because the filament operates at a temperature considerably higher than that required for recrystallization of tungsten, grain growth in a filament of pure tungsten is excessive, and the filament is so weak that it sags under the pull of gravity, and breaks easily. To obviate this difficulty, thoria is added to the tungsten to inhibit grain growth; the resultant smaller grain structure produces a stronger metal.

Summary of Recrystallization and Grain Growth

1. Recrystallization starts at a lower temperature than does crystallization, and is completed within a narrower temperature range.

2. The final grain size after recrystallization will be smaller:

    a. the greater the degree of prior strain hardening

    b. the lower the temperature above that required for complete recrystallization

    c. the shorter the time at temperature

    d. the faster the rate of heating to temperature

    e. the more insoluble impurities present (within limits)

    f. the more finely divided the impurities

Note: Impurities in solid solution have a negligible effect.

3. Grain growth in the solid state may occur in:

    a. plastically deformed metals

    b. compressed powders

    c. electrolytically deposited metals

    d. alloys in which a new phase is formed

4. During the process of grain growth the larger grains grow at the expense of the smaller grains, which eventually disappear. Grain growth is the result of migration at the grain boundary.

5. Grain growth occurs more rapidly at high temperatures. At a temperature near the melting point, a few seconds often suffice to pro-

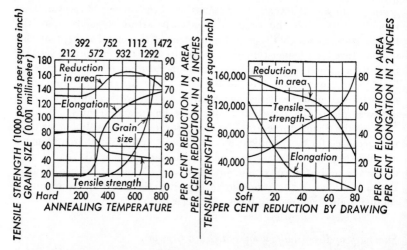

Figure 13. Effects of cold drawing and of annealing on high (66-34) brass.
From *Metals Handbook*, 1939 ed., courtesy American Society for Metals.

duce grains, which may not change further on extended exposure.

6. Germination causes the formation of abnormally large grains because of certain conditions of nonuniformity. During germination, preferential growth of some grains is favored by their large size.

7. The grain size in cast metals that undergo no change in the solid state, or that have not been worked mechanically, cannot be changed appreciably by heating, even at temperatures approaching the melting point.

Hot Working

Hot working, such as hot rolling or hot forging, is work carried out at a temperature at or above the recrystallization temperature of the metal in question. Although the grains are being deformed constantly and being broken up constantly, new ones are forming constantly to take their places. The result of hot working is therefore the same as that of cold working followed by annealing. Some metals, such as cadmium, lead, tin, and zinc, are not hardened by working at room temperature because their recrystallization temperatures are at or below room temperature, and the work done on them at room temperature is therefore actually hot working.

Questions

1. Define the following terms:

   a. Plastic deformation   e. Hot working
   b. Slip   f. Work hardening
   c. Slip interference   g. Recrystallization
   d. Cold working

2. Describe the difference between elastic and plastic deformation.
3. Describe the progress of slip.
4. What are the chief characteristics of metallic deformation?
5. What is the basis of hardness in metals?
6. Give four practical applications of the use of work hardening.
7. How may metal be softened after work hardening?
8. Describe the process of recrystallization and grain growth.
9. Name several factors which affect grain growth.
10. Explain the difference between hot working and cold working, and their effects upon metal.
11. How do dislocations contribute to the softness of pure metals?
12. How can dislocations be removed?
13. How is the strength of a metal affected by dislocations?

3 / Alloys and Constitutional Diagrams

Because most pure metals are too soft for the greater number of industrial purposes, it is frequently necessary to combine two or more metals to form an *alloy*. The internal structure that results when such a mixture solidifies from the molten state is of great practical importance with respect to the uses to which the alloy is to be put, and the study of this structure is one of the most important branches of metallurgy. A *constitutional diagram*, also called an *equilibrium diagram*, is a graph which shows the relationships between the relative amounts of solid and liquid portions of all constituents over a range of temperatures, under conditions of equilibrium. Constitutional diagrams are plotted from data obtained from a series of *cooling curves* of alloys that contain different proportions of the constituent metals.

A study of its constitutional diagram provides information with respect to the normal structure and many of the physical and chemical properties of an alloy. Because a constitutional diagram represents a theoretical state of equilibrium which exists only under ideal conditions, the information obtained from it is to be considered as only approximate.

Cooling Curves

The thermal changes that occur as a pure metal or alloy is cooled from a temperature at which it is molten to a temperature below that at which it has become solid are plotted as a graph, with time as abscissas and temperatures as ordinates. An investigation carried out to obtain information of this type is known as *thermal analysis*. To obtain data for a cooling curve, a pyrometer is inserted in the molten mass, and is read at intervals as the mass is cooled slowly to a temperature below that at which it is completely solidified.

The cooling curve of a pure metal falls smoothly until the solidification temperature is reached, when it turns and remains horizontal for a period of time, and finally resumes its smooth descent as the solid metal cools. The horizontal section represents the period during which the metal is changing from the liquid state to the solid state, and the latent heat of solidification is being dissipated.

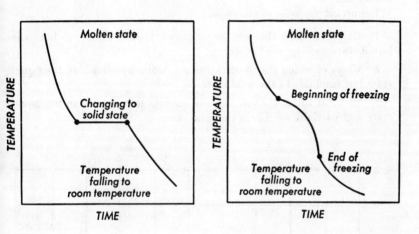

Figure 14. Cooling curve of a pure metal.

Figure 15. Typical cooling curve of a binary alloy, the component metals of which form a solid solution.

Some special alloys produce cooling curves of the same type, but solidification of most alloys does not occur at a single temperature, but extends over a range of temperatures. A second change of direction of the curve takes place at the temperature at which solidification is complete.

Constitutional Diagrams

From each of the cooling curves determined for a series of alloys containing different proportions of two metals, are read off the temperatures at which the initial evolutions of heat occurred. These temperatures are plotted as ordinates, with the corresponding percentage compositions as abscissas. The resulting curve is the locus of temperatures at which each alloy starts to solidify, and is called the *liquidus*. Similarly, the temperatures at which the second pause in the steady drop of temperature occurs on the cooling curve of each alloy become ordinates for another line on the constitutional diagram. This line is the locus of temperatures at which each alloy has completed solidification, and is called the *solidus*. Constitutional diagrams can be computed for alloys that consist of more than two pure metals, but they are much more complicated than are those for binary alloys.

Important varieties of alloys are:

1. Alloys of which the constituents are mutually soluble in both the liquid state and the solid state.

2. Alloys of which the constituents are mutually soluble in the liquid state but insoluble in the solid state.

3. Alloys of which the constituents are mutually soluble in the liquid state and partially soluble in the solid state.

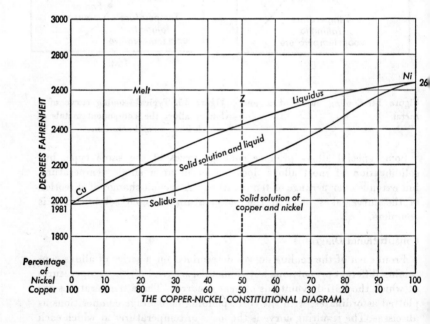

Figure 16. Constitutional diagram of copper-nickel alloys.

From *Metals Handbook*, 1948, courtesy American Society for Metals.

Alloys of Metals Mutually Soluble in Both the Liquid State and the Solid State

Since the different substances present in an alloy can dissolve in one another and form a homogeneous mixture when the alloy is melted, such melted alloys are classified under the general heading of solutions. For example, if the two metals that compose an alloy are soluble in each other in both the liquid and the solid state, this means that the metals remain as completely dissolved in each other *after* solidification as before. The solution formed after the alloy solidifies is known as a *solid solution*. A solid solution can be formed in one of two ways:

1. If the atoms of the solute and solvent are of the same size, the solute atoms can replace some of the solvent atoms. The result is a *substitutional* solid solution. Monel metal, a copper-nickel alloy, is an example of a substitutional solid solution.

2. If the solute and solvent atoms are not of the same size, the solute atoms may occupy the spaces between the solvent atoms. This will produce an *interstitial* solid solution. Iron and carbon are two elements which will combine in an interstitial solution.

Figure 16 is the constitutional diagram for the alloys of copper and nickel; the melting point of pure copper is 1981°F, and that of pure nickel is 2646°F. At all temperatures above the liquidus, alloys of every composition are liquid; at all temperatures below the solidus, alloys of every composition are solid. The area between the liquidus and the solidus represents a range in which the constitution is a mushy mass of both liquid and solid metal.

Alloys of Metals Mutually Soluble in the Liquid State but Insoluble in the Solid State

Figure 17, the constitutional diagram of cadmium-bismuth alloys is illustrative of alloys of this type. Pure cadmium melts at 610°F, and pure bismuth at 520°F. When bismuth is added to a mass of cadmium, the melting point becomes progressively lower than that of pure cadmium; conversely, when cadmium is added to a mass of bismuth, the melting point becomes progressively lower than that of pure bismuth. The melting point, approached from either direction, reaches a common minimum value of 291°F for an alloy consisting of 60 per cent bismuth and 40 per cent cadmium. The alloy which has this composition is known as the *eutectic* alloy, and 291°F is the eutectic temperature. In solid form, the eutectic alloy consists of an intimate mixture of fine crystals of nearly pure cadmium and nearly pure bismuth. The temperature remains constant while solidification of this or any other eutectic alloy is taking place; in this respect a eutectic alloy behaves like a pure metal.

The lines CE and EB form the liquidus, and the horizontal line AED is the solidus. In the field above the liquidus, only the liquid phase consisting of the two mutually soluble molten metals is present. In the field contained within the triangle CEA, the mixed liquid and solid cadmium are present; in the field contained within the triangle BED, the mixed liquid and solid bismuth are present. In the field below the solidus and

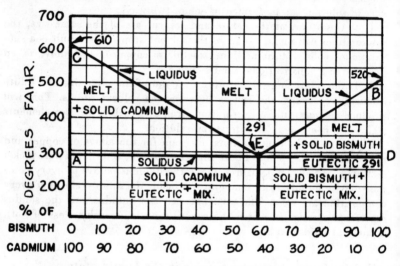

Figure 17. Constitutional diagram of cadmium-bismuth alloys.

at the left of the perpendicular dropped from E, solid cadmium and the eutectic mixture of cadmium and bismuth are present; in the field at the right of the perpendicular, solid bismuth and the eutectic mixture are present.

The information shown in the diagram makes it possible to compute the approximate relative amounts of any constituent for any selected temperature by the *law of horizontals*. The composition at any stated temperature is that indicated at the point of intersection of the horizontal line with the liquidus. Two examples of possible computations are shown, in each case for a molten alloy which consists of 80 pounds of cadmium and 20 pounds of bismuth:

1. The melt is cooled from any point above the liquidus to the eutectic temperature, 291°F. As the alloy cools, cadmium crystallizes out, and the remaining liquid accordingly becomes progressively richer in bismuth until, when the eutectic temperature is reached, the remaining liquid contains the entire original 20 pounds of bismuth plus enough cadmium to satisfy the eutectic proportion of 40 per cent cadmium and 60 per cent bismuth. The amount of cadmium x that remains to solidify with the bismuth at the eutectic point is therefore calculated by the proportion:

$$x/20 = 40/60$$
and $$x = 13.3 \text{ pounds}$$

The total amount of eutectic is therefore the sum of the 20 pounds of bismuth and the 13.3 pounds of cadmium, or 33.3 pounds of metal. The amount of cadmium that crystallizes from the melt before the temperature drops to 291°F is the original 80 pounds less the 13.3 pounds which remained to solidify in the eutectic mixture, or 66.7 pounds.

2. The melt is cooled from any point above the liquidus to 400°F. The horizontal line for 400°F intersects the liquidus at the point that corresponds to a composition of 61 per cent cadmium and 39 per cent bismuth. The amount of cadmium that still remains molten at this temperature is therefore found by the proportion:

$$x/20 = 61/39$$
and $$x = 31.3 \text{ pounds}$$

The amount of cadmium that has solidified is therefore 80 pounds less 31.3 pounds, or 48.7 pounds.

Alloys of Metals Mutually Soluble in the Liquid State and Partially Soluble in the Solid State

A typical constitutional diagram of alloys of this type is the lead-tin diagram, shown in Figure 18. In this diagram, *AEC* is the liquidus, and *ABEDC* is the solidus.

These metals form two types of solid solutions:

1. Solutions of tin in lead, which are designated *alpha* solutions, and which lie in the field at the left of *B*. The point *B* corresponds to a composition of the alpha solid solution of about 19.5 per cent tin and 80.5 per cent lead, which represents the maximum solubility of tin in lead in the solid state.

2. Solutions of lead in tin, which are designated *beta* solutions, and which lie in the field at the right of *D*. The point *D* corresponds to a

composition of the solid beta solution of about 2.5 per cent lead and
97.5 per cent tin, which represents the maximum solubility of lead in tin
in the solid state.

The eutectic formed at point *E* consists of a mixture of alpha and
beta solid solutions. The field *ABE* contains both liquid and solid, but
in this case the solid is not a pure metal, but the alpha solid solution;
similarly, the field *CDE* contains liquid and solid beta solution. The field
above the liquidus contains only liquid; the field below the solidus contains only alpha and beta solid solutions.

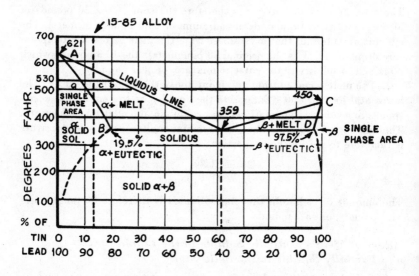

Figure 18. Constitutional diagram of lead-tin alloys.

The amount of phase or constituent present at any stated temperature
can be determined by the so-called *lever-arm principle*; this is illustrated
by the example of a mixture which contains 15 per cent tin and 85 per
cent lead. As a molten mixture of this composition is cooled from a temperature above the liquidus, alpha solid solution starts to crystallize out
at a temperature of about 555°F and continues to crystallize as the temperature is lowered further. At about 530°F, the composition of the alpha
solid solution is indicated by the intersection *a* of the 530°F horizontal

line with the line AB ; it is about 9 per cent tin and 91 per cent lead. The composition of the liquid phase is shown by the intersection b of the same temperature line with the line AE; it is about 22 per cent tin and 78 per cent lead. The perpendicular which represents the original melt of 15 per cent tin and 85 per cent lead intersects the 530°F line at the point c; at this temperature the percentage of alpha solid solution to total melt is bc/ab, and the percentage of liquid to total metal is ac/ab. The ratio bc/ab is 0.54 and the ratio ac/ab is 0.46. If the original melt contained 100 pounds of metal, therefore, 54 pounds are contained in the alpha solid solution and 46 pounds are contained in the liquid. The 54 pounds of alpha solid solution are composed of:

$$0.09 \times 54 = \quad 4.9 \text{ pounds of tin}$$
$$0.91 \times 54 = 49.1 \text{ pounds of lead}$$

and the 46 pounds of liquid are composed of:

$$0.22 \times 46 = 10.1 \text{ pounds of tin}$$
$$0.78 \times 46 = 35.9 \text{ pounds of lead}$$

Cored Crystals

When an alloy in the liquid state is cooled rapidly, the metal that is present in greatest concentration begins to crystallize out first, and diffusion throughout the mass is not sufficiently rapid to establish equilibrium in fact. As a result, the composition differs materially from that which the equilibrium diagram indicates. Because those portions which solidify first differ in composition from the portions which solidify later, the resulting crystals and dendrites are nonuniform. For example, dendrites of a nickel-copper alloy that is cooled too rapidly to permit complete diffusion have a heterogeneous structure that is richest in nickel at the center. A crystal which has such a nonuniform structure is called a *cored crystal*.

Questions

1 Define the following terms:

   a. Constitutional diagram
   b. Equilibrium
   c. Binary alloy
   d. Eutectic alloy
   e. Solidus
   f. Liquidus
   g. Solid solution
   h. Single-phase field

2 Draw typical cooling curves of a pure metal and of an alloy, and discuss the difference between them.

3 Draw and label the cooling curve of an alloy of two metals that are mutually soluble in both the liquid state and the solid state. What conclusions can be drawn from a study of the curve?

4 Draw and label the constitutional diagram typical of alloys of two metals that are mutually soluble in the liquid state but insoluble in the solid state. Explain how the diagram was derived.

5 Draw and label the constitutional diagram typical of alloys of two metals that are mutually soluble in the liquid state and partially soluble in the solid state.

6 Referring to one of the curves of the three preceding questions, describe and explain what occurs as an alloy of any chosen composition cools and solidifies.

7 Under what conditions does an alloy of eutectic composition behave like a pure metal? Explain.

8 State the law of horizontals and describe its application.

9 Determine by the law of horizontals the composition of a lead-tin solder that melts at 210°F.

10 What is a cored crystal? How does it form?

4 / Structures of Iron and Steel in the Solid State

Allotropy and Transformation Points

The existence of an element in more than one form of space lattice structure is called *allotropy*. The term is sometimes extended to apply to the existence of a compound in more than one form, but the term *polymorphism* is preferred in that case. Allotropy is in many respects a valuable property in an industrial metal, but relatively few metals are allotropic. Iron may exist in any one of three allotropic forms, depending upon its temperature.

The temperatures at which changes of structure take place in a metal as it is heated from room temperature to its melting point are called *transformation points* or *critical points*. As a metal is cooled from the molten state to room temperature, the changes occur in reverse order, usually at temperatures a few degrees lower than the respective points observed in heating.

Critical Points and Structures of Pure Iron

The form of iron that exists at room temperature is composed of grains that have body-centered cubic space lattice structures, and is known as *alpha iron*. Alpha iron is soft, ductile, and magnetic. When heated above 1420°F, alpha iron loses its magnetic properties, but retains its body-centered cubic structure. This form of iron has sometimes been considered a different allotropic form because it differs from the alpha iron existing at room temperature in being nonmagnetic, and has been called *beta iron*. Because no change of lattice structure is involved, however, it is customary to consider it nonmagnetic alpha iron.

At about 1670°F the crystalline structure changes to the face-centered

33

cubic form, and the iron assumes the allotropic form known as *gamma iron*. At this point a considerable absorption of heat occurs, and the volume of the iron contracts. Gamma iron is nonmagnetic and is of slightly greater density than the forms that have body-centered cubic structures.

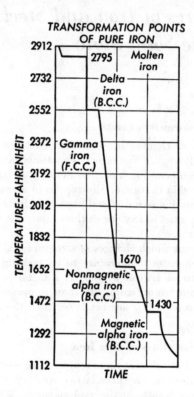

Figure 19. Cooling curve of pure iron.

At the third critical point, 2552°F, the face-centered cubic lattice reverts to the body-centered form, and the iron again becomes magnetic; this form is known as *delta iron*.

At 2795°F a final pause takes place in the rise of temperature as the iron absorbs the heat required for fusion and enters the liquid state.

Critical Points and Structures of Steel

Plain carbon steel is an alloy of iron and carbon. Other alloying elements may be added to produce steels for special purposes. The structure of

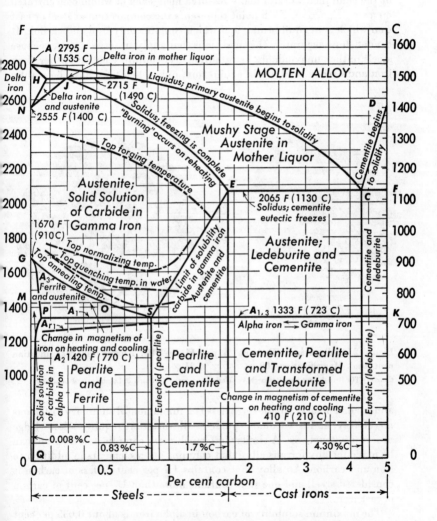

Figure 20. Iron-carbon constitutional diagram.

steel is determined by the amounts and nature of alloying materials, and by the rate of cooling from the molten state.

The *iron-carbon constitutional diagram* shown in Figure 20 is the result of the combined work of many research men, each of whom concentrated on a single portion. Each point represents the composition of steel or cast iron which is in equilibrium, and which contains only iron and carbon. It cannot show accurately the conditions that apply to actual steels, because steels, either unavoidably or intentionally, contain elements other than iron and carbon, and are in equilibrium only when they have been fully

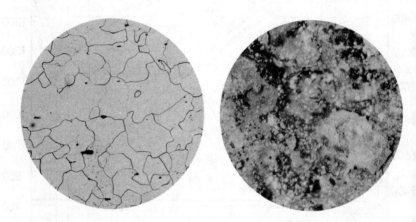

Figure 21A. Pure iron. Etchant 5 per cent nital. Magnification 100 ×.
Courtesy The British Cast Iron Research Association.

Figure 21B. Rusted iron. Magnification 21 ×.
Courtesy Bausch & Lomb Optical Company.

annealed. The diagram nevertheless is a fundamental summary of ferrous metallurgy, and is a valuable guide in all work that requires a knowledge of ferrous structures. A similar but much more complicated diagram can be made for any ferrous alloy that contains other elements in addition to iron and carbon. An alloy that contains 1.7 per cent or less of carbon is considered steel, and one that contains more than 1.7 per cent of carbon is cast iron.

The maximum solubility of carbon in alpha iron is about 0.035 per cent at 1333°F, and it drops to less than 0.01 per cent at room temperature.

At room temperature under conditions of equilibrium, therefore, any carbon present in excess of that small amount must exist in a form other than that of a solute in a solid solution. This additional carbon is present in chemical combination as iron carbide (Fe_3C), which is called *cementite*. Cementite contains 6.68 per cent of carbon by weight, is hard and brittle, and is magnetic at room temperature.

A solid solution of which alpha iron is the solvent is called *ferrite*. Because of the extremely small amount of carbon which it can contain in solid solution, ferrite in a steel that contains only iron and carbon

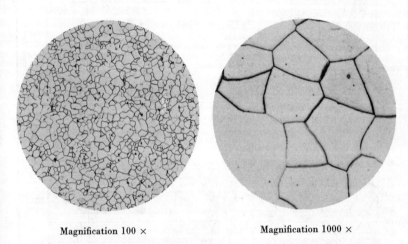

Magnification 100 ×   Magnification 1000 ×

Figure 22. Ferrite.

Courtesy United States Steel Corporation.

may be considered substantially pure iron, and it is called pure iron in some texts. Because the ferrite of an alloy steel may contain in solid solution appreciable amounts of other elements, however, it is better to use the term only in its exact meaning: *a solid solution of any element in alpha iron.*

The various temperatures at which pauses occur in the rise or fall of temperature when steel is heated from room temperature or cooled from the molten state are called *arrest* points; these are identical with the critical points. The arrest points obtained on heating are designated *Ac*,

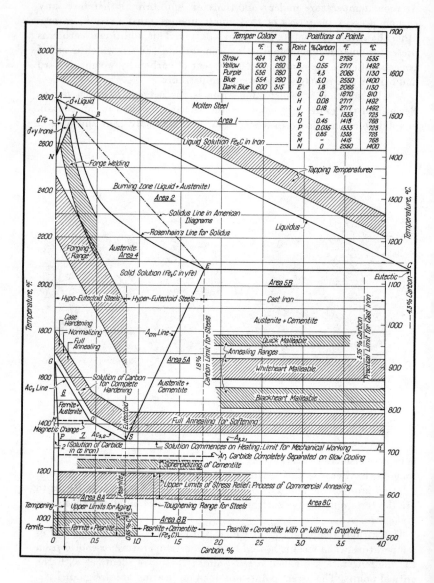

Figure 23. Iron-carbon constitutional diagram. (Modified for students and shopmen by R. Whitfield.)

From *Metal Progress*, courtesy American Society for Metals.

and those obtained on cooling are designated *Ar*. The suffixes *c* and *r* are respectively from *chauffage* (French for *heating*) and *refroidissement* (French for *cooling*).

The first critical point on heating occurs at 410°F, the temperature at which cementite loses its magnetic properties. The magnetic properties of steel are accordingly reduced at this point, although there is no change in crystal structure. This point is at the same temperature on heating or cooling, and is designated A_0. It has little importance in metallurgy.

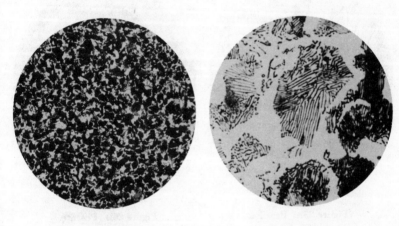

Magnification 100 ×   Magnification 1000 ×

Figure 24. Hypoeutectoid steel. Nital etch.

Courtesy Bethlehem Steel Company.

The first critical point at which a change of structure occurs is at 1333°F, the *decalescence* point, designated Ac_1. Here a considerable amount of heat is absorbed by the steel as the structure changes in part to the face-centered cubic form.

Above 1333°F, the temperatures at which further changes occur vary considerably with the composition of the steel, as is indicated on the diagram. For steels that contain less than about 0.45 per cent carbon, for example, there is a critical point designated A_2, which corresponds to the loss of the remaining magnetic properties of the steel. This is the same temperature at which magnetic alpha iron changes to the nonmagnetic form. The A_2 temperature is the same for heating or cooling.

The next critical point Ac_3, which lies on the line GOS, marks the change of all alpha iron of the steel to the gamma form. The critical point which corresponds to the change of the iron from the gamma form to the delta form is of no practical importance in metallurgy.

Magnification 1000 ×

Figure 25A. Pearlite.

Courtesy United States Steel Corporation.

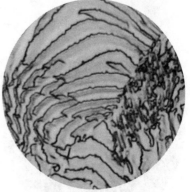

Magnification 2400 ×

Figure 25B. Pearlite.

Courtesy Bausch and Lomb Optical Company.

When steel is cooled, the reverse transformations take place. The most noticeable difference between the temperature of an Ac point and that of the corresponding Ar point is that between Ac_1 and Ar_1. The Ar_1 point for a 0.2 per cent plain carbon steel, for example, is at about 1260°F, or about 73 degrees below the corresponding Ac_1 point. At the Ar_1 point, known as the *recalescence* point, the evolution of heat is so considerable that the steel sometimes can be seen to redden. The range between the recalescence point and the decalescence point is often called the *critical range*.

Steels are classified with respect to carbon content. A steel containing 0.83 per cent of carbon is called *eutectoid steel* because its composition corresponds to the eutectoid point on the iron-carbon diagram. Like all steels when they are at room temperature and in a state of equilibrium, eutectoid steel consists of a mixture of ferrite and cementite, but in this case the particles of ferrite and the particles of cementite form thin plates or lamellae, which alternate throughout the mass to form a characteristic structure known as *pearlite*.

A steel that contains less than 0.83 per cent of carbon is called a *hypoeutectoid steel*. Its structure consists of a small amount of pearlite with an excess of ferrite, which collects at the grain boundaries. Because ferrite is in excess, hypoeutectoid steel is softer and more ductile than eutectoid steel.

Figure 26. High carbon steel. Etchant 4 per cent picral. Magnification 100 ×.
Courtesy The British Cast Iron Research Association.

A *hypereutectoid steel* is one that contains more than 0.83 per cent of carbon. It has a structure of pearlite with an excess of cementite that forms a network at the grain boundaries. The excess of cementite makes hypereutectoid steel harder, more brittle, and less ductile than eutectoid steel.

At temperatures above the A_3 point, the structure of steel is that of a solid solution with gamma iron as the solvent. This solid solution, whether the solute consists of iron carbide only, or of any number of other elements, is known as *austenite*. Austenite exists at room temperature in plain carbon steels only if they contain more than 0.9 per cent of carbon and have been quenched rapidly from about 1630°F. Special alloy steels which contain 18 per cent of chromium and 8 per cent of nickel show austenitic structures when less drastic quenches are employed.

When hypoeutectoid steel is heated to 1333°F the pearlite content changes to austenite; as heating is continued the excess ferrite is absorbed in the austenite grains, until at the temperature that corresponds to the line *GOS*, the entire mass is of austenitic structure.

It is definitely established that the carbon atoms in austenite occupy interstitial positions in the F.C.C. lattice, causing the parameter of the

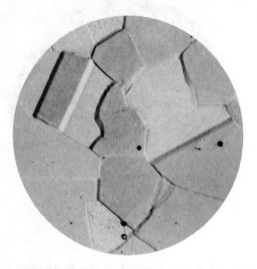

Figure 27.   Austenite.   Magnification 500 ×.
Courtesy United States Steel Corporation.

lattice to increase progressively with the carbon content. This leads one to infer that the carbon atoms make room for themselves in the interstitial pockets among the iron atoms, which otherwise are closely packed.

With a hypereutectoid steel the pearlite changes to austenite at 1333°F just as it does in a hypoeutectoid steel, and the excess cementite goes into solution as the temperature is raised further, until at the temperature which corresponds to the line *SE* the entire structure becomes austenitic.

Structures of Cast Iron

Molten cast iron starts to solidify at the temperature determined by the *ABC* line, and completes solidification at the *ECF* line of the diagram. Point *C*, which corresponds to a composition of 4.3 per cent carbon and 95.7 per cent iron, is the eutectic point of the iron-carbon diagram. The eutectic mixture that solidifies at that point consists of an austenite which contains 1.7 per cent of carbon in solid solution, and cementite; this eutectic is called *ledeburite*. Alloys that contain more than 1.7 per cent and less than 4.3 per cent of carbon consist of a mixture of austenite, cementite, and ledeburite at temperatures between 2065°F and 1333°F. Below 1333°F, cast iron consists of pearlite and cementite.

Questions

1 Define the following terms:

   a. Allotropy   e. Critical range
   b. Critical point   f. Recalescence
   c. Arrest point   g. Ledeburite
   d. Transformation point

2 What is the iron-carbon constitutional diagram? In what way is it of value to a metallurgist?

3 Draw a simplified iron-carbon diagram and explain the changes which occur in the structure of an iron-carbon alloy of any chosen composition as it is heated from room temperature to 2600°F.

4 Define the following terms:

   a. Austenite   d. Pearlite
   b. Ferrite   e. Alpha iron
   c. Cementite   f. Gamma iron

5 What experimental evidence shows that the structures of iron and steel change at certain definite temperatures?

6 Define the following terms:

   a. Eutectoid steel
   b. Hypoeutectoid steel
   c. Hypereutectoid steel

7 What are the properties of alpha iron?

8 What is the recalescence point?

9 What facts show that iron is allotropic?

10 What is the relation of the critical points to heating and cooling?

5 / Iron Ore and Production of Pig Iron

Next to aluminum, the most commonly occurring metal is iron, which constitutes about 5 per cent of the earth's crust. It occurs in rocks, soil, plants, and the blood streams of human beings and animals. Its extraction is commercially profitable only from those mineral formations that are high in iron content and low in objectionable impurities. Because it is chemically active it is not present in nature in the free state, but exists only in chemical combination with other elements. The forms that are used for commercial production of iron are principally the oxides.

Varieties of Ore

An *ore* is a natural mineral deposit from which a useful metal can be extracted at a profit. The iron ore mined in greatest quantity in the United States is *hematite*, which is essentially ferric oxide. Pure ferric oxide is 70 per cent iron, but the hematite supplied to the blast furnace contains about 50 per cent of iron, and impurities including silica, alumina, manganese, moisture, and small amounts of sulfur and phosphorus. Other important iron ore minerals are: *limonite*, a hydrated ferric oxide which has the chemical composition $2 Fe_2O_3 \cdot 3 H_2O$, and which in the pure state contains 59.8 per cent of iron; and *magnetite*, ferrosoferric oxide, which has the chemical composition Fe_3O_4 and which in the pure state contains 72.4 per cent of iron.

Sources of Iron Ore

The principal ore in the Lake Superior district is hematite, some of which is partially hydrated and therefore approaches the composition of limonite. Lake Superior ore as shipped runs about 52 per cent iron, 8.2 per cent silica, 0.8 per cent manganese, 0.09 per cent phosphorus, and 11 per cent moisture. High-grade hematite is mined also in Wyoming. Most of the hematite mined in New York, New Jersey, and Pennsylvania, and the small amounts mined in California, New Mexico, and Utah, require concentration for economical smelting. Most of the ore in Alabama is also hematite, but with a lime content which makes it to some extent self-fluxing; it runs about 36 per cent iron, 15 per cent lime, and 0.3 per cent phosphorus. Limonite is mined in Alabama, Georgia, Missouri, Texas, and some other southern states, but the total tonnage is small compared with that of hematite. Most limonite ores require washing to remove clay and other impurities before shipping.

The ores of the Birmingham, Alabama, district are primary deposits which contain only about 35 to 40 per cent iron. Ordinarily it would not pay to work a deposit so low in iron, but the proximity of deposits of coal and of limestone makes it possible to smelt the ore near the mines. The resulting saving in cost of transportation to a distant smelter, and the fact that much of the ore contains almost enough lime to make it self-fluxing in the blast furnace, make it possible to work these mines profitably.

Ores are grouped according to relative content of iron, phosphorus, silica, manganese, and moisture. The principal classifications are bessemer and nonbessemer grades; bessemer ore contains not more than 0.045 per cent phosphorus. The Mesabi Range produces both bessemer and non-bessemer ores, which are divided into grades according to the amount of iron impurities in each.

Taconite

The really big reserves of iron ore in the Lake Superior region are found in the country rock called *taconite*. Geologically, taconite is the source of the present ore bodies. Enormous quantities of these taconites contain between 25 and 35 per cent metallic iron. The iron exists chiefly as fine grains of hematite and magnetite embedded in a matrix of silica. Taconite is very difficult to drill and blast, but easy to grind. Magnetic separation of the magnetite from the silica particles is a simple, efficient operation. However, separation of the hematite from the silica particles is more difficult. Hematite is classed as nonmagnetic and therefore responds weakly to a magnetic field. Hence, a chemical method known as *froth flotation* is employed; in this process the powdered taconite ore is introduced into a soapy aqueous solution, and air is blown through the solution to create frothy bubbles. The iron oxide particles attach themselves to the frothy bubbles, while the silica remains in the underlying water. When the froth is removed completely, the concentrate contains 63 per cent iron.

In one operation, where magnetite predominates in the taconite, the rock is ground and the magnetite is separated by magnetic means. The final concentrate, which contains 63 to 65 per cent iron and about 8 per cent silica, is formed into balls or pellets, and is heated to about 2000°F in a continuous shaft-type furnace before being charged into the blast furnace in the same manner as iron ore.

Treatment of Ore

Most of the ore from open-pit mines requires treatment to make it commercially usable because it is not sufficiently uniform in size, or because it contains too much lean rock or too much sand to be suitable for use in blast furnaces. In the Hall-Rust mine, for example, about 35 per cent of the ore requires screening and crushing; lean rock is removed during the crushing operation.

Natural concentration of the ore mined at the western end of the Mesabi Range is incomplete in that the silica is loosened but is not dissolved. These so-called *wash ores* therefore require further concentration: the removal of the remaining sand particles is accomplished by washing the ores with water in agitators; the average output of the largest concentrating unit is more than 2500 tons per hour.

Beneficiation

Beneficiation means treating a low-grade ore so that most of the waste rock and valueless portion is separated from the fractional residue of the valuable mineral contained in the ore. Beneficiation reduces the silica content of the ore to 9 per cent or less, and increases the iron content to 60 per cent or more. The average output of the largest concentration unit is more than 2500 tons per hour.

Sintering

A high percentage of the iron ore presently mined in the Lake Superior region must be sintered (agglomerated) before being charged into the blast furnace. Blast-furnace flue dust is high in iron content, and must be sintered before being reintroduced into the blast furnace.

During sintering, a mixture of finely ground ore and about 10 per cent of the ore's weight of powdered coal or coke is ignited, and air is sucked through the mixture to promote combustion. Because of the high temperature reached, the iron fuses, forming a strong cellular mass. This mass is suitable for use in the blast furnace. Sintering may be used also to reduce materially the sulfur content of ores.

Pig Iron

Pig iron is obtained by reduction smelting in the blast furnace. The process includes chemical reduction of the iron compounds present in iron ore, and mechanical separation of the resulting iron from the slag that also is formed. The iron ore is reduced by bringing carbon monoxide into contact with it at a high temperature. Some of the manganese and silicon and all of the phosphorus contained in the ore are also reduced and enter the iron as impurities. Oxides such as those of aluminum, calcium, and magnesium, which are not reduced, enter the slag.

Operation of a Blast Furnace

The Charge

The raw materials used in charging a blast furnace are iron ore, coke, flux, and air. In blast furnaces in the northern United States, approximately 2 tons of ore, 1 ton of coke, $\frac{1}{2}$ ton of flux (usually limestone), and 4 tons of air are required for each ton of pig iron produced. The coke serves both as fuel and as reducing agent, and some of it combines with the iron. The limestone, in addition to reacting with the gangue to form slag, reacts with sulfur in the ore and carries that impurity into the slag.

The Smelting Process

The charge is loaded at the top of the blast furnace where the temperature is between 300°F and 400°F. At this temperature the ascending stream of carbon monoxide resulting from the combustion of coke starts to react with the descending charge, reducing the iron content of the ore partly to iron and partly to ferrous oxide. At the same time some of the carbon monoxide is cooled by the charge to a temperature at which it forms carbon dioxide and free carbon in the form of a fine powder or soot. Some of the free carbon penetrates a short distance into the porous ore and some deposits on the surface of the ore and on the walls of the furnace. At approximately the middle of the stack some of the carbon that was deposited in or on the lumps of ore reduces to metallic iron any ferrous oxide which remains in the furnace. The remainder of the carbon is dissolved by the iron. This dissolved carbon, by lowering the melting point of the iron, converts it from the solid state to a spongy mass.

The temperature in this section of the stack is sufficiently high to decompose the limestone into lime and carbon dioxide. The carbon dioxide reacts with the coke to form carbon monoxide, and the lime combines with the acid gangue to form a slag.

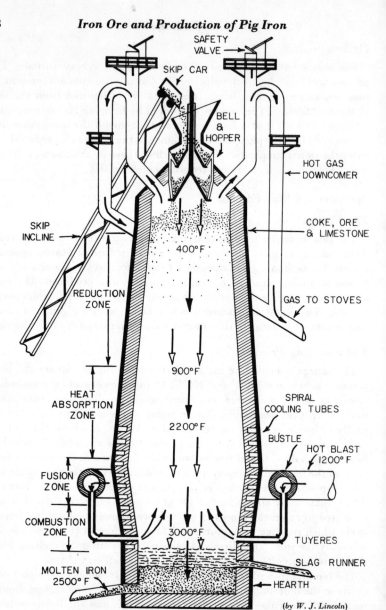

Figure 28. Cutaway diagram of blast furnace.

(by W. J. Lincoln)

As the partially reduced ore descends, it encounters increasingly high temperatures and increasing concentration of carbon monoxide, which accelerate the reactions. Finally the rate of reaction becomes sufficient to reduce all ferric oxide to ferrous oxide and metallic iron. In about eight hours from the time it was loaded, each portion of charge reaches the top of the bosh. At this stage the slag still is pasty and contains a considerable amount of ferrous oxide which is not reduced until it reaches the hearth.

At the tuyeres the oxygen of the hot blast reacts with coke to form carbon dioxide, which reacts at once with more coke and is reduced to carbon monoxide.

The final reactions take place in the region between the tuyeres and the bottom of the furnace. Here the last of the ferrous oxide is reduced to iron, the iron and slag both become liquid, the oxides of silicon, manganese, and phosphorus are reduced, and the elements enter the iron; the lime combines with the coke ash and gangue to complete the formation of liquid slag. The molten iron and slag trickle into the hearth, where the lighter and insoluble slag floats on the surface. Coke extends practically to the bottom of the hearth and is in contact with the molten pig iron for several hours before tapping takes place; as a result the iron becomes saturated with carbon.

Tapping

Approximately 250 tons of pig iron are drawn from the furnace every 6 hours. The slag is tapped more frequently than is the pig iron, to prevent it from rising to the level of the tuyeres and clogging them. If the slag is to be solidified into granular form, it is allowed to flow to concrete basin granulating pits where it passes through a stream of water. Slag is sometimes used to make cement, for ballast, or sometimes as a fertilizer.

If pig iron is to be shipped, it is run into molds on a moving casting machine and cooled. If it is to be used in the plant in which it is produced, it is stored in molten condition in a vessel known as a *mixer*, until it is required in one of the steel furnaces. Iron from several blast furnaces is run into the mixer, which has a capacity of about 1500 tons; this system provides greater uniformity in the quality of iron.

Composition of Pig Iron

Pig iron is a hard, brittle, and impure form of iron, used as an intermediate product in the making of commercial grades of iron and steel.

Low contents of sulfur and phosphorus are desirable because presence of either in iron or steel increases brittleness. The average composition of pig iron is:

Carbon—per cent	3.50	to	4.25
Silicon—per cent	1.25		1.25
Manganese—per cent	0.90	to	2.50
Sulfur—per cent	0.04		0.04
Phosphorus—per cent	0.06	to	3.00
Iron—per cent (by difference)	94.25	to	88.96
	100.00		100.00

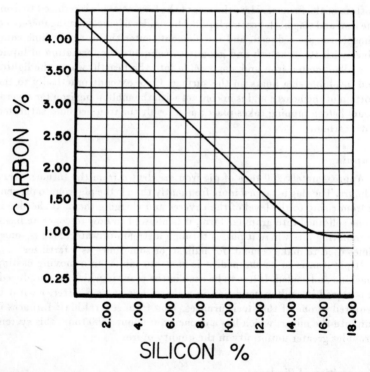

Figure 29. Relation of silicon and carbon content in pig iron. The carbon level gradually decreases as silicon content is increased.

Courtesy Union Carbon and Carbide Corporation.

Questions

1 What is ore?
2 What iron ore is the chief source of iron in the United States, and what is its composition as supplied to the blast furnace?
3 Discuss treatment of ore from open-pit mines to make it commercially usable.
4 How is ore graded to meet specified requirements?
5 What are the chief impurities in iron ore?
6 Why can the ores of the Birmingham, Alabama, district be worked profitably although they contain only between 35 and 40 per cent of iron?
7 What is the difference between bessemer and nonbessemer ore?
8 Discuss treatment of taconite to make it commercially usable.
9 Explain the purpose of beneficiation.
10 Explain sintering.
11 Define the following terms:

a. Smelting   d. Pig iron
b. Immiscible   e. Gangue
c. Slag

12 Name the solid raw materials which compose the charge of a blast furnace, and explain the part played by each in the making of pig iron.
13 Describe the changes which occur in the raw materials as they descend in the blast furnace.
14 What is the principal reducing agent in the blast furnace?
15 Name the products of combustion which leave the furnace through the down-comer, and describe how they are cleaned.
16 Name three uses of:

a. Pig iron
b. Slag

17 What is the average composition of pig iron?
18 What effect on steel has each of the following elements:

a. Carbon   c. Manganese
b. Silicon   d. Phosphorus

6 / Chemistry Involved in the Metallurgy of Iron and Steel

While the net result of the various reactions involved in the making of fully deoxidized steel from iron ore is chemical reduction, intermediate reduction and oxidation reactions are required to eliminate the various impurities that would impair the quality of the steel. These reactions are carried out variously in acidic or basic environments as required by different impurities.

Acids and Bases

In metallurgical terminology an *acid* is the oxide of a nonmetal, and a *base* is the oxide of a metal. The acid oxides chiefly used in metallurgy are silica (SiO_2) (sand) and phosphorus pentoxide (P_2O_5). The basic oxides most commonly used are ferrous oxide (FeO), lime (CaO), and manganese monoxide (MnO). Reactions of acids with bases result in the formation of compounds that constitute slags. Typical examples are:

$$CaO + SiO_2 \rightarrow CaSiO_3$$
$$FeO + SiO_2 \rightarrow FeSiO_3$$
$$MnO + SiO_2 \rightarrow MnSiO_3$$

Oxidation and Reduction

Other fundamental reactions of metallurgical refining processes are oxidation-reduction reactions. Oxidation, in the restricted sense in which the term is used in a discussion of smelting processes, is the combination of an element with oxygen to form an oxide, or the combination of an oxide with more oxygen to form a higher oxide. Reduction is exactly the reverse of oxidation; oxygen may be removed from an oxide to form the element, or a part of the oxygen may be removed from a higher oxide to form a lower oxide of the element.

Oxidation reactions can take place between elements and oxygen directly:

$$C + O_2 \rightarrow CO_2$$
$$2\,Fe + O_2 \rightarrow 2\,FeO$$
$$2\,Mn + O_2 \rightarrow 2\,MnO$$
$$Si + O_2 \rightarrow SiO_2$$

Reduction seldom is accomplished in metallurgical practice by direct disintegration of an oxide into the element and free oxygen. Instead it is necessary to use an intermediate reducing agent that takes oxygen from the oxide which is to be reduced, and is thereby itself oxidized. Reduction and oxidation are thus complementary simultaneous reactions. The most important reducing agents used in ferrous metallurgy are carbon (C), aluminum (Al), silicon (Si), and manganese (Mn). Typical oxidation-reduction reactions of metallurgical processes are:

$$FeO + C \rightarrow Fe + CO$$
$$2\,FeO + Si \rightarrow 2\,Fe + SiO_2$$
$$FeO + Mn \rightarrow Fe + MnO$$

In these reactions ferrous oxide is reduced to metallic iron, while carbon, silicon, and manganese are oxidized to carbon monoxide, silicon dioxide (silica), and manganese monoxide, respectively. Other typical reactions involve the reduction of ferric oxide by carbon monoxide:

$$Fe_2O_3 + CO \rightarrow 2\,FeO + CO_2$$
$$2\,Fe_2O_3 + 8\,CO \rightarrow 7\,CO_2 + 4\,Fe + C$$

When raw materials charged into a steel furnace are heated to about the melting point, iron, silicon, and manganese react with some of the oxygen of the furnace atmosphere. Because iron comprises the greater part of the charge it is the first to oxidize:

$$4\,Fe + 3\,O_2 \rightarrow 2\,Fe_2O_3$$

The ferric oxide mixes with the molten charge and is reduced to ferrous oxide:

$$Fe_2O_3 + Fe \rightarrow 3\,FeO$$

The ferrous oxide dissolves in the molten metal and oxidizes some of the silicon and manganese, and the reduced metallic iron again becomes part of the bath:

$$Si + 2FeO \rightarrow SiO_2 + 2Fe$$
$$Mn + FeO \rightarrow MnO + Fe$$

Slags

A flux that is added to the charge reacts with the oxides of iron, silicon, and manganese to form the viscous material known as *slag*. Slag is immiscible with and lighter than the molten metal and therefore rises to the top.

A slag is acid, basic, or neutral, depending upon the type of flux added. If ferrous oxide or similar oxides are present, the slag has an oxidizing action. A basic oxidizing slag, for example, contains an excess of lime and an oxidizing agent, usually ferrous oxide.

Because of the important functions performed by slag, it is sometimes said that a steelmaker makes slag rather than steel. Those functions are:

Control of the composition of the metal by dissolving the various oxides.

Protection of the metal on which it floats. The blanket of slag prevents excessive oxidation of the metal by the furnace gases, and the loss of heat that would result from such oxidation. It serves also as thermal insulation and protects the metal from overheating.

Fluxes

The temperatures usually attained in metallurgical furnaces are not sufficiently high to melt the *gangue*, a claylike mixture of oxides of aluminum and silicon, which is found as an impurity in iron ore. It is therefore necessary to add a *flux*, the function of which is to react with the gangue and to reduce its melting point to such a value that the entire charge melts and produces a slag sufficiently fluid to be handled readily.

A flux may be acid, basic, or neutral. The only commonly used acid flux is silica, which reacts with any basic material present to form a silicate slag. The basic fluxes most frequently used are limestone, which is essentially calcium carbonate, and dolomite, a mixture of calcium carbonate and magnesium carbonate. These fluxes break down into the oxides of the metals, with evolution of carbon dioxide:

$$CaCO_3 \rightarrow CaO + CO_2$$
$$MgCO_3 \rightarrow MgO + CO_2$$

Both limestone and dolomite have the dual effect of removing some impurities and making the slag more fluid. Fluorspar (CaF_2) is a neutral flux that is used only to increase the fluidity of slag. It is one of the more important raw materials used in steelmaking, and is used in the manufacture of all kinds of steel. It increases the fluidity of the slag without change in chemical properties. Because of the increased fluidity, impurities are more quickly eliminated from the molten steel. Fluorspar also permits quicker transference of heat to the metal.

Any basic or neutral slag should be low in silica, alumina, and sulfur compounds.

Questions

1 Define the following terms as used in metallurgy:

    a. Acid   c. Oxidation

    b. Base   d. Reduction

2 Name the principal acid oxides used in metallurgy.
3 Name the principal basic oxides used in metallurgy.
4 What is slag, and why is it important in metallurgical processes?
5 Why is it stated sometimes that a steelmaker makes slag rather than steel?
6 What is a flux, and what is its function in steelmaking?
7 What is the commonly used acid flux?
8 What are the commonly used basic fluxes?
9 Write chemical reactions for the formation of a slag.
10 Explain why oxidation and reduction are complementary simultaneous reactions.

7 / Cast Iron and Wrought Iron

Cast iron is a term that designates a series of alloys of iron and carbon that contain more than 1.7 per cent carbon, together with various quantities of silicon, manganese, phosphorus, and sulfur. The amounts of the different elements found in most commercial cast irons are:

Carbon	2.50 to	3.75 per cent
Silicon	0.50	3.00
Manganese	0.40	1.00
Phosphorus	0.12	1.10
Sulfur	0.01	0.18

The American Society for Testing Materials Tentative Specification A196-42T defines cast iron as: "Essentially an alloy of iron, carbon, and silicon, in which the carbon is present in excess of the amount which can be retained in solid solution in austenite at the eutectic temperature. When cast iron contains a specially added element or elements in sufficient amounts to produce a measurable modification of the physical properties of the section under consideration, it is called alloy cast iron. Silicon, manganese, sulfur, and phosphorus, as normally obtained from raw materials, are not considered as alloy additions."

Gray Cast Iron

Gray cast iron is the form of iron most widely used for castings. Its matrix is composed of pearlite with minor amounts of ferrite. If the iron contains an appreciable amount of phosphorus, the structure also includes *steadite*, a hard and brittle substance containing about 10 per cent phosphorus, and consisting of a mixture of iron and iron phosphide. Steadite usually is present in the grain boundaries of gray cast iron, and appears under microscopic examination as a fine structureless area.

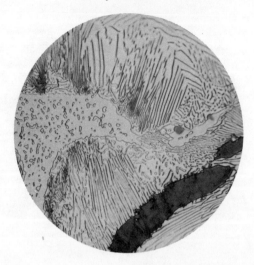

Figure 30. Gray iron. Etchant 4 per cent picral. Magnification 600 ×.
Courtesy The British Cast Iron Research Association.

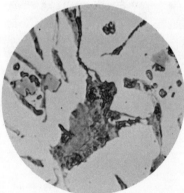

Figure 31. Gray Iron, showing random graphite flakes, unetched. Magnification 100×.
Courtesy International Nickel Company, Inc.

Figure 32. Graphite in gray cast iron. Magnification 500×.

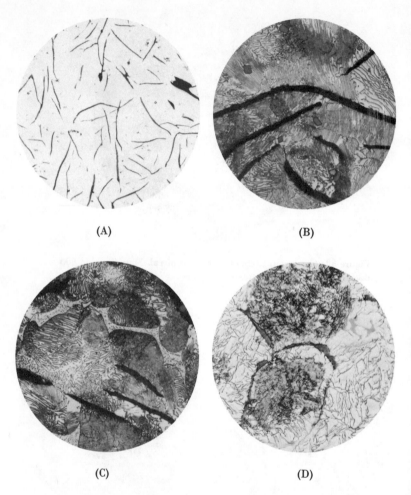

(A)   (B)

(C)   (D)

Figure 33. Gray cast iron.

(A)   Showing distribution of graphite; unetched; magnification 100 ×.

(B)   Showing graphite flakes and detail of pearlite matrix; etchant 2 per cent nital; magnification 500 ×.

(C)   Steadite in gray iron containing phosphorus; etchant 2 per cent nital; magnification 500 ×.

(D)   Alloy gray cast iron; etchant 2 per cent nital; magnification 500 ×.

Courtesy International Nickel Co., Inc.

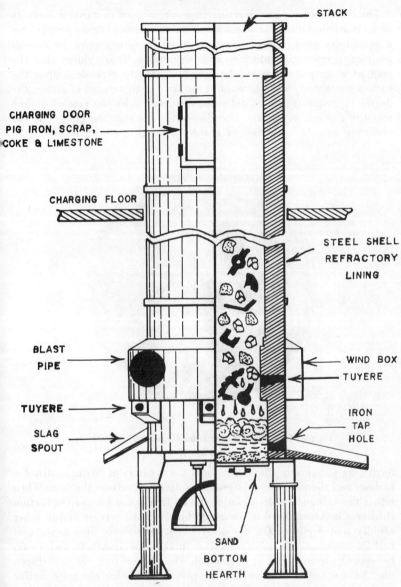

Figure 34. Cutaway diagram of cupola furnace.

The characteristic gray fracture from which gray cast iron derives its name is attributable to the presence in its structure of flakes of graphite, a practically pure form of free carbon; graphite is formed by decomposition of cementite into iron and free carbon. It is evident that the amount of graphite that can be formed is largely dependent upon the carbon content of the cast iron, but for any stated amount of carbon the degree of graphitization is determined primarily by the rate of cooling and the content of silicon. The slower the cooling, the greater is the tendency toward formation of graphite. Because silicon is soluble in

Figure 35. Pearlite in cast iron. Magnification 2000 ×.
Courtesy Bausch & Lomb Optical Company.

ferrite, its presence in the iron reduces the capacity of ferrite to dissolve carbon, and thereby promotes graphitization and softens the iron. When either the carbon or silicon content of an iron is too low for the section thickness involved, hard iron carbide forms at the corners and in other rapidly cooled places. On the other hand, excessively high content of carbon or silicon in heavier sections makes the iron soft and weak, because the iron will be open grained. The graphitizing effect of silicon may be assisted or impeded by the presence of other elements. Sulfur tends to form iron sulfide, which impedes the formation of graphite by stabilizing cementite. If manganese is present to about 0.35 per cent in excess of the theoretical amount required to combine with sulfur, manganese sulfide is formed instead of iron sulfide, and the stabilizing effect of iron sulfide is removed. A considerable excess of manganese, however, inhibits graphite formation because it combines with carbon to form manganese carbide, leaving correspondingly less free carbon to exist in the form of graphite. As an alloying element, manganese imparts density and high strength. A manganese-sulfur ratio of 6 to 1 is suggested. Alloying elements such as aluminum, nickel, and titanium are soluble in ferrite, and therefore promote graphitization in the same manner as does silicon; other elements such as chromium, tungsten, and vanadium form carbides and therefore inhibit graphitization by holding carbon in combination. A high phosphorus content weakens cast iron and increases its fluidity in the molten state, but has little effect on graphitization.

The properties of gray cast iron vary over a wide range depending upon the method of making, heat treatment, and composition. Alloy cast irons are available for a wide range of special purposes. In general, gray cast iron is characterized by its power to damp vibrations and by the wear resistance imparted by the lubricating effect of graphite; both properties make gray cast iron a useful material for the construction of machinery. Gray cast iron of suitable composition is readily machinable and is an economical material of which to make many metal parts used in various industries.

Production of Gray Cast Iron in the Cupola Furnace

Gray cast iron is made by melting pig iron, scrap iron, scrap steel, or a combination of all, in a cupola furnace or *cupola*, as it usually is known. The composition of the charge is calculated to produce the desired composition of gray iron. Although the function of a cupola is remelting rather than refining, there is some loss of silicon and manganese, and there are some changes in the contents of other elements in the metal during the operation. Because a cupola is a continuous melting unit, the uniformity of its product is not equal to that obtainable in a steel furnace, where metal is melted in batches and adjustments in composition must be made before tapping.

Comparison of Cupola and Blast Furnace

Cupola	*Blast Furnace*
Shell lined with firebrick	Shell lined with firebrick
Taphole and slag hole	Taphole and slag hole

Cupola	*Blast Furnace*
Tuyeres	Tuyeres
Wind box	Bustle pipe
Drop bottom	Solid bottom
Charged at the side	Charged at the top
No stoves or dust catchers	Stoves, dust catchers, gas washers
Simple construction	Complicated construction
Charged with pig iron or scrap	Charged with ore
Coke in contact with metal	Coke in contact with metal
Slightly oxidizing	Strongly reducing
Primarily remelting	Primarily smelting
Low temperature operation	High temperature operation
Low fuel consumption	High fuel consumption
Intermittent or continuous operation	Continuous operation
Air may be cold or preheated	Air is preheated
Composition of product adjustable to some extent after tapping	Composition of product not adjustable after tapping

Figure 36. Malleable iron melting furnace.
Courtesy Whiting Corporation.

White Cast Iron

White cast iron normally is so low in silicon plus carbon that during and after solidification no carbon is precipitated as graphite. All carbon therefore exists in combination with iron as iron carbide, and the structure consists of pearlite and cementite. A similar structure can be obtained on a section of gray cast iron by cooling it rapidly; this is called *chilled cast iron.* White cast iron is-hard, brittle, and almost impossible to machine. It is used to a limited extent in applications which require these properties, such as plowshares, car wheels, chilled rolls, dies, and grinding balls.

The greater part of all white cast iron produced is intended for conversion into malleable cast iron. For that purpose carbon and silicon must be present in such amounts that the cementite is decomposed into free

Figure 37. White iron. Nital etch. Magnification 100 ×.
Courtesy National Malleable and Steel Castings Company.

carbon and iron during the malleableizing process. Some white cast iron is made in cupolas, and some is made in air furnaces from iron and steel scrap and pig iron. From the known carbon and silicon contents of each of the materials in the furnace charge, the total amount of each is com-. puted. Because the charge is not in direct contact with fuel, the carbon content of the metal is not increased during the operation.

The compositions of white cast iron suitable for conversion into the two grades of malleable cast iron covered by the American Society for

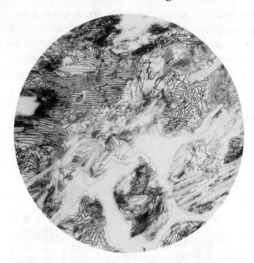

Figure 38. Hard white iron as cast.   Magnification 500 ×.
Courtesy General Electric Company.

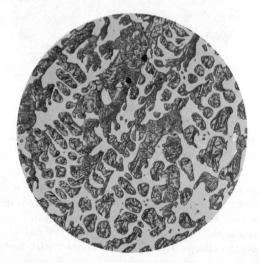

Figure 39. Martensitic white iron (nickel hard). Etchant 4 per cent picral. Magnification 100 ×.
Courtesy The British Cast Iron Research Association.

Testing Materials Specification A47-33 have been given in the *Cast Metals Handbook* of the American Foundrymen's Association:

	Grade 35018			Grade 32510		
Carbon—per cent	1.75	to	2.30	2.25	to	2.70
Silicon—per cent	0.85	to	1.20	0.80	to	1.10
Manganese—per cent	less than	0.40		less than	0.40	
Phosphorus—per cent	less than	0.20		less than	0.20	
Sulfur—per cent	less than	0.12		0.07	to	0.15

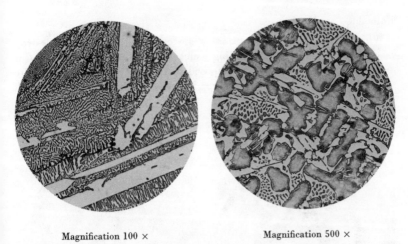

Magnification 100 ×   Magnification 500 ×

Figure 40. Two photomicrographs of hypereutectic white iron. Etchant 4 per cent picral.
Courtesy The British Cast Iron Research Association.

The Air Furnace

The air furnace consists of a shallow hearth covered by an arched roof designed to reflect the heat of combustion upon the charge.

The fuel, which usually is pulverized bituminous coal, is fed by a burner; an air blower supplies the amount of air required for combustion of the fuel, and during a portion of the operation supplies an excess of air for oxidizing some of the elements in the melt.

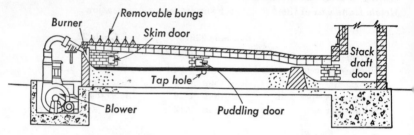

Figure 41. Cutaway diagram of air melting furnace.
Courtesy Whiting Corporation.

Malleable Cast Iron

White iron castings after cleaning are converted into malleable iron castings by a process of annealing which requires several days. By this means the iron carbide is decomposed and the pearlite and cementite structure characteristic of white iron is converted into a mixture which

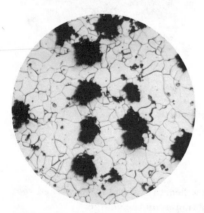

Figure 42. Good malleable iron. Magnification 100 ×.
Courtesy General Electric Company.

Figure 43. Malleable iron. Nital etch. Magnification 100 ×.
Courtesy National Malleable and Steel Castings Company.

consists substantially of graphite and ferrite. Graphite in malleable cast iron is in the form of nodules, called *temper carbon*, which under microscopic examination are readily distinguishable from the graphite flakes in gray cast iron. Good malleable iron contains practically no combined carbon.

When castings are to be malleableized in a batch-type oven, they are placed by hand in cast-iron pots called *rings* or *saggers*, and often are packed with sand or crushed slag. The pots are sealed with fireclay to protect the castings from oxidation by contact with furnace gases.

When the oven is full it is sealed and the temperature is raised very slowly to between 1550°F and 1600°F. That temperature is maintained for a period of from 40 to 60 hours and then is reduced as slowly as possible to 1400°F. The temperature is held in the critical range (1380°F to 1280°F) for about 20 hours, and then is decreased to 1200°F. At that time the doors of the oven are opened, and the pots are allowed to cool to handling temperature, after which the castings are dumped.

At times the castings may warp after annealing. Since they are capable of absorbing shock at this time, they can be straightened by blows with a hammer. *Other than this, they cannot be worked.* The name malleable cast iron is misleading; it is not readily workable, but has greater toughness after heat treatment because its structure is essentially a mechanical mixture of commercially pure iron and graphite nodules. Malleableizing seeks to transform the casting from a hard, brittle structure into a structure that is soft and tough.

The total time that elapses from the time the castings are packed to the time they are dumped may be as long as 90 hours.

In addition to the batch method of malleableizing, there are continuous processes in use; the castings are moved through the oven at practically uniform speed, either in cars or on continuous conveyors or rollers.

Some castings are malleableized by *short-cycle* annealing, which is completed in a period of from 15 to 60 hours. The white iron chosen for such a process usually has high silicon content and sometimes contains alloying elements which promote rapid graphitization. The process is further accelerated by use of a small and well-insulated oven that has low thermal lag, and which retains a temperature of nearly 1200°F when each new batch of castings is loaded, thereby saving several hours of heating. The castings should be of such shape that no packing is required, and they should be relatively small, so that the mass that must be heated is kept to a low value.

Malleable cast iron is more ductile and more resistant to shock than is gray cast iron, and therefore is suitable for many more applications. It is used in large quantities for such materials as pipe and pipe fittings, and in the automotive and many other industries.

Ductile Cast Iron

It has long been recognized that the properties of irons are influenced significantly by the presence, shape, and distribution of graphite. The graphite in ordinary grades of cast iron is in the form of thin flat flakes that are distributed through a matrix of pearlite and some ferrite. This flake graphite is largely responsible for the excellent machinability, high damping capacity, good wear resistance, and excellent foundry properties of gray iron. On the other hand, the graphite flakes are largely responsible also for greatly limiting the mechanical properties of gray cast iron and its poor toughness and limited tensile strength.

In malleable iron, however, the graphite has a nodular shape and is distributed through a *ferritic* matrix, not in a pearlitic malleable matrix. Malleable irons have relatively higher mechanical properties; one of the principal reasons is the presence of nodular graphite. For this reason, metallurgists have long recognized that if the graphitic carbon in as-cast gray iron could be produced in the form of nodules or spheres instead of in flakes, considerable improvement in mechanical properties would result. The reason is that when the graphitic carbon is in the form of nodules or spheres, irons have higher strength and better toughness, because these particles interrupt the continuity of the matrix much less than does the same amount of carbon in flake form. As a result of this knowledge, a new engineering material described as *ductile cast iron* has been developed.

Ductile cast iron is a high carbon ferrous product containing graphite in the form of spheroids. The spheroid in ductile iron is a single polycrystalline particle, whereas a nodule of graphite in malleable iron is composed of an aggregate of fine flakes. Ductile cast iron closes the gap between cast iron and steel. It has all the advantages of cast iron from the process viewpoint: low melting point, good fluidity and castability, ready machinability, and low cost, plus the additional advantages of high yield strength, high elasticity, and a substantial amount of ductility. Thus it is suited for many applications hitherto considered beyond the scope of cast iron.

The spheroidal graphite structure is obtained by introducing into the molten iron in the ladle, shortly before casting, a small effective amount of magnesium in the form of a magnesium-containing agent such as a

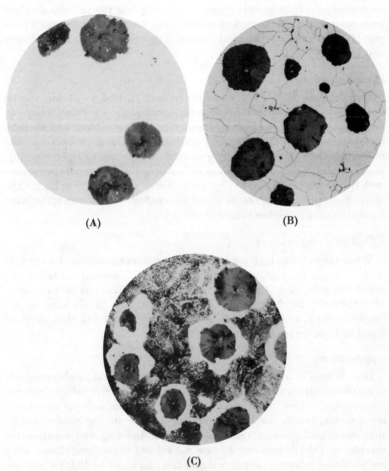

Figure 44. Ductile iron.

(A) Containing random graphite spheroids; unetched; magnification 100 ×.

(B) Ferritic; etched; magnification 250 ×.

(C) Pearlitic; etched; magnification 250 ×.

Courtesy International Nickel Co., Inc.

nickel-magnesium alloy that contains 50 to 80 per cent nickel. Common cast iron compositions melted in the cupola or in other furnaces can be used. The spheroidal graphite in ductile iron is·formed as cast, whereas continued heat treatment is required to precipitate the nodular graphite from as-cast carbide in malleable iron. It is the spheroidal shape of the graphite that removes the weakening and embrittling effects of ordinary cast iron.

Heat Treatment

Ductile irons have been successfully heat treated to obtain special properties without noticeable effect on the graphitic structure. They can be annealed readily to give a completely ferritic matrix and thereby increase ductility. They respond also to quenching, tempering, and normalizing much as do ordinary gray irons. By quenching and tempering, improvement in tensile strength to higher than 150,000 pounds per square inch (psi) has been obtained; by normalizing, which involves heating to above the critical temperature, and air-cooling, tensile strengths have been increased to higher than 120,000 psi.

Effect of Temperature

When subjected to high temperatures, gray iron containing flake graphite is susceptible to growth and relatively rapid internal oxidation. To avoid this growth, it is necessary to obtain a fine graphite texture, or to alloy the iron with chromium. It has been found that ductile irons are not nearly so susceptible to growth as is plain gray iron, and that they are equal or superior to chromium alloy irons.

Applications

Ductile iron is used by the automotive, agricultural implement, and railroad industries. It is used also for machine tools, crankshafts, pumps, compressors, valves, and the like; for rolling mill rolls, railroad car wheels, ingot molds, textile machinery, pipe products requiring good strength, toughness, and pressure tightness; for road-building and construction machinery. Ductile iron castings are not subject to size limitations. They are produced commercially in weights from 2 ounces to 100,000 pounds, with section thicknesses from 0.10 inch to 48 inches.

Wrought Iron

Wrought iron was known long before the beginning of recorded history, and was the form of iron used industrially until the invention of the modern processes for making steel. Wrought iron has been replaced to a great extent by steel. It is defined by the American Society for Testing Materials as: "A ferrous material, aggregated from a solidifying mass of pasty particles of highly refined metallic iron with which, without subsequent fusion, is incorporated a minutely and uniformly distributed quantity of slag." The iron is exceptionally pure and is low in carbon, while the slag consists of iron silicate, an inert, noncorroding, glasslike material. Iron and slag form an intimate mechanical mixture that differs materially from that of an alloy, such as steel.

Figure 45. Wrought iron.
Courtesy A. M. Byers Company.

Wrought iron now is produced economically from pig iron. Pig iron is melted in a furnace lined with ferrous oxide, and mill scale or other form of iron oxide is added. Under the basic conditions maintained by the ferrous oxide lining, practically the entire carbon content of the pig iron is oxidized and removed. Most of the other impurities also are oxidized, and the silicon reacts to form the iron silicate slag. The furnace used is called a *puddling furnace* because the melt is stirred or *puddled* to accelerate oxidation.

As the process approaches completion, the heat of reaction becomes insufficient to keep the iron in molten condition. The refined iron gathers into small sticky lumps that collect into a spongy ball saturated with liquid slag. The ball, which weighs from 400 to 500 pounds, is removed from the furnace, put through a squeezer to remove as much surplus slag as possible, and formed into compact *blooms*, which are rolled at once into rough flat sections known as *muck bars*. In order to obtain more uniform distribution of slag through the iron, the muck bars are cut into lengths, piled, heated to welding temperature, and rolled.

The slag, which constitutes from 1 per cent to 3 per cent of the total weight of wrought iron, is distributed through the iron in fibers. The fibrous structure, which is similar in appearance to that of hickory, is quite different from the crystalline structure of steel. Table 2 shows a comparison of the constituents of wrought iron and two types of soft steel suitable for making welded pipe.

The presence of slag fibers, of which there may be 250,000 or more per square inch of cross section, is responsible for many of the desirable attributes of wrought iron. Perhaps the most important function served by the fibers is that of a mechanical barrier which, by confining corrosion largely to the surface, protects the iron from pitting. Wrought iron is therefore a good material from which to make water pipes. Its fibrous structure gives wrought iron excellent resistance to shock and vibration, making it particularly suitable for the manufacture of such products as railroad drawbars, air-brake pipes, and engine bolts. Wrought iron is welded readily by any of the commonly used welding processes, the presence of slag making it self-fluxing.

Wrought iron may be formed by hot or cold methods, and is readily machinable. Threads cut on wrought iron are sharp and clean because the chips crumble and clear the dies instead of forming long spirals.

TABLE 2. PERCENTAGE COMPOSITION OF WROUGHT IRON AND STEEL

	Wrought Iron	*Acid Bessemer Steel*	*Open-Hearth Steel*
Carbon—per cent	0.02	0.07	0.10
Manganese—per cent	0.03	0.35	0.40
Phosphorus—per cent	0.12	0.10	0.03
Sulfur—per cent	0.02	0.05	0.03
Silicon—per cent	0.15	0.02	0.02
	0.34	0.59	0.58
Slag—per cent	3.00	—	—

Questions

1 Define the following terms:

   a. Cast iron   c. Graphitizer
   b. Steadite   d. Chilled iron

2 What are the properties of gray cast iron, and how does it differ from white cast iron? State four uses of each.

3 What is the function of silicon in the making of gray cast iron?

4 Compare the cupola furnace with the blast furnace.

5 Describe in detail the process of converting white iron castings into malleable iron castings.

6 Compare the properties of ductile cast iron with those of malleable cast iron and gray cast iron.

7 Discuss how the properties of cast iron can be changed.

8 How is the spheroidal graphite structure in ductile cast iron obtained?

9 Why does wrought iron resist corrosion?

10 State five properties of wrought iron.

11 What happens to the cementite of white iron during the malleableizing process?

FLOW SHEET OF MANUFACTURE OF CAST IRON

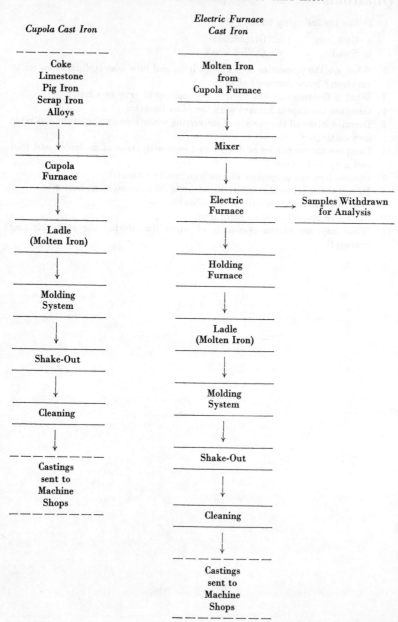

Cupola Cast Iron

———————

Coke
Limestone
Pig Iron
Scrap Iron
Alloys

———————

↓

Cupola
Furnace

↓

Ladle
(Molten Iron)

↓

Molding
System

↓

Shake-Out

↓

Cleaning

↓

Castings
sent to
Machine
Shops

———————

*Electric Furnace
Cast Iron*

———————

Molten Iron
from
Cupola Furnace

↓

Mixer

↓

Electric
Furnace → Samples Withdrawn
for Analysis

↓

Holding
Furnace

↓

Ladle
(Molten Iron)

↓

Molding
System

↓

Shake-Out

↓

Cleaning

↓

Castings
sent to
Machine
Shops

———————

Courtesy Ford Motor Company.

8 / Steel

Steel, the most versatile and most important of all industrial metals, is produced in many types, each designed to serve most effectively in one or more applications. Defined in terms of its constituents, steel is an alloy of iron with not more than 1.7 per cent carbon. Professor Stoughton of Lehigh University defines steel in terms of its properties: "Steel is an iron-carbon alloy which is malleable at least in some range of temperature and in addition is either: (a) cast into an initially malleable mass, or (b) is capable of being hardened by sudden cooling, or (c) is both so cast and so capable of hardening." The malleability of steel distinguishes it from cast iron and pig iron; initial malleability when cast distinguishes steel from malleable cast iron; hardenability by rapid cooling distinguishes steel from wrought iron.

The characteristics that are largely responsible for the versatility of steel are the range of hardness and the changes in its other physical properties that are obtained both by variation of the amount of carbon in its composition and by heat treatment. A high-carbon steel, cooled rapidly from a high temperature by quenching in water or other liquid, may be hard enough to scratch glass, while an unhardened steel of low carbon content is soft enough to be scratched by a needle. A hard steel can be made so brittle as to shatter under a blow from a hammer, whereas a soft steel can be made so ductile that it can be drawn into a wire much finer than a human hair. The tensile strength of steel varies over the wide range of from 40,000 to 500,000 pounds per square inch (psi).

Many steels adapted to specific purposes have been developed by adding to plain carbon steel any one, or often more than one, of various alloying metals. In this way it is possible to produce a steel to meet almost any required specification.

Until the middle of the nineteenth century steel was rare and too expensive to be used widely in industry. At about that time, William Kelly in Kentucky, and Sir Henry Bessemer in England, independently discovered that by blowing a blast of cold air through molten iron a useful form of steel could be produced at moderate cost.

Development of the Siemens-Martin open-hearth process commenced in about 1860; after the development of the basic process, use of this

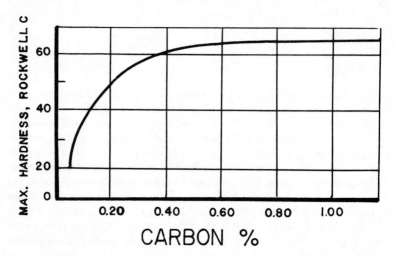

Figure 46. Effect of carbon content on maximum hardness of steels.
Courtesy E. F. Houghton and Company.

type of furnace increased rapidly, until at the present time a large proportion of all steel made in the United States is made in basic open-hearth furnaces.

Development of the electric arc furnace for making steel started in the early years of the twentieth century. Its use has increased rapidly, particularly in the making of the finest grades of tool and alloy steels, for which it is especially well adapted.

Reasons for Selecting a Steel

1. Cost of producing the part
   a. Annealing characteristics
   b. Machinability
   c. Ease of fabrication at mill

   d. Ease of control to meet composition limits

   e. Comparison with other steels

   f. Price

2. How it meets engineering requirements

   a. Manner in which it will stand up in service

   b. How it will stand up under the loads imposed on it

   c. How slowly the steel can be quenched and still obtain maximum strength (every steel is dependent on its carbon content for hardness).

Plain Carbon Steel

A plain carbon steel is a steel in which carbon is the only alloying element added to the iron to control its physical properties. Addition of a relatively small amount of carbon, followed by heating and quenching, changes pure iron from a soft, ductile material to a hard, strong one. Plain carbon steels contain from 0.08 to 1.7 per cent carbon, and small amounts of other elements, particularly sulfur, silicon, manganese, and phosphorus.

The carbon content controls the hardness, strength, and ductility of steel. The higher the carbon content, the harder is the steel, and therefore the greater is its resistance to plastic deformation by bending, scratching, or penetration. Conversely, reduced carbon content results in an increasingly ductile steel which can be deformed readily by cold work such as forging or drawing. Tensile strength increases directly with increase in carbon content up to 0.9 per cent, and then falls off slightly. When ability to hold a cutting edge and resistance to wear are required, the carbon content is increased to more than 0.9 per cent, at a sacrifice of tensile strength and toughness. Increase of carbon content has the following effects:

Increased hardness and wear resistance

Decreased tendency to warping and cracking

Decreased ductility and malleability, both hot and cold

Decreased toughness and resistance to corrosion

Lowered welding and forging temperatures

Lowered transformation points on cooling

Carbon steels are classified broadly into machining steels, forging steels, and tool steels. Machining steels usually are supplied cold rolled. Because they have a carbon content of only between 0.1 and 0.2 per cent, they are used widely for carburizing purposes. Forging steels contain between 0.3 and 0.6 per cent carbon, and can be heat treated to increase hardness and strength. Tool steels are those that contain more than 0.6 per cent carbon and nominally 0.25 per cent silicon and 0.25 per cent manganese. Because they do not retain their hardness at high temperatures, tools made of plain carbon tool steels fail in service if speed of machining is too great or if cuts are too deep. They can, however, be reworked and rehardened a number of times. Steels of this type can be given a high surface hardness over a soft, tough core.

Effect of Other Elements in Plain Carbon Steel

The sulfur content usually is kept below 0.05 per cent, although amounts as high as 0.12 per cent are now believed to cause little trouble in most cases. Steels which contain more than 0.12 per cent sulfur tend to crack and tear during rolling. This condition, known as *hot shortness*, is caused by the presence of iron sulfide, which melts during forging and rolling. The trouble is overcome by adding sufficient manganese to combine with the sulfur to produce manganese sulfide, which instead of melting, becomes plastic and forms fibers when the steel is rolled. In steel which is intended for screw cutting, the presence of these fibers is beneficial, since, as do the slag fibers in wrought iron, they cause the chips to break off cleanly, keeping the edge of the cutting tool free.

Another adverse influence exerted by sulfur is the production of surface imperfections during fabrication. These are expensive to remove, and the sulfur segregates in such a manner that mechanical properties of the steel vary from the center to the outside of an ingot. This explains why automobile fenders and hoods cannot be made from steel that contains more than 0.031 per cent sulfur. About 85 per cent of the sulfur encountered during steelmaking comes from the coke, because none of the chemically combined organic sulfur can be removed by washing. Therefore the most logical point to reduce the sulfur is at the blast-furnace stage.

Plain carbon steels usually contain from 0.6 to 0.9 per cent manganese. Its presence in steel increases tensile strength, and in refining operations it is an essential constituent because of its action as a deoxidizer and desulfurizer. The content of phosphorus customarily is kept below 0.05 per cent because a greater proportion tends to produce a coarse grain structure which makes steel brittle when cold; this condition is called *cold shortness*. Phosphorus also has an effect somewhat like that of carbon in increasing the hardness of steel. It has some beneficial effects, in that it improves the machining qualities, and in low-carbon steel improves resistance to corrosion.

Silicon increases the tensile strength of steel. In steels intended for use in magnetic circuits, silicon is an important element because it reduces hysteresis losses. In smelting operations, silicon is important as a deoxidizer.

Oxygen in Steelmaking

Oxygen is used either to speed combustion of the fuel or to oxidize the carbon in the charge.

Distinct advantages gained by the use of accelerated oxidation are: (1) heats require less time, (2) a higher bath temperature increases the fluidity of the slag, (3) less fuel is required, (4) less bank erosion occurs because there is less time of exposure to high-oxide slags. The principal disadvantage of the use of oxygen is the splashing of the slag, which may cause the roof and the front wall of the furnace to erode.

The most common use of oxygen for decarburization is in making low-carbon steels. There is a normal slowing of carbon elimination in the neighborhood of a carbon content of 0.12 per cent. At this low carbon content, the elimination of carbon is accelerated by the use of oxygen.

The Acid Process

Steel is made by an acid process only if the raw materials to be used are low in phosphorus and sulfur, because those elements are not removed by this process. The acid Bessemer process is the acid process in widest use in the United States. The initial reaction is oxidation of the iron content of the charge to ferrous oxide by the oxygen of air. Subsequently the silicon and manganese react with the ferrous oxide, forming metallic iron and oxides of silicon and manganese. The oxides of silicon and manganese enter the slag; finally any remaining ferrous oxide is reduced to metallic iron by reaction with the carbon. Neither the phosphorus nor the sulfur is oxidized.

The Basic Process

Use of a basic process is necessary for making steel from raw materials that contain appreciable amounts of phosphorus and sulfur; the basic open-hearth process is typical. Oxidation of the undesired elements takes place in substantially the same manner as in the acid process except that the charge includes ferric oxide ore which also has an oxidizing effect. Limestone is added to the charge to keep it basic, and reacts with sulfur and phosphorus to form calcium sulfide and calcium phosphate, respectively, both of which enter the slag.

Comparison of Steels

The open-hearth process is more expensive in operation than is the Bessemer process, but the steel produced is of better quality for most purposes because the open-hearth process permits more complete de-oxidation and better control over the reactions, with resulting better elimination of inclusions such as oxides and sulfides. Open-hearth steel contains fewer blowholes because less hydrogen, oxygen, and nitrogen are trapped in it. The loss of iron in open-hearth smelting is less than in the Bessemer process.

Bessemer steel has advantages over open-hearth steel for some uses. The excellent machining qualities, imparted mainly by the sulfur content but partly by the nitrogen content, make this steel particularly suitable for such machine parts as screws. The high phosphorus content imparts greater stiffness for the same carbon content than is obtainable in open-hearth steel; this characteristic is important in the steel sheets intended for making tin plate used in the manufacture of cans. Bessemer steel is better suited for welding, and stronger welded pipe can be made from bessemer than from open-hearth steel. Bessemer steel is considerably more affected by cold working than is open-hearth steel, an attribute which may be favorable or unfavorable, depending upon the use for which the steel is intended.

Questions

1 Define steel in terms of its constituents.
2 What is the most important element, other than iron, in steel, and what properties does it confer on steel?
3 Define the following terms:

    a. Plain carbon steel   d. Ductility
    b. Hot shortness   e. Tensile strength
    c. Cold shortness

4 What are the effects of increased carbon content of a steel?
5 What advantages are gained by use of plain carbon steels for tools?
6 What effect has each of the following elements on a plain carbon steel:

    a. Sulfur
    b. Manganese
    c. Silicon

7 Compare the acid process and basic process of making steel.
8 State the differences between machining, forging, and tool steels.

9 | Bessemer Steel

The Bessemer process was the first by which steel could be made in commercially useful quantities, and by which its composition could be controlled as required. With the development of the process, steel began to supplant wrought iron as the most important metal of industry.

The Bessemer process can be operated either as an acid process or as a basic process. The acid process does not reduce the content of phosphorus or sulfur, and is not suitable for refining pig iron of high phosphorus content. However, it is used in the United States because of the availability of ores of low phosphorus content.

Composition of the Charge

Successful operation of a bessemer converter requires a supply of molten pig iron which usually is supplied from a mixer in which it is kept molten after having been tapped from the blast furnace. The iron in the mixer is kept at a temperature of about 2200°F to 2400°F, which is the most favorable range for charging the converter. The use of a mixer conserves the heat of the iron from the time it is received from the blast furnace, and maintains great uniformity of composition.

The composition of iron must be held within certain limits if it is to be converted to steel by the acid Bessemer process. Because neither phosphorus nor sulfur is eliminated in this process, the proportions of those elements in the finished steel are somewhat higher than in the pig iron, as a result of loss of iron and other elements during the process. For that reason, the content of phosphorus should not exceed 0.10 per cent, and sulfur should not exceed 0.05 per cent. Manganese in excess of about 0.45 per cent tends to produce a watery slag with resultant slopping, and a poorer grade of steel. Carbon in pig iron suitable for use as bessemer

81

raw material should be between 4.0 and 4.5 per cent, and silicon should be between 1.00 and 1.75 per cent.

The content of silicon is particularly important; an excess causes high temperatures in the converter, resulting in the formation in the belly of the converter of solid masses of slag known as *kidneys* or *salamanders*. If the blow is too hot as the result of the presence of too much silicon, it can be cooled by the addition of scrap metal low in silicon, or by blowing steam through the metal. An inadequate amount of silicon, on the other hand, causes a cold blow and produces a slag with high content

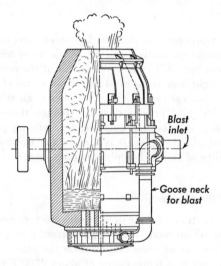

Figure 47. Fifteen-ton Bessemer converter.
Courtesy Jones and Laughlin Steel Corporation.

of ferrous oxide, which makes it erosive. A cold blow can be corrected by side blowing, which consists of tipping the converter so that some of the tuyeres are above the surface of the metal. The air from these tuyeres, in passing over the bath instead of through the molten metal, causes the carbon monoxide to burn inside the converter instead of at the mouth; this results in an increase in the temperature of the melt.

The reactions which take place in the converter during the blow are:

1. Oxidation of some of the iron to ferrous oxide by the oxygen of the air blown from the tuyeres through the metal

2. Reaction of the ferrous oxide with silicon and manganese to form their respective oxides, and to reduce the ferrous oxide to iron

3. Combination of ferrous oxide with silicon oxide to form iron silicate, and combination of manganese oxide with silicon oxide to form manganese silicate

4. Reaction of ferrous oxide and carbon to form iron and carbon monoxide

At the end of the blow the converter is tilted and the steel is poured into a ladle, where any required additions are made. Addition of manganese in some form is necessary to deoxidize the ferrous oxide that remains in the metal at the end of the blow, and thus convert it to metallic iron. Spiegeleisen, which contains about 8 per cent manganese, originally was used for the purpose, but it is now more economical to use ferromanganese, which contains about 80 per cent manganese. Because the greater part of the carbon is removed from the metal during the blow, it is necessary to recarburize it to the extent required for the grade of steel which is being made. A part of the required carbon is obtained from the ferromanganese, which contains about 6 per cent carbon. The remaining carbon requirement sometimes is obtained by adding carbon in the form of coke dust, but usually it is more economical to add molten pig iron.

The time required for the blow depends mainly upon the silicon content of the iron used, but usually it is between 10 and 20 minutes. The speed with which steel can be made is one of the advantages of the Bessemer process. This is particularly important when relatively small amounts of several different varieties of steel are required.

Basic Oxygen Process (Basic Bessemer)

Description of the Furnace

The basic oxygen steelmaking process is comparatively new. It is carried out in a cylindrical furnace lined with basic refractories. The refractory lining is approximately 38 inches thick, and usually consists of three sections: a permanent lining of burned magnesite brick, a working lining of pressed tar-dolomite brick, and a layer of rammed mixture of tar-dolomite between the permanent lining and the working lining. The life of the lining is estimated as between 300 and 400 heats. About 5 days are required to reline the furnace. The bottom of the furnace remains in good condition indefinitely, but the body lining wears evenly on the side walls in the reaction zone. The furnace differs from the conventional converter in that it has no tuyeres, wind box, blast pipe, or removable bottom. The rated capacity of the furnace is 54 tons of steel per heat; each heat requires about 37 minutes. It is estimated that the capital

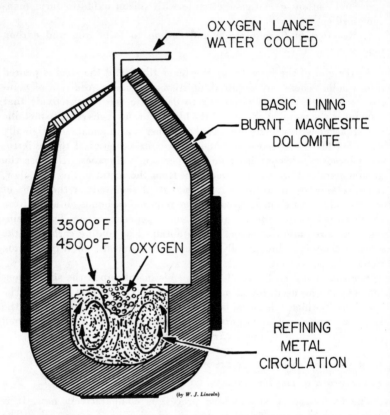

(by W. J. Lincoln)

Figure 48. Cutaway diagram of basic oxygen furnace (tilting modified converter) in original "LD" (Linz-Donowitz) basic oxygen process (Linz, Austria).

investment is only $15.00 per annual ingot ton, as compared with at least $40.00 per annual ingot ton for new open-hearth facilities.

Operation

Charging is similar to practice with conventional converters. It begins with the tilting of the converter to receive the scrap. The scrap amounts

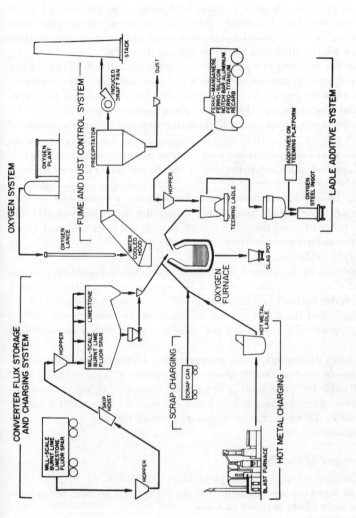

Figure 49. Flow diagram of basic oxygen process.
Courtesy Jones and Laughlin Steel Corporation.

to 10–15 tons per heat, or about 25–30 per cent of all metallics used. It is very important to note here that the principal source of the unwanted alloys in open-hearth steel is purchased scrap; the presence of these alloys decreases ductility. Since a lower ratio of scrap is used in the basic oxygen process, steel made by this process has a lower alloy content than has open-hearth steel, and therefore greater ductility.

When the charging of scrap has been completed, about 40 tons of hot metal and small quantities of roll scale and lime are added. The converter is then tipped upright, and a water-cooled lance is lowered to a predetermined height above the surface of the molten metal. The tip of the lance directs a vertical jet of high-purity oxygen, under a pressure of 100–150 psi, at the surface of the molten bath. The thermochemical reactions that take place refine iron to steel. First, ferrous oxide (FeO) is formed, part of which enters the slag, while the remainder diffuses through the bath. Carbon monoxide (CO) gas forms immediately and starts a vigorous boiling action throughout the bath. As a result, the refining actions are accelerated. Since the temperature approaches closely the boiling point of iron, the solubility of the oxygen increases, and accelerates the diffusion of the ferrous oxide throughout the bath, producing more rapid oxidation.

Completion of the process is indicated by a clearly visible drop in the flame at the mouth of the converter. The oxygen lance is withdrawn, and the converter is tilted to a horizontal position to permit skimming off the fluid slag. After the slag has been skimmed off, the converter is rotated to the opposite direction, and the steel is poured into a conventional ladle.

The gases driven out of the converter are drawn by fans into water-cooled hoods located directly above the converter, at a rate of more than 200,000 cubic feet per minute. The gases are sprayed with water to remove the heavier particles and to lower the temperature from 3000°F to less than 500°F. Electrostatic precipitators recover the fine dust, which is sintered and reused.

Advantages of the Process

1. Charging is simple, and requires about 3 minutes.

2. The scrap can be obtained from the plant itself because the process requires only 25–30 per cent of scrap.

3. The use of dust precipitators keeps the surrounding area extremely clean.

4. The process is not dependent on the chemical analysis of the iron.

5. Since the steel produced is low in nitrogen, phosphorus, and sulfur, it has high purity and ductility.

Chemistry of the Process

Carbon is eliminated as in the open-hearth process.

Manganese content is usually higher than in the open-hearth process; the percentage is closely related to the amount of manganese in the molten iron used to charge.

Phosphorus content is approximately 0.01 per cent, which is equivalent to that in good open-hearth practice.

Sulfur removal is as good or better than in the open-hearth process, because of more vigorous action of the bath, higher operating temperature, and elimination of fuel as a source of sulfur.

Nitrogen content is 0.004 per cent or less, since the refining agent contains virtually no nitrogen.

Oxygen content is usually about 0.04 per cent, which is less than that of normal open-hearth steel.

Questions

1 What changes in composition take place in the conversion of pig iron into steel?
2 What is the fundamental chemical principle of the Bessemer process?
3 What causes the rise in the temperature of the steel during the blow?
4 What undesired elements are not eliminated in the Bessemer process?
5 Why is it necessary to deoxidize and recarburize at the end of the blow?
6 State three advantages and three disadvantages of the Bessemer process.
7 Why has the tonnage of steel produced by the Bessemer process decreased in the past decade?
8 State five uses of Bessemer steel.
9 Why is sulfur added sometimes to Bessemer steel?
10 Describe the basic oxygen process for steelmaking.
11 What are the advantages of the basic oxygen process?
12 Explain the chemistry of the basic oxygen process.

10 / The Open-Hearth Process

While Kelly and Bessemer were devoting their efforts to perfecting their process to a point at which steel could be produced on a commercial scale, another group of metallurgists was endeavoring to manufacture steel by the open-hearth process. This process was destined eventually to outstrip the Bessemer process with respect to annual production. The pioneer of the open-hearth group was an English metallurgist named Heath, who attempted to make steel in a reverberatory puddling furnace which was heated by radiation reflected from the roof and sides. Heath was unsuccessful because the available heat input was insufficient to keep the refined metal in the molten condition and it became pasty. Little progress was made toward finding a method of supplying the required additional heat until Siemens invented the regenerative system in 1860; by applying it to the puddling furnace he made the open-hearth process workable.

There are two methods of making steel in the open-hearth furnace, the acid and the basic. The methods differ with respects to the furnace lining and slag, the materials charged in the furnace, and in some of the chemical reactions which take place during the refining of the metal.

About 60 per cent of the charge for the acid open-hearth furnace consists of scrap which is low in phosphorus and sulfur; the pig iron charged is also low in phosphorus. No other material is included in the initial charge, but small amounts of limestone are added later to thin down the acid slag, and additions of iron ore are made as required to control the carbon content of the melt. Typical maximum tolerances for the undesired constituents in the acid open-hearth process are: 2 per cent silicon, 2 per cent manganese, 0.030 to 0.035 per cent sulfur, and 0.030 to 0.035 per cent phosphorus.

In 1875 Sidney Thomas, another English metallurgist, discovered that phosphorus could be removed from the iron if a basic lining were used in the Bessemer converter, and if limestone or burned lime were added before and during the blow to keep the slag basic. This discovery soon led to the use of a basic lining in the open-hearth process. Under normal operating conditions the greater portion of the charge consists of scrap and pig iron which are charged in about equal amounts, with iron ore and limestone making up the balance.

About 90 per cent of all steel produced in the United States is made by the open-hearth process, and the preponderant part of all open-hearth steel is made by the basic process.

Operation of a Basic Open-Hearth Furnace

Charging

Before charging, the taphole is closed with dolomite, and partial heat is applied; the direction of flow of the burning gases is reversed every 20 minutes to insure even distribution of heat in the furnace.

The bottom of the hearth is spread evenly with a small amount of quick-melting scrap in order to prevent the limestone from fusing to the furnace bottom. This is followed by a layer of limestone lumps, and full heat is applied. The limestone is charged before the major portion of the metal charge to permit the full effect of the lime boil to be secured. If it were placed at the top of the charge, it would act as thermal insulation, and would be the first constituent to enter the slag, making it too viscous to permit flushing the preliminary or runoff slag. The limestone is followed by a layer of iron ore and finally by a layer of scrap or of scrap and cold pig iron.

Melting Down

The charge is heated until the scrap becomes hot and partially fused, which requires about two hours. Sufficient molten pig iron to make up the required balance of the charge then is drawn from the mixer into a transfer ladle which is transported by an overhead electric traveling crane to a position in front of the furnace door.

The purification of the metal takes place in three stages: the ore boil, the lime boil, and the working period.

Ore Boil

During the *ore boil* the oxygen of the oxides in the iron ore and in the slag combines with the carbon in the pig iron to generate carbon

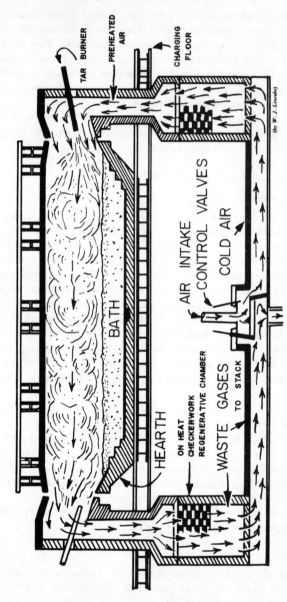

Figure 50. Cutaway diagram of open-hearth furnace and chambers.

monoxide, which burns above the bath. The oxygen also oxidizes the silicon, manganese, and phosphorus which are contained in the iron in small amounts. The oxides of those elements enter the slag, increasing its volume and retarding the rate of heat transfer from the flame to the metal. At the end of about two hours virtually all of the silicon and the greater part of the manganese have been oxidized. When excess silicon is present, it is necessary to use a high percentage of limestone to prevent the acid-forming silicon from attacking the basic refractories of the open-hearth furnace. In a recently developed method of reducing the silicon content of the hot metal, mill scale is added to produce very rapid oxidation of the silicon. This reaction is similar to that taking place in the first minute of the Bessemer blow.

The steady oxidation of the carbon gradually fills the slag with tiny bubbles of carbon monoxide, causing it to froth and rise. This slag, which is known as *flushoff* or *runoff slag*, forms soon after the molten pig iron is added, and is drawn off through a hole at the rear of the furnace and into a slag pot. The flushoff slag removes a portion of the silicon, manganese, and phosphorus; therefore a smaller amount of limestone than otherwise would be required is sufficient to retain the oxides of silicon and phosphorus in the tapping slag.

The ore boil continues for several hours longer, during which time the scrap is melted almost completely. The carbon content of the metal gradually decreases, the temperature of the bath rises, and evolution of carbon dioxide resulting from calcination of the limestone (lime boil) becomes predominant.

Lime Boil

The *lime boil* lasts between two and three hours and takes place in the presence of a strongly reducing slag. During this stage the carbon dioxide released from the limestone on the bottom of the hearth causes the bath to bubble violently; this forces some of the metal to break through the slag covering and to become exposed to the oxidizing action of the flame. During the same period lime rises to the surface and forms a more basic slag which is capable of retaining phosphorus and sulfur.

Elimination of Sulfur in the Open-Hearth Process

Notwithstanding the time required to refine open-hearth steel and the high temperature encountered, the slag basicity and fluidity are the most important factors in the removal of sulfur. Because of the limitations in the chemistry involved, it is uneconomical to reduce the sulfur content of the tapped steel to below the average level of the sulfur content of the metallics charged into the furnace.

Working Period

The working period starts at the end of the lime boil. At this stage practically all impurities have been eliminated except carbon, which has been reduced to about 0.5 per cent above the value at which the steel is to be tapped. Any required alloy additions are made at this time. Skillful management is required during the final operations, which last an hour or more.

Because reduction of the carbon content raises the melting point of the metal, it is necessary to increase the temperature of the furnace correspondingly in order that the steel may be sufficiently liquid at the end of the heat to be tapped and cast into ingots.

Further reduction in carbon content of the steel must be kept under closest control and this requires careful regulation of the properties of the slag. If carbon is removed too rapidly there is danger that the metal may not be hot enough to tap; in such a case it is necessary to *pig up*— that is, to add pig iron to increase the carbon content of the metal. On the other hand, if carbon is removed too slowly it is necessary to *ore down* —to add ore to hasten oxidation of the carbon. To maintain this control, the carbon content and temperature of the metal and the composition of the slag must be determined at intervals.

Tests for Carbon Content

By modern methods of testing, the carbon content of a heat of liquid steel may be determined within two minutes from the time the sample is taken from the furnace. This permits the furnace operator to make adjustments required to hold the carbon content of the heat within a close range. Three tests are made for carbon content: a fracture test, a carbometer test, and a chemical analysis.

For the fracture test a specimen of metal is drawn off, cooled rapidly in water, and broken with a sledge hammer. The appearance of the metal exposed in the fracture permits a fairly accurate estimate of carbon content and shows whether the steel contains phosphorus, which is a matter of particular concern at the end of the lime boil.

The carbometer measures the magnetic properties of the steel; these properties bear a definite relationship to the carbon content. The test is accurate to ± 0.03 per cent and requires only two minutes from the time the sample is poured into a spoon and solidified, to the finished analysis.

Chemical analysis by combustion provides a check on the other tests and also shows the amounts of any alloy metals. Such a test is accurate to within ± 0.005 per cent and requires between 10 and 12 minutes, of which 5 minutes are consumed in fusing the sample to make certain that it is completely oxidized.

Slag Tests

Tests of the slag are no less important than are the tests for carbon, because it is the oxidizing power of the slag, as represented by its content of iron oxide, which results in elimination of carbon from the metal, and it is the basic character of the slag, as determined by its content of lime, which prevents phosphorus from re-entering the metal. An excess of iron oxide in the slag shows that the metal contains too much oxygen, which may cause *inclusions* to form. These are foreign substances such as sulfur and oxygen which are held in the steel, and which must be kept at a minimum in most grades of steel.

The *pancake* test is a rapid method for determining visually the iron oxide content and the basic quality of freshly poured slag. A small sample of slag is poured into a shallow spoon, where it takes the shape of a small pancake. From the color, concavity, regularity of shape, and luster, when compared with standard samples, an experienced operator can estimate accurately both the basicity and oxidizing power of the slag. If a more strongly oxidizing slag is required, it is customary to add limestone or mill scale, a pure oxide of iron which forms on the surface of steel as it cools and which is removed when the steel is rolled in the rolling mill.

Temperature Measurements

The temperature of the bath is measured by one of several methods at various times during the heat.

In one method a long steel bar is inserted into the bath and is moved back and forth until the portion immersed in the metal melts off; the bar is then withdrawn and inspected. If the end of the bar is pointed, a cold bath is indicated; if nicks appear near the end of the bar, the bath is too hot; a clean square end indicates that the temperature is correct.

In a second method a spoonful of molten steel is poured slowly from the spoon. Since the length of time the metal will remain molten depends upon its temperature, it is possible to judge the temperature by observing how the metal flows and by the amount of solidified metal which remains in the spoon.

Various types of pyrometers are used to measure the temperature of the steel either in the bath or as it is being tapped into the ladle or poured into ingot molds.

Tapping

When the tests for carbon content, composition of slag, and temperature have shown all three to be satisfactory, and after additions have been made to control deoxidation of the steel, the heat that has been in the furnace for eight hours or more is ready to tap.

The taphole is opened and the molten steel flows down a spout or *runner* into a large ladle of just sufficient capacity to hold the heat of steel. When the ladle is full, the slag, which is tapped with the steel, overflows into a slag pot. While the metal is flowing down the runner, various ladle additions are made to adjust its composition. One of the most common is ferromanganese, which contains 80 per cent manganese, 12 per cent iron, 6.5 per cent carbon, and 1 per cent silicon; its addition serves to increase the content of manganese and carbon in the steel and to deoxidize it. Anthracite coal often is thrown into the ladle to increase the carbon content. Aluminum sometimes is added to control deoxidation and grain size.

Duplex Operation

The open-hearth furnace is used also for additional refining of steel which has been given initial treatment in a Bessemer converter. The molten metal from the converter is transferred to the open-hearth furnace, where the phosphorus and more of the carbon content are removed by oxidation. The heat is finished by conventional open-hearth procedure. This method of duplex operation effects considerable saving of time as compared with refining the ore entirely in the open-hearth furnace.

Questions

1 Explain the order and manner in which raw materials are charged into an open-hearth furnace.
2 Describe and explain:

    a. Ore boil
    b. Lime boil
    c. Working period

3 What are the functions of the slag blanket?
4 Name the order in which the impurities are removed.
5 What element is eliminated from open-hearth steel which remains in Bessemer steel, and why is it possible to eliminate that element?
6 Describe the tests for:

   a. Carbon
   b. Slag
   c. Temperature

7 What are the properties of basic open-hearth steel?
8 State five uses for basic open-hearth steel.

11 / The Electric Arc Furnace

Electric arc furnaces may be of the direct-arc variety in which the arc passes between a carbon electrode or electrodes and the metal which is being refined, or of the indirect-arc type in which the arc is formed between carbon electrodes above the metal. In the direct-arc furnace the metal is melted and kept molten by direct contact with the arc; in the indirect-arc furnace the metal is heated by radiation. The efficiency of a direct-arc furnace is higher than that of an indirect-arc furnace. One disadvantage of a direct-arc furnace is that the temperature of the metal at the points where the arc strikes is materially higher than the average temperature throughout the melt.

As with the Bessemer and open-hearth processes, steel can be made in the electric furnace by either acid or basic methods; steel that is to be poured directly into molds to make steel castings usually is made by the acid process, while steel that is to be made into ingots ordinarily is made by the basic process. Arc furnaces vary in size from a laboratory model with a capacity of 25 pounds to those with capacities between 50 and 200 tons now being used in refining steel in quantity.

Advantages of the Electric Arc Furnace

Furnaces in which fuel is used to obtain the heat necessary for melting the charge of course require oxygen to support combustion, and the atmosphere is necessarily an oxidizing one. Since an electric furnace requires no fuel, the atmosphere can be made neutral or reducing if desired. No undesired sulfur is added to the melt in an electric furnace as it is in a furnace which burns a fuel containing sulfur, such as coal, coke, oil, or gas. It is possible to reduce the sulfur content of steel in an electric furnace because a strong basic and deoxidizing slag can be maintained. Since such a reducing slag contains practically no ferrous or manganese oxides as do the oxidizing slags of the open-hearth furnace, it increases the stability of the sulfides, such as calcium sulfide, that are insoluble in the metal. The reactions in the electric furnace that aid in control of the sulfur content of the steel are:

$$2\,CaO + CaC_2 + 3\,FeS \rightarrow 3\,CaS + 2\,CO + 3\,Fe$$
$$2\,CaO + CaC_2 + 3\,MnS \rightarrow 3\,CaS + 2\,CO + 3\,Mn$$

Nearly all electric furnaces are constructed so as to tilt in order to facilitate the removal of one slag and the substitution of another of a different type, as, for example, use of an oxidizing slag followed by a reducing slag for more complete purification of the metal. Use of a reducing slag permits addition to the metal of alloying elements which are easily oxidized, and of which considerable amounts would be lost by oxidation if the slag were oxidizing in character. Elimination of sulfur is promoted by a reducing slag.

Because extremely high temperatures can be attained in an electric arc furnace, the metal is more fluid and therefore retains fewer inclusions and is less subject to segregation. The higher temperature also makes possible the addition of a considerably higher proportion of alloy metals and the use of alloy metal of higher melting points than could be handled in the lower temperatures of the open-hearth furnace. Accurate control of temperature is possible in the arc furnace, and the temperature can be held reasonably constant at any desired value.

Operation of an Electric Arc Furnace

Charging

The metallic portion of a cold charge for an arc furnace is steel scrap of known composition, and consists of light and heavy material of such average density that charging can be completed in a single operation. A typical charge might be stated as about 35 per cent heavy, 40 per cent medium, and 25 per cent light scrap.

Production of high-grade steel requires the presence in the bath of an excess of carbon in order that a vigorous ore boil may be secured by reaction with the iron ore. Unless the scrap is high in carbon, this excess, in the form of anthracite coal or broken electrodes, is charged with the scrap.

Limestone sometimes is added with the charge, but because that procedure retards melting, most operators prefer to add it only after the charge is fairly well melted. It must be added soon enough to prevent corrosion of the banks of the hearth by impurities in the metal. Since an excess of limestone increases the electric resistance of the bath and therefore retards the process, it is customary to use as little as possible to provide the required basicity; the amount seldom exceeds 5 per cent.

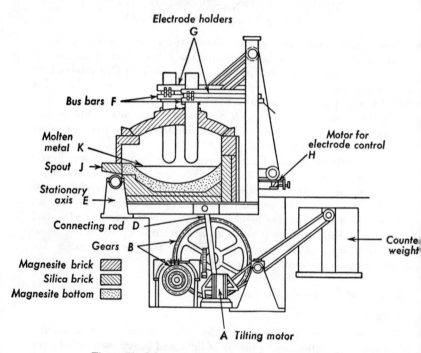

Figure 51. Schematic diagram of Héroult arc furnace.
From *Metals Handbook*, 1948, courtesy American Society for Metals.

Oxidation Period

The reactions are similar to those which take place in the open-hearth furnace except that oxidation is less because no air is being blown through the arc furnace. The silicon, manganese, and phosphorus oxidize at relatively low temperatures and form a slag with the lime. As oxidation proceeds, the temperature is raised to about 2900°F to expedite oxidation of carbon and to make the slag sufficiently fluid for removal of solid inclusions.

Deoxidation Period

After removal of the oxidizing slag any alloy metals and any additional carbon required are added, and the current is increased to raise the temperature to a value at which the second slag is melted quickly. The reducing slag is formed by adding to the bath fluorspar and lime, with any carbon required to react with the lime to form calcium carbide; sometimes combined silicon in the form of silica sand or ferrosilicon also is added. All reducible oxides in the bath are reduced by the calcium carbide to metal which is returned to the steel. The slag also removes sulfur by converting it to calcium sulfide. It is often necessary to add more coke at intervals to maintain the calcium carbide slag.

Addition of Alloying Elements

Highly refined steel sometimes is melted in an arc furnace for the purpose of adding alloying elements in percentages which would be too high to handle in an open-hearth furnace, or alloying elements of which the melting points are higher than the temperatures obtainable in an open-hearth furnace.

The affinity of an alloy for oxygen determines whether it should be added in the electric arc furnace in metallic form or as a ferroalloy. For example, alloys of nickel, copper, and molybdenum are added early during the oxidizing period because they do not oxidize. Oxidizable alloys or metals such as ferromanganese, ferrochromium, ferrosilicon, and aluminum must be added under the reducing slag to avoid any loss as a result of oxidation. However, various mills disagree as to best practice. Some mills add all alloys except ladle additions before the reducing slag is made up.

Tapping

When analysis indicates that the composition of the steel is correct, a final deoxidizer such as ferrosilicon is added and the temperature is raised to about 3000°F. Aluminum may be added for control of grain size. The furnace then is tilted rapidly to permit the steel to run into the ladle before the slag appears. The slag is poured on top of the steel to protect it from oxidation until it is poured into ingots.

Duplex Operation

As a matter of economy, an arc furnace sometimes is used only for final refining of steel that has been refined partially in a Bessemer converter or an open-hearth furnace, or occasionally for the refining of pig

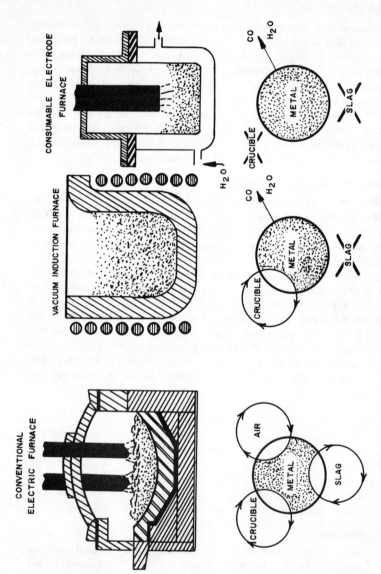

Figure 52. Schematic representation of methods of vacuum melting.
Adapted from *Metal Progress*, courtesy American Society for Metals.

iron. In any case, the metal is supplied in molten condition to the arc furnace, making unnecessary the expensive melting process.

Slagging materials are added immediately in order to produce a slag over the metal as quickly as possible. Pig iron from a blast furnace requires addition of iron ore or mill scale for oxidation of the large carbon content; steel from a Bessemer converter usually requires addition of coke or broken electrodes for recarburization. Steel from an open-hearth furnace usually is free of phosphorus and requires only treatment with reducing slag, although treatments with both oxidizing and reducing slags are more generally used.

Vacuum Induction Process

This recently developed process improves the impact properties, fatigue properties, and stress-rupture properties of a metal at elevated temperatures by permitting very close control of the chemistry of composition. The detrimental gases such as oxygen, nitrogen, and hydrogen are removed from the molten metal during the processing, thus permitting addition of alloys containing large amounts of such reactive elements as titanium, aluminum, and silicon. Certain alloys can be made by this process that can be made in no other way. Impurities are prevented from entering the metal, resulting in greater cleanliness of the metal and improved properties. (*See also* Chapter 25.)

In the vacuum induction process the entire charge is melted in a basic-lined high-frequency induction furnace. The complete cycle—melting and pouring—is carried out under a vacuum of from 5 to 20 microns.

Questions

1 What are the advantages of the electric arc furnace?
2 Name the raw materials used in making basic electric steel.
3 Why is an excess of carbon required?
4 Describe the melting process.
5 Describe the oxidizing period.
6 Describe in detail the deoxidizing period.
7 Why are alloying elements not lost during electric steelmaking?
8 What is duplex steel?
9 Explain the importance of the vacuum induction process.

12 / Ingot Practice

By far the largest proportion of all steel made is cast into ingots for later fabrication by appropriate shaping operations. After molten steel has been tapped into a ladle it is poured into ingot molds, the capacities of which cover a range from several hundred pounds to many tons. The design of the mold and the details of pouring steel into it have important effects upon the quality of the steel.

Segregation

Molten steel, like any other alloy in molten condition, is a nearly homogeneous liquid. If cooling during solidification were sufficiently slow to maintain conditions of equilibrium, the composition of the portion of steel solidifying at each successively lower temperature, and correspondingly the composition of the residual melt, would change constantly, as is the case with every alloy. As the iron-carbon diagram shows, the last portion to solidify of a steel containing iron and carbon only would have a concentration of carbon higher than that of portions which had solidified earlier, and higher than the content of carbon in the original melt. The same principle holds true to greater or less degree for the other alloying elements, and this varying concentration or *segregation* is particularly troublesome with carbon, sulfur, and phosphorus.

To impede segregation and to keep the steel as homogeneous as possible, measures are taken to cool it rapidly below the temperature of solidification, so that thermal equilibrium cannot be reached. This is accomplished through use of ingot molds with thick walls which conduct the heat away and thus chill the steel when it is poured in. The shape and size of the mold also are such as to promote rapid cooling. In general, the use of a small ingot mold is favorable to prevention of segregation, and the finest steels are cast in small ingots.

Dendrites

Formation of large dendrites is unfavorable to the production of steel of adequate strength, and therefore precautions are taken to keep them small. They extend from the walls of the mold at approximately 90 degrees, and where a corner exists, as at the bottom of the mold, their line of interference forms a plane of weakness, called a *plane of cleavage*, at an angle of about 45 degrees with walls and bottom. This effect is avoided by use of ingot molds with rounded corners.

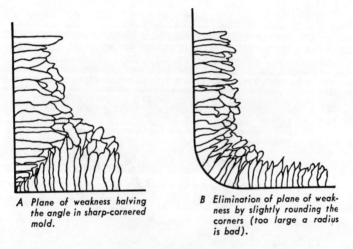

A *Plane of weakness halving the angle in sharp-cornered mold.*

B *Elimination of plane of weakness by slightly rounding the corners (too large a radius is bad).*

Figure 53. Sharp-cornered and round-cornered molds.

Reproduced by permission from D. M. Liddell and G. E. Doan, *The Principles of Metallurgy*, copyright 1933 by the McGraw-Hill Book Co., Inc., New York.

When steel is poured, the first crystals which are formed on the surface of the mold are small and of practically the same composition as that of the melt; these are called *chill crystals*. The crystals which form subsequently tend to build large dendrites if cooling is slow, but if cooling is sufficiently rapid, large numbers of equiaxed crystals of no definite orientation are formed. Rapid cooling is therefore important for production of favorable crystal size as well as for reduction of segregation.

Ingot Molds

Ingot molds are tapered containers made of cast iron; some are big-end-down and some are big-end-up. They are made in various types of cross section, including squares or rectangles with rounded corners, polygons, circles, and fluted patterns. When molten steel is poured into a mold, that portion next to the wall of the mold solidifies rapidly and produces a shell of solid steel which contracts and shrinks away from the wall of the mold; this contraction is sufficient to permit removal of the ingot from the mold after solidification is complete. As the metal contracts, a conical shrinkage cavity known as *pipe* forms at the center of the ingot near the top; this portion of the ingot is discarded because it cannot be rolled subsequently into sound steel.

Figure 54. Ingot mold. Etchant 4 per cent picral. Magnification 60 ×.
Courtesy The British Cast Iron Research Association.

Big-end-up molds are used for the finest steels, because the shape permits the molten steel to solidify from the bottom, with the result that fewer gas bubbles and inclusions of solid matter remain in the usable part of the ingot, and the length of pipe is less.

A *shrinkhead casing*, often called a *hot top*, is a reservoir generally used on big-end-up ingot molds; it supplies additional molten steel to the upper portion of the ingot as solidification progresses, and so reduces the extent of the pipe. A shrinkhead casing is made of a refractory material which serves as thermal insulation to maintain the steel at a high temperature. One type of casing is fixed on the top of the mold and remains there until the ingot is ready for stripping. Another variety has an outside diameter smaller than the inside diameter of the mold, and is held on top of the mold by supporting blocks; when the ingot mold has been filled, the blocks are removed and the shrinkhead casing drops with the steel as the ingot contracts.

Deoxidation Practice

After steel has been tapped into a ladle it still contains dissolved free oxygen and combined oxygen in the form of iron oxide. Unless some deoxidant is added, free oxygen comes out of solution and iron oxide reacts with carbon to form carbon monoxide during the period of solidification of an ingot, with resulting evolution of gas. Steels of highest quality,

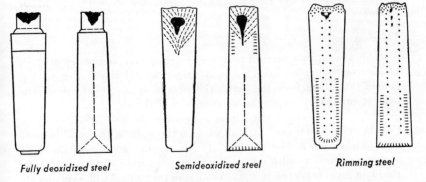

Fully deoxidized steel   Semideoxidized steel   Rimming steel

Figure 55. Condition of steel in ingot molds.
Courtesy Gathmann Engineering Company.

particularly tool steels, forging steels, and nearly all alloy steels, must be as free as possible from the blowholes caused by the bubbles of evolved gas. By use of suitable slags in the furnace and addition of proper deoxidants in the furnace or in the ladle or both, it is possible to produce steel which is almost completely deoxidized. Because such steel solidifies in the ingot mold with little or no evolution of gas bubbles, it is called *killed* steel. Completely deoxidized steel shows only slight segregation and has a favorable physical structure with relatively few large dendrites, which are broken up by subsequent shaping work. Fully deoxidized steel is poured into big-end-up ingot molds provided with shrinkhead casings, to keep to a minimum the portion of each ingot which must be discarded.

Semideoxidized or *semikilled* steel is less homogeneous in structure than is completely deoxidized steel, and contains more inclusions and blowholes. It is partially deoxidized in the ladle by the addition of manganese. Steel of this kind is used largely for heavy structural shapes and plates because it is cheaper to produce than is killed steel.

A *rimmed* steel is one to which relatively little deoxidant is added and in which consequently a considerable volume of gas is evolved and rises to the surface as the ingot solidifies. An ingot of rimmed steel has a surface shell of solid metal of low carbon content, an intermediate section which contains blowholes of various sizes, and a core of metal in which the contents of carbon, sulfur, and phosphorus are high. A rimmed steel usually is soft, with a carbon content of about 0.20 per cent or less. Rimmed steel has a good surface finish and a high degree of ductility which makes it suitable for making thin sheets, tinplate, wire, and similar products.

Teeming

Molds are set in a vertical position on heavy cast iron plates or *stools* which usually are mounted on small rail cars, and the molten steel is poured into them from the ladle; this operation is called *teeming*. The ladle is carried by an overhead crane, and if the ingots are to be top-poured, each in turn is filled from a spout in the bottom of the ladle. Care must be taken not to splash steel on the walls of the mold because any splashed steel oxidizes promptly and usually adheres to the ingot as a surface defect called a *scab*. To facilitate pouring without splashing, the ingot mold is necked-in at the bottom so that a pool will form quickly and cushion the stream of molten steel. The temperature at which steel is teemed is of importance, because a temperature which is either too high or too low may cause the formation of surface defects. To avoid possibility of splashing, ingot molds sometimes are filled from the bottom; the molten steel rises from a channel into which it is poured through an opening called a *gate*. The inside surface of an ingot mold should be smooth and free from crevices into which molten steel can flow. Various washes sometimes are applied to the walls of a mold to prevent any splashed metal from adhering, but they are unnecessary if temperature of both molds and steel, and the rate of teeming, are kept at the proper values.

Stripping and Preheating

After solidification is complete, an ingot is *stripped*. Big-end-up ingots are either pushed or pulled from the molds according to the type of

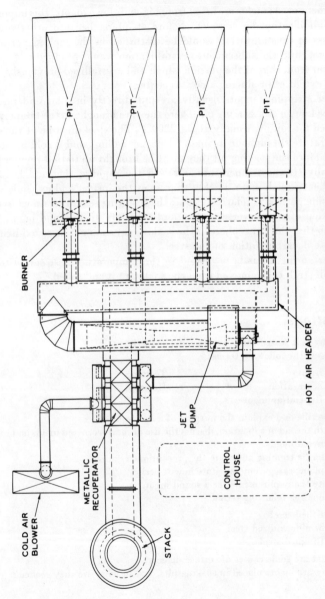

Figure 56. Schematic diagram of soaking pit with jet-pump recuperative system.
Courtesy Surface Combustion Corporation.

equipment used; molds are lifted from big-end-down ingots. It is important that solidification be complete before stripping because otherwise the progress of crystallization would be disturbed in the molten portion of the ingot, and the subsequent crystallization would produce a structure different from that of the portion which had crystallized previously, even to the extent of producing a porous center.

After stripping, ingots usually are put directly into an underground furnace known as a *soaking pit*. There they are brought to the temperature required for rolling, usually about 2200°F. The steel must be at the same temperature throughout its mass for proper rolling, and if cold ingots are charged into the soaking pit they must be brought up to that temperature gradually. If the temperature is raised too rapidly, the outside of the ingot heats much more quickly than does the inside, and expands while the inside portion is relatively cool; the resulting severe strain causes the formation of a transverse fissure, and therefore an unsound ingot. Care is taken that the atmosphere of the soaking pit shall not be so oxidizing as to cause decarburization of the steel.

After an ingot has been heated to the temperature best suited to the grade of steel, it is removed from the soaking pit and rolled.

Questions

1 Why is it stated that "steel begins in the ingot"?
2 Define the following terms:

    a. Ingot   d. Teeming
    b. Segregation   e. Hot top
    c. Big-end-up ingot

3 Describe and explain the purpose of a hot top.
4 With the aid of a diagram, discuss the use of sharp-cornered molds and round-cornered molds.
5 Describe teeming and state the precautions required.
6 Explain segregation and state how it occurs.
7 State the requirements for a sound ingot.
8 Define the following terms:

    a. Killed steel
    b. Semideoxidized steel
    c. Rimmed steel

9 What are some causes of surface defects?
10 Why are ingots placed in soaking pits, and in what are they soaked?

13 | Working, Shaping, and Joining of Iron and Steel

Hot Rolling

Hot rolling to sections of various sizes and of uniform cross section is the first step in the preparation of steel for use. An important effect of hot rolling is to break up the dendrites and large grains of an ingot into smaller grains, and to flatten them and elongate them in the direction of rolling. This results in a fiberlike structure which gives steel different physical properties in longitudinal and transverse directions. Inclusions such as oxides, silicates, and sulfides also are flattened and elongated to form *stringers*. Blowholes are reduced in size or are closed entirely by rolling.

Steel is rolled between heavy chilled, cast, or forged steel rolls which revolve in opposite directions, and by successive passes gradually reduce the thickness to the desired value. Portions of the ingot which include pipe or other bad defects are removed and discarded early in the rolling process.

Cold Rolling

Because hot rolling of sheets less than 0.05 inch in thickness is not economical, thin sheets are produced by cold rolling.

Hot-rolled strip is pickled by passing it through a bath of dilute acid, and is washed successively in cold water, hot water, and steam, to remove the surface scale of oxide formed during hot rolling. It then is given a light coat of oil to prevent reformation of oxide, and is cold rolled to the desired thickness, usually in three-high tandem mills. The result is a thin sheet of steel with smooth surfaces. Cold rolling increases the tensile strength and yield strength of steel, but hardens it excessively. To restore ductility and to make it suitable for deep drawing, cold-rolled sheet is softened by annealing; this is done in a nonoxidizing atmosphere to prevent surface oxidation.

Tinplate used for making cans for food is sheet steel of light gage, coated with pure tin. Proper adhesion of the tin requires particularly

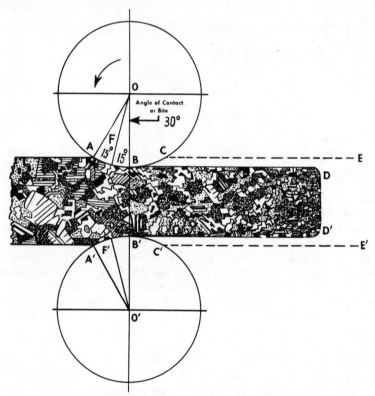

Figure 57. Action of rolls on a piece of plastic metal.

From *Metal Quality*, courtesy Drop Forging Association.

careful preparation of the sheet steel. In addition to the cold-rolling operations ordinarily performed on sheet steel, that intended for tin-plating is etched to assist adhesion and is given additional treatment in pickling baths to provide a surface which can be wetted by molten tin. The sheet is passed through molten tin, which is spread evenly by rolls.

The tin forms on the steel sheet a thin layer of iron-tin alloy that is covered by a thick layer of pure tin; the result is a firmly bonded surface of pure tin.

Wire and Tubular Products

Wire drawing consists of drawing a rod through a series of successively smaller dies until it conforms to a predetermined shape. During the drawing operation, plastic deformation takes place.

Lubricants must be used during metal drawing to permit easy passage of the metal through the dies; to permit the dies to resist the high

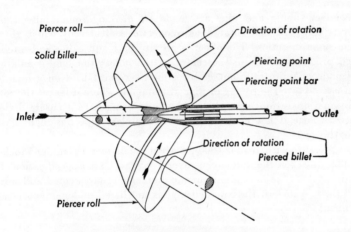

Figure 58. Method of producing seamless tubing.
Courtesy Bethlehem Steel Company.

temperatures reached; to prevent the dies from welding to the metal; and to permit rapid production of smooth-finished work. The lubricants must remain stable under the high pressures and high temperatures encountered during drawing; they should be slippery enough to reduce the coefficient of friction, thereby minimizing the heat generated; they must adhere to the moving surfaces and spread evenly over them; they must be corrosion-resistant. Drawing compounds are applied by brushing, swabbing, or spraying. Lubricants used include mineral oils containing special additives, lubricating greases, soaps, water-soluble pastes, and pastes compounded from water-soluble and oil-soluble compounds.

Butt-welded pipe is made from sheet steel strips known as *skelp*. In the most modern mills, butt-welded pipe is made by a continuous process from skelp supplied in coils of considerable size. The skelp is shaped into pipe and is welded as it passes through a series of six pairs of rolls of appropriate shape.

Lap-welded pipe differs from butt-welded pipe in that the edges are beveled and overlapped before welding. Reheated skelp is shaped first into cylindrical form in a bending machine by grooving rolls, the grooves of which determine the outside diameter, and is then heated in a furnace to welding temperature. From the furnace it is passed through a pair of grooved rolls that press it against a bullet-shaped anvil and thus weld the seam.

Seamless tubing and pipe are made by piercing round billets. The billet is squeezed and rotated between two conical rolls which rotate in the same direction. The effect is an alternate squeezing and bulging which produces a hole in the center of the billet; the hole is enlarged and made of uniform diameter by a piercing point held on a mandrel and inserted into the tubing or pipe at the point where it leaves the rolls. Additional operations smooth the tube and bring it to accurate dimensions. Because it has no weld or seam, seamless pipe has high tensile strength.

The pressure withstood is 1000 psi when the diameter is under 5 inches, and 800 psi when the diameter is over 5 inches. Under specifications for pressure tubes, the maximum outside diameter of lap-welded and seamless tube is 4 inches. Butt-welded pipe is not listed under pressure-tube specifications.

Forging

Forging is a development of the craft of the blacksmith who forged hot iron by hammering it into desired shapes; modern forging includes shaping of hot metal by either hammering or pressing. Forging retains and even improves the desirable qualities which were imparted to steel in the rolling mill. Many forging operations are conducted by use of closed impression dies into which plastic hot metal is forced by the impact of a hammer or by pressure.

The simplest type of impact forging equipment is the board drop hammer. The lower portion of the die is fixed to an anvil, while the upper portion is positioned accurately above it and is fixed to a ram which in turn is attached to maple boards which lie between a set of mechanically driven rolls. When pressure is applied to a pedal, the rolls squeeze against the boards and revolve in opposite directions, thus raising the boards, ram, and upper portion of the die. If pressure on the pedal is maintained, the rolls separate when the ram has reached the highest point of its stroke, and the ram drops by gravity, hammering the upper portion of the die into the work, and thus forcing the hot metal into both portions of the die. The hammer continues to operate in this fashion as long as pressure on the pedal is maintained; if pressure is released when the hammer has reached the top of the stroke, it is held there by clamps until released by another pressure of the pedal. A gravity drop hammer operates at a normal velocity of about 14 feet per second.

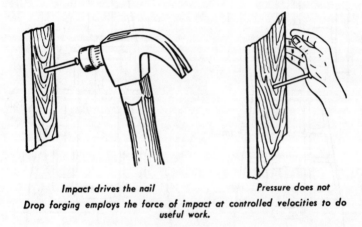

Impact drives the nail   Pressure does not

Drop forging employs the force of impact at controlled velocities to do useful work.

Figure 59. Impact and pressure.

Courtesy Chambersburg Engineering Company.

A similar piece of apparatus is the steam or air drop hammer, which is operated by steam or air pressure. The hammer is connected to a piston working in a cylinder at the top of the frame so that pressure can be applied on the downstroke, increasing the impact over that which could be secured by gravity alone. A hammer of this kind can develop a useful velocity as high as 30 feet per second, and therefore exerts greater forging effect than can a gravity drop hammer.

Pressure forging is accomplished by squeezing instead of by hammering. Small forgings are made in closed impression dies in vertical presses operated by mechanical pressure; a similar piece of equipment is the

IMPACT CONTROLLED FOR FORGING

$$\text{ENERGY AVAILABLE} = \frac{WV^2}{2g}$$

| GRAVITY DROP | ACCELERATED DROP |

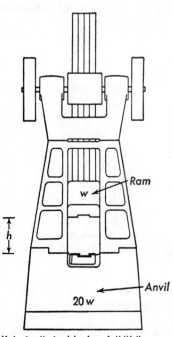

Velocity limited by free fall "h."
Forging effect influenced by ratio of anvil weight to ram weight.
Normal velocity approximately 14 feet per second.

Velocity accelerated by steam or air.
Anvils made higher ratios when not in one piece.
Normal useful velocity approximately 30 feet per second.

Figure 60. Drop hammers.
Courtesy Chambersburg Engineering Company.

Figure 61. Double-acting steam drop hammer.
Courtesy Chambersburg Engineering Company.

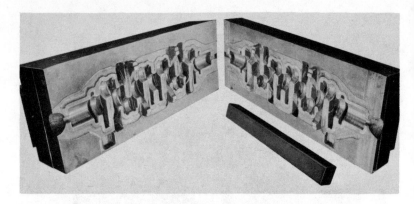

Figure 62. Dies for forging crankshaft for Allison aircraft engine.
From *Metal Quality*, courtesy Drop Forging Association.

horizontal forging machine which produces large or small forgings. The largest pieces, such as shafts for steamships and hydroelectric machinery, are forged directly from ingots in a hydraulic press in which extremely high pressure can be exerted.

A large proportion of sheet steel is shaped into finished articles by pressing between dies, a process that is actually cold forging. Spinning is a shaping method in which metal is held in a revolving chuck and is shaped over a form by pressure of a tool held against it. Spinning, bending, and stamping of pieces in a cutting die all may be considered pressing operations.

Figure 63. Forged crankshaft for Allison aircraft engine.
From *Metal Quality*, courtesy Drop Forging Association.

Cold Extrusion of Steel

Cold extrusion can be carried out on most grades of steel with carbon content up to 0.6 per cent, when the steel has been stabilized by full annealing or spheroidization. Such steels have comparatively low deformation pressures and less tendency to work harden, whereas steels of higher carbon content work harden rapidly, limiting the amount of cold

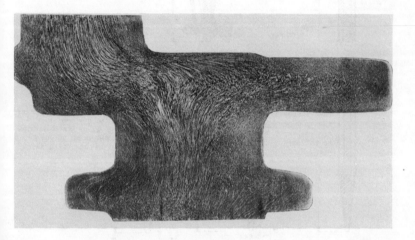

Figure 64. Macro etch illustration of cross section from Allison aircraft-engine crankshaft.

From *Metal Quality*, courtesy Drop Forging Association.

work that is feasible in a single operation. Cold extrusion is carried out by either backward or forward processes.

Backward Extrusion

A descending punch strikes a cut or sized slug contained in a closed die, forcing the displaced metal to flow up around the punch in a direction opposed to the direction of punch travel.

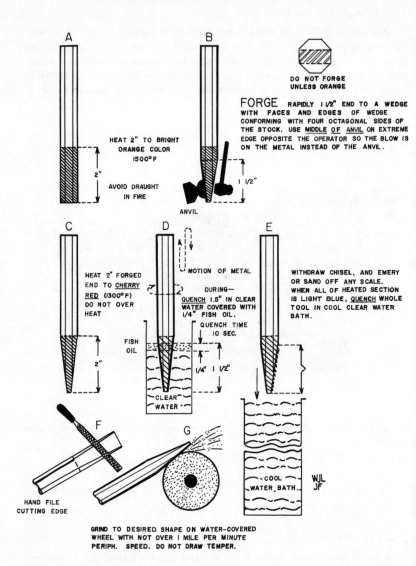

Figure 65. Forging, hardening, and tempering (with flat cold chisel using $\frac{1}{2}$ in. by 6 in. octagonal stock; carbon 0.80 to 0.90).

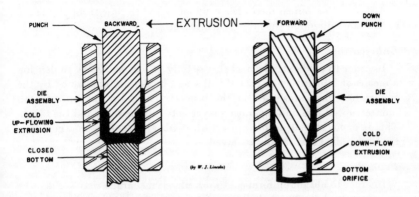

Figure 66. Two basic cold-extrusion movements.

Figure 67. Die after extruding one bar without glass lubricant.

Courtesy Vanadium Corporation of America.

Figure 68. Die after extruding 32 bars using glass lubricant.

Courtesy Vanadium Corporation of America.

Forward Extrusion

A descending punch forces the confined metal to flow through the die in the same direction that the punch travels.

Lubrication

In order to cold extrude steel successfully, it was necessary to develop a means of lubrication so that the dies and holder are unaffected by the tremendous pressures used in the process. The lubricant provides an uninterrupted film and separating layer between the tools and the blank. This is a metal soap formed integrally with a phosphate coating. It is readily applied and readily removed.

Procedure

In order to obtain uniform cold work physical characteristics, the piece is first rough shaped to such dimensions that the final cold work operation, or a series of cold work operations on different areas, will bring the part close to finish tolerances and uniform hardness. The starting blank may be solid or hollow, but it must be thoroughly cleaned of scale, surface oxides, grease, and soil, so that bare metal is exposed. The work is usually process annealed, between $1150°F$ and $1250°F$, after each cold operation. This results in no grain growth, but does result in a recrystallization, which develops a fine-grain structure ideally suited for further heavy cold working. The finished part is stress relieved at a temperature well below the critical temperature, to prevent development of stress cracks in service.

Advantages of Cold Extrusion

1. It permits use of low-carbon steels to obtain the mechanical properties that are ordinarily obtainable only through use of high-alloy steels when conventional shaping methods are used.

2. It is useful for shaping other carbon steels, many alloy steels, stainless steels, and special steels.

3. Very little scrap is lost in the process.

4. Because of the fine finish and close tolerances obtained on the finished part, little trimming or final machining is required.

5. Fewer operations are required than in conventional shaping methods.

6. Cold extrusion improves mechanical properties without the need for heat treating.

7. Tools can be changed rapidly.

8. Complex parts can be produced quickly.

9. Since the metal follows the shape of the part, the extruded part can be made stronger and of smaller cross section than a part produced by machining.

Hot Extrusion of Steel

Although the extrusion process dates from 1797 when Joseph Bramah of England patented a press for the manufacture of lead pipe, the application of extrusion to steel dates from after World War I. In Germany,

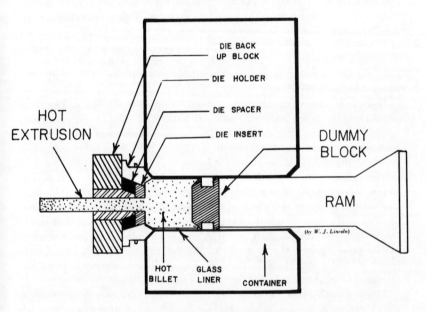

Figure 69. Hot extrusions (Ugine-Sejournet Process).

Mannesman developed an extrusion process for the manufacture of steel pipe, utilizing a vertical mechanical press to pierce and extrude in one motion. For lubrication of the billets and dies, a mixture of tallow and salt was employed. In 1937, steels were extruded in England, employing the conventional horizontal hydraulic extrusion presses as are used in the nonferrous field, and the same type lubricants for the billets and dies as were used by Mannesman. After World War II the use of glass as a lubricant for the extrusion of steel was patented by the Ugine-Sejournet group.

Materials which have been extruded by the hot extrusion processes, as applied to steel, are: carbon and low alloy steels, corrosion- and heat-resistant steels, high-temperature steels and alloys, titanium and its alloys, zirconium and its alloys, and a number of the "exotic" materials used in the field of nuclear energy.

In the hot extrusion of steels, the billets are not scalped (extruded in such a manner that a thin shell of metal representing the outer surface of the billet is left in the container of the press) as they are in the non-ferrous extruding process. For this reason, the billets must have careful surface preparation prior to heating for extrusion. This preparation can be accomplished either by turning or grinding. The surface-conditioned billet of the proper diameter for the particular press tooling is cut to a length yielding a multiple of the desired length, or weight, of product. Billets for solid shapes are solid, but billets for tubing and pipe are either drilled or pierced before extruding.

Heating is done with salt baths, by induction, or in a furnace. The most popular methods are induction and salt baths because of their speed and surface protection from oxidation. The temperatures used vary with the materials to be extruded and may be as high as 2350°F for some corrosion-resistant materials.

The lubrication of the billets varies according to the system of hot extrusion employed. When using the French (or glass lubrication) system, the billet is rolled over a sheet of fiber glass or a layer of powdered glass, and, as in the case of tubular extrusions, powdered glass is placed inside the hole in the hot billet. A quantity of glass is placed in the container of the press against the face of the die and the hot lubricated billet is inserted in the container. If the tallow and salt lubrication is used, the lubricant is swabbed into the container and on the die.

The basic difference in the hot extrusion of steel tubular products from the extrusion of solids is in the press setup. When extruding tubular prod-ucts, a mandrel is attached to, or inserted in, the ram or stem of the extrusion press. When the press closes, this mandrel passes through the hole in the billet and through the die opening so that the die has an annu-lar opening to form the tube or pipe. When extruding a solid shape, the mandrel is omitted, and the solid ram, or stem, of the press pushes the solid billet through the die. The full cycle of the extrusion process as applied to steel is approximately as follows:

1. Lubricate die and container or die and billet.
2. Load billet into container.

3. Close press.

4. Extrude by the use of high pressure in the main ram cylinder to develop a total force on the main ram that will vary with the press capacity to as high as 12,000 tons. (The press force is sufficient to overcome the friction of the metal and to accomplish the deformation required to form the extruded product.)

5. After the press has reached the end of its stroke and has left a short section of the original billet, or discard, the ram is retracted and the press is opened.

6. The discard is separated from the extruded product by sawing or shearing, and the press is cleared for the next billet.

After extrusion, the product, whether a solid bar or a tubular section, is usually cleaned to remove the lubricant. When glass has been used, it is removed by either a hydrofluoric acid bath or a molten caustic bath. After cleaning, the extrusion is annealed and descaled if the material requires it. After inspection, the extrusion may be further worked, usually by cold drawing, etc., and straightened.

A number of advantages of the hot extrusion process have been found.

1. Shapes not possible to produce by rolling can be made by extrusion.

2. Shapes can be produced that will eliminate machining operations and conserve material.

3. It lends itself to the production of small orders of rollable shapes, as changes from one size or shape to another can be accomplished with virtually no·break in production.

4. Materials difficult or impossible to roll can be extruded.

5. The directionality phenomenon of mechanical properties is minimized with extruded products.

Casting

Casting is a method of shaping metal by pouring it while in the molten state into prepared molds that usually are made of sand when iron or steel is to be cast. Neither hot nor cold working is involved in the process.

A casting has approximately the shape and dimensions required in the finished article, and needs only a moderate amount of machining to complete its shaping. An article of intricate shape therefore can be cast and finished at reasonable cost, whereas the number of operations that would be required in forging the same article would make the cost prohibitive; some shapes are so intricate that forgings would be impractical in any event. Pieces of great size usually are formed by casting, also for reasons of economy.

Cast metal lacks the desirable directional properties, and thus has less resistance to repeated stresses than has forged metal. It usually contains a greater number of blowholes and other discontinuities, and is more porous than forged metal. A steel casting of a suitably chosen grade of steel, properly heat treated after casting, however, has strength and toughness approaching those of a forging. Steel to be used for casting is melted in some foundries in open-hearth furnaces, but more often in electric arc furnaces in order that no oxidation shall occur. Electric induction furnaces also are coming into use for the purpose.

Continuous Casting

The continuous casting process will eliminate ingot casting, soaking pits, and the blooming mill because the tapped metal will be cast directly into a casting, equivalent in cross-sectional area to a bloom. Other advantages of the continuous casting method over conventional casting methods are:

1. It produces a sounder steel because the heat withdrawal pattern is obtained by applying regulated cooling below the mold while the section still has a liquid core.

2. As a result of fast cooling, the casting has a fine and uniform crystalline structure which has a minimum of ingotism and segregation.

3. The mold used during continuous casting approximates in behavior the big-end-up mold used to cast the best quality steels. Since the continuous process supplies an infinite or constant hot top for progressive feeding, it contributes to the elimination of pipe.

4. The surface of the continuously cast ingot is superior to the surface of an ingot obtained by conventional practice in respect to checks, scabs, and entrapped slag. The process offers less opportunity for dirt or other foreign matter to enter the casting.

5. Continuously cast steel ingots have no taper and may be as long as desired.

6. A small ingot is generally conceded to give the best quality product. By their nature, continuously cast steel ingots are small in cross section.

The Mold

The most important factor to be considered in continuous casting is the mold. It should have high heat conductivity, and the mold metal should not be wetted by the liquid steel, that is, the liquid steel in the mold must show a strong negative meniscus (like mercury in a thermometer). The surface of the mold must remain smooth at all times during the casting operation. Since any oxide present during the casting will promote the wetting of the mold by the steel, it is necessary to add a small amount of hydrocarbon into the mold to combine with and thus eliminate free oxygen above the metal pool.

Shortly after the shell of the billet has been formed, the billet shrinks away from the mold. As a result, the metal is in direct contact with the surface of the mold for only a few inches, and it is only within this short

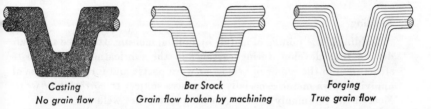

| Casting | Bar Stock | Forging |
| No grain flow | Grain flow broken by machining | True grain flow |

Figure 70. Comparison of grain in a casting, a machined bar stock, and a forging.
Courtesy Chambersburg Engineering Company.

distance that it can lose heat to the mold by direct contact. While in contact with the mold the steel must gather structural strength as rapidly as possible, thus enabling it to withstand the action of relative movement past the mold surface. This is the critical period for the formation of the surface because an initially well-formed billet will remain so unless damaged later on. Proper cooling of the mold is essential. The more rapid the casting rate, the better the surface.

Influence of Slag

Continuous casting requires a slag-free steel because any slag that enters with the metal will float on the surface of the liquid pool and eventually find its way down between the mold wall and the casting, thereby delaying freezing and causing shrinkage cracks large enough to ruin the billet.

Procedure

To eliminate the slag the steel is poured first from the electric induction furnace into a *tundish*, designed to drain out the slag, and thence into the water-cooled mold. (As a rule the tundish is heated.) The slower the flow of the metal through the tundish, the better the elimination of the slag. Argon, a nonreacting gas, is introduced above the liquid pool in the mold to prevent oxidation, since oxidation products cause wetting and fouling of the mold surface. Argon was selected because of its high density and because it forms no compound with iron.

Below the water-cooled mold the casting passes through an insulated chamber which controls the speed of further cooling. Below the insulated chamber is the withdrawal mechanism, pinch rolls, which regulate the speed of movement of the steel billet. The billet is now cut to specified length by an oxyacetylene torch, and the cut-off section of the billet is lowered to a horizontal position by a cradle arrangement.

Welding

Welding is the joining of metals by mutual melting. It is used to repair defects that develop during casting; in the fabrication of composite structures; in the repair of broken or worn parts; and for the purpose of applying extra metal, especially to surfaces subject to corrosion or wear. Most of the commonly used engineering metals are weldable.

The techniques of both process metallurgy and physical metallurgy are involved in welding. An example from process metallurgy is thermit welding. If nickel oxide is added to the thermit mixture of iron oxide and aluminum, it is reduced to metallic nickel, which alloys with the molten metal contained in the crucible. The post heat treatment of a welded article is an example of the use of physical metallurgy in welding; it may either soften or harden the welded article.

Welding involves very complex metallurgical reactions because of the speed with which it takes place and the exceedingly high temperatures reached. During welding there are usually greater and more violent temperature changes than in any other fabricating process. The peak temperature is highly localized, resulting in a sharp temperature gradient between the molten weld metal and the base metal only a fraction of an inch away.

There are three groups of weldable steel:

1. The low-carbon low-alloy steels that contain less than 0.15 per cent carbon are readily weldable.

2. Steels that contain 0.35–0.50 per cent carbon, and low-alloy steels that contain 0.15–0.30 per cent carbon, can be welded with proper precautions.

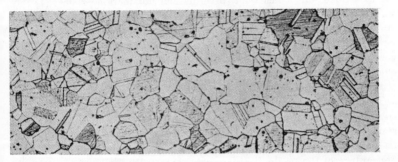

Figure 71. Structure of alpha brass base metal. All the grains are of the same composition.

From *Braze Welding of Iron and Steel*, 1957, courtesy International Acetylene Association.

3. The so-called high-carbon steels that contain more than 0.50 per cent carbon, and alloy steels that contain more than 0.30 per cent carbon, can be welded with difficulty, because of their high carbon content.

The making of a weld involves (1) the melting of the edges of the joint and of any filler metal that is used, and (2) the heating and the subsequent cooling of the zone of the base metal adjacent to the weld. Melting is followed by solidification, which forms a single weld structure known as a *weldment*.

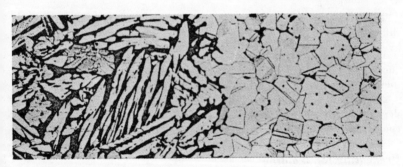

Figure 72. Welded joint between alpha brass base metal (*right*) and alpha plus beta weld metal.

From *Braze Welding of Iron and Steel*, 1957, courtesy International Acetylene Association.

The welded joint consists of three distinct but merging zones: (1) the weld metal, which is that portion of the metal that has been in the molten state, and consists of a mixture of filler material that has been deposited and the base metal that melted during welding; (2) the heat-affected zone, which did not melt but was heated to a temperature high enough to change its original microstructure and properties; and (3) the base metal that remained unaffected by the welding cycle.

To make a proper weld, the following weld joint properties must be considered: (1) the mechanical properties of strength, ductility, impact resistance, and elasticity; (2) the physical properties of electric conductivity and thermal conductivity; (3) corrosion resistance; and (4) the appearance of the product.

The physical properties of a welded joint may or may not be the same as those of the base metal, for the following reasons:

1. When filler metal is used, it may or may not be of the same composition as the base metal. Alloy filler metal generally is used where a tensile strength exceeding 70,000 psi is required. When high-alloy steels are to be welded, the filler metal should have as nearly as possible the same composition as the base metal. When no filler metal is used, as in resistance welding, the properties of the welded joint vary, depending on the design of the joint, the heat of welding, etc.

2. Since the deposited weld metal acts as a casting, its microstructure differs from that of the base metal.

3. The microstructure of the base metal adjacent to the weld is altered by the heat developed during welding.

The heat generated during welding affects the welded joint in two different ways:

1. The microstructure determines such properties as strength, ductility, shock resistance, and corrosion resistance. These properties depend on metal composition, temperature reached, and cooling rate.

2. The amount of distortion and shrinkage stresses caused by the heat depends on the shape of the joint, the rate at which it was heated and cooled, and the time it was held at maximum temperature.

Preheating for Welding

Preheating the workpiece before welding reduces the differential between the temperature of the base metal and that of the weld. *General preheat* refers to preheating the entire workpiece; *local preheat* refers to preheating only the vicinity of the weld. General preheat is the preferred method because it minimizes localized stresses.

The principal benefits derived from preheating are those that result from slower cooling of the weld zone, particularly at the low temperatures in the martensitic formation range. Slow cooling thus (1) promotes completion of the austenite transformation at high temperatures, (2) permits hydrogen to diffuse out of the weld joint, and (3) lowers the internal stresses set up during cooling.

The effects of preheating are:

1. It eliminates or decreases the danger of cracking by lowering the temperature differences between the unaffected base metal and the weld metal.

2. It minimizes hard zones adjacent to the weld because it minimizes formation of martensite.

3. It holds shrinkage stresses to a minimum.

4. It decreases distortion.

5. It increases the rate of diffusion and escape of hydrogen from steel.

Plastic Welding

The earliest form of plastic welding was hand forging on an anvil; the smith heated to the welding temperature the pieces to be welded, and welded them together by hammering. The same result can be accomplished mechanically, but both the hand process and the mechanical processes of *forge welding* are too slow and too expensive for use as modern production methods. A more modern form of plastic welding is *resistance welding*. The heat required for bringing to a plastic condition the edges to be welded is secured by passing an electric current through them. In *butt welding* by the resistance method, each piece of metal to be welded is clamped firmly to a copper electrode to keep the resistance low at those points, and the edges to be welded are pressed together. The greatest electric resistance in the circuit is at the area of contact, and the greatest heating effect therefore takes place there. When the metal at the edges becomes plastic, the pressure causes the edges to weld together. In *spot welding* the pressure is applied by two opposed electrodes, and welding occurs at the spot of contact. *Seam welding* is similar to spot welding; pressure is applied by rollers which serve also as electrodes, and the weld is in the form of a seam at the line of contact. Both spot welding and seam welding are rapid processes and can be adapted to automatic methods; they are used for the most part for welding sheet and other thin sections.

Fusion Welding

The source of heat for fusion may be burning gas, an electric arc, or thermit.

Gas welding usually is done with a torch to which oxygen and either acetylene or hydrogen are supplied under pressure. The edges to be joined are melted by the flame of the torch, and additional metal is flowed into the joint from a filler rod. The proportions of the two gases are adjusted to give a neutral flame for most welding operations.

In *arc welding* the arc is struck between two electrodes or between one electrode and the metals which are to be joined. A carbon rod may be used as one electrode, and any additional metal required may be supplied from a filler rod which is melted by the arc as welding proceeds. When a metal rod is used as an electrode, it acts also as a filler rod. Metal welding rods are of various compositions and sometimes are coated with materials which have a beneficial effect upon the weld. An indirect process of arc welding utilizes the heat liberated by formation of molecular hydrogen from atomic hydrogen. Molecular hydrogen is passed through an arc formed between two tungsten electrodes, where it is decomposed into the atomic form. Liberation of intense heat produced at the surface of the work by the recombination of these hydrogen atoms into molecular hydrogen results in welding the edges. The high temperature and nonoxidizing atmosphere make this an excellent method of welding, but the considerable expense limits its use.

Thermit welding utilizes the exothermic reaction which takes place between ferrosoferric oxide and aluminum to provide both the necessary heat and the iron required as a filler. Both the iron oxide and the aluminum used are in finely divided form, and when the mixture is ignited, this reaction takes place:

$$3\,Fe_3O_4 + 8\,Al \rightarrow 9\,Fe + 4\,Al_2O_3$$

The process is used principally for repairing large parts; a sand mold is prepared around the broken section and the thermit mixture is placed in a crucible directly above. When the thermit is ignited, the molten iron produced flows into the mold and welds the break together.

Postwelding Heat Treatment

Cast, rolled, extruded, and wrought metal articles are benefited by postwelding heat treatment. Because of the chemistry involved, this treatment is particularly desirable for castings. The treatment insures crack-free welds before the metal enters service, and insures satisfactory metallurgical properties that can withstand service conditions.

The minimum postwelding heat treatment should be a stress relief, which consists of heating to 1100–1250°F to remove virtually all residual stresses next to the weld. A more important factor is that it reduces hardness of the heat-affected zone to the point where machining or chipping presents no problem.

Advantages of Weldments

1. Fabrication and assembly are simplified.
2. Finishing costs are low.
3. Weldments permit flexibility and freedom in design.
4. By increasing strength, decreasing weight or size, improving appearance, and producing a uniform product, the quality of products is raised by use of weldments.
5. Dissimilar materials and different types of metal forms can be joined in the same product.
6. Welding can be used for such delicate parts as wire, and for such large parts as railroad cars, ships, bridges, tanks, pipelines, automobile bodies, and the like.
7. Virtually all common metal forms such as bar stock, extrusions, castings, stampings, forgings, and rolled stock can be welded, as well as the most widely used plate and sheet.

The principal adverse effects that may result from the welding of a casting are: (1) quench hardness; (2) thermal stresses; (3) hot cracks; and (4) cold cracks (cracks that develop well below the critical temperature). These adverse effects are mitigated by selection of welding methods best suited to the type of steel or iron to be welded, by various additions to metal welding rods used in arc welding, and sometimes by relief of stresses by subsequent annealing procedures.

Welding of Dissimilar Metals

Dissimilar metals are used when different sections of a unit must have different properties that cannot be obtained by use of a single metal. Irrespective of the method used to weld dissimilar metals, they must be preheated to at least 200°F. The composition of the joint will vary according to the composition of the component parts.

Testing Weldments

Mechanical testing and metallography play important roles in the science and practice of welding. Microscopic examination discloses any changes in the microstructure that have been caused by the heating effects in the weld metal and in the heat-affected zones of the base metal. This information applied to the condition of the weld joint permits the welder to avoid high stresses in the joint.

Corrosion of Welded Joints

If the weld metal is of the same or very similar composition to that of the base metal, the corrosion resistance of the joint is virtually the same as that of the base metal. If the weld metal differs chemically or metallurgically from the base metal, corrosion may occur at the weld or adjacent to it.

Welding Failures

Welding failures may be caused by (1) rapid change of physical conditions during welding and cooling, and (2) residual stresses caused by localized heating in the martensitic structure that exists after cooling. (Martensite has low ductility and therefore poor ability to withstand the stresses. It forms in the base metal and thereby causes much weld cracking.)

The most serious defect is the cracking that may occur in the weld metal and in the adjacent base metal. *Hot cracking* occurs at high temperatures because the solidified weld metal or the heat-affected base metal cannot withstand even relatively low-internal stresses at elevated temperatures. *Cold cracking*, which occurs at temperatures well below the critical temperature, is caused by hydrogen dissolved in the austenite. When the hydrogen is released, the unstable austenite transforms at ordinary temperatures.

Cracking is also caused by improper manipulation of the welding rod or electrode and the welding puddle, because there may not be proper plastic flow during the heating and cooling operations. Cold laps can occur if the base metal is not brought to the fusion temperature. Flux or oxide may be included in the weld. Other causes of cracking are the presence of stress raisers and locked-in stresses. Locked-in stresses are caused by rapid expansion of the heated base metal around the weld during the heating period. Upon cooling, the weldment does not return to its original dimensions, and the shrinkage of the solidified weld metal causes internal stresses to develop. If the internal stresses are high enough, they may cause hot cracking or permanent distortion and may also lead to formation and propagation of cold cracks. Internal weld stresses can be eliminated by any heat treatment that is appropriate to the condition, shape, and material involved.

The factors that determine the ability of a steel to be welded without cracking are:

Composition. Composition is important in determining the final hardness and brittleness of a welded joint. In a high-carbon steel the formation of martensite is promoted; that is the purpose of ordinary heat treatment for tempering and the like, and it explains why high-carbon steels are not easily weldable. The primary effect of such carbide-forming elements as molybdenum, manganese, vanadium, and chromium is to promote formation of martensite and thus to improve the ability of the steel to withstand abrasion and impact. The higher the carbon and alloy contents of a steel, the more readily it hardens in the heated zone, and therefore the slower must be the cooling to prevent cracking.

Maximum temperature in welding and time held at that temperature. These factors determine the amount of austenite that will be formed and therefore the amount of change in microstructure that will occur.

Rate of cooling. The rate of cooling exerts the greatest influence on the structures formed in the heat-affected zone. Because of the speed at which welding is carried out, the cooling rates are rapid, and as a result, the microstructure formed is martensite. The rate of cooling depends on the size and shape of the article being welded, the difference in temperature between the weld joint and the base metal, and the total heat input.

Hydrogen entrapment. Hydrogen may be introduced into the base metal from the coatings of welding rods, producing a supersaturated solution. (A *supersaturated solution* is a solution that holds a greater concentration of solute than it can normally hold at a given temperature; if the equilibrium of the solution is disturbed, as, for example, by lowering the temperature, the excess solute [in this case hydrogen] precipitates out of solution.) Since hydrogen is more soluble in molten steel than in solid steel, it will attempt to escape from the supersaturated solution during the cooling period. If it is trapped within a discontinuity in the metal and within a hardened area, cracking may result.

As a result of the welding operation, cracks are likely to occur in the heat-affected zone of the base metal and in the weld metal. Cracking occurs less frequently in the weld metal than in the base metal, because most weld metals have low carbon content and therefore are unaffected by the rapid cooling that occurs after welding.

Questions

1   What is the difference between hot rolling and cold rolling, and when is each used?
2   How is wire made?
3   What is tinplate, and how is it made?
4   Describe briefly the methods used for making:

   a. Butt-welded pipe
   b. Lap-welded pipe
   c. Seamless pipe

5   Define forging, and name several methods.
6   Compare rolling and forging.
7   State five advantages of cold extrusion.
8   State three advantages of hot extrusion.
9   Explain the welding operation; include the metallurgy involved.
10   Explain the necessity for preheating and postheating a welded article.

14 / Principles of Heat Treatment
of Steel

At the beginning of the twentieth century, engineers believed that the heavier the structure, the better was the design. However, when speed became an important factor with the introduction of automotive

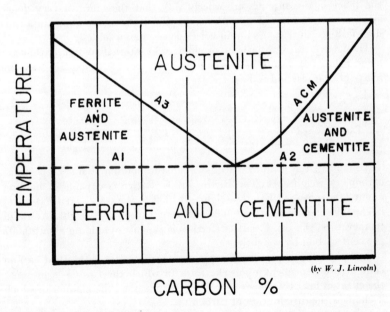

(by W. J. Lincoln)

Figure 73. Schematic iron-carbon diagram showing areas important in heat treatment.

equipment, the airplane, and the like, excess weight meant poor design and loss of power. To develop the high strength with less weight that is demanded of modern metals, the engineer turned to the metallurgist for help. The metallurgist met these requirements by the application of careful heat treatment methods and greater care in fabricating processes.

According to the definition accepted by the American Society of Metals, the American Society for Testing Materials, and the Society of Automotive Engineers, *heat treatment* is "an operation, or combination of operations, involving the heating and cooling of a metal or an alloy in the solid state for the purpose of obtaining desirable conditions or properties." Heat treatment therefore includes both the hardening of steel by any of the various processes, and softening it by full or partial annealing.

Theory of Heat Treatment

The concept of solid metals as crystalline substances is essential for an elementary understanding of the theory of heat treatment. In the solid state the atoms of the metallic elements in the crystal are so packed into the space lattice in such an orderly way that they form a very dense structure. In the liquid state, however, the atoms move about in a random fashion, so that the liquid is less dense than a solid.

The lattice transformations, or changes in internal structure, that are listed below occur only in iron and make it possible to explain why the alloys of iron respond to heat treatment.

Structure	*Temperature Range*			*Name*
Body-centered cubic (B.C.C.)	2552	to	2795°F	Delta
Face-centered cubic (F.C.C.)	1670	to	2552°F	Gamma
Body-centered cubic (B.C.C.)	Room	to	1670°F	Alpha

At 1330°F, the F.C.C. lattice is capable of holding about 0.8 per cent carbon by weight. At 2066°F, the F.C.C. lattice is capable of holding about 2.0 per cent carbon by weight. At 1330°F, the B.C.C. lattice (ferrite) is capable of holding about 0.03 per cent carbon by weight. At room temperature, the B.C.C. lattice (ferrite) is capable of holding about 0.007 per cent carbon by weight.

The physical properties of a plain carbon steel of a specified carbon content are dependent upon the form in which the carbon is present; the effect of heat treatment therefore depends upon the manner in which it changes the distribution of carbon.

For a hypoeutectoid or eutectoid steel, the first step in any heat treating operation, whether for the purpose of softening or of hardening, is

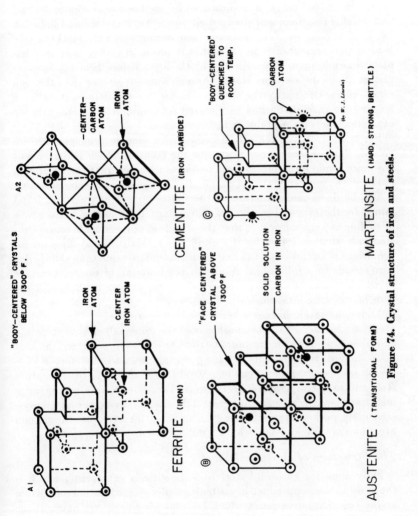

Figure 74. Crystal structure of iron and steels.

conversion of the steel to a solid solution consisting of homogeneous austenite. That is accomplished by heating uniformly to a temperature above the critical range, as represented by the line GS in Figure 20, and maintaining that temperature until all carbon has dissolved and diffusion has become complete. The length of time during which the steel must be held at that temperature in order that diffusion of carbon may be complete depends upon the structure of the steel before heating, because carbon atoms diffuse more slowly through some structures than through some others. Ordinarily the steel is not held at this temperature for a longer time than that which is required for complete diffusion, because of the tendency of the grains to become coarser.

For a steel of hypereutectoid composition similar principles are involved, but the steel is not heated to the point at which homogeneous austenite would form, but to some temperature above that corresponding to line PSK, but below that corresponding to line SE of Figure 20.

The hardness and structure of the finished piece of steel are determined by the rate and method of its cooling. A *full anneal* is the result of cooling at such a low rate that the structure of the steel becomes substantially that indicated by the iron-carbon constitutional diagram for conditions of equilibrium. Cooling at any rate more rapid than that which corresponds to a full anneal results in the lowering of the temperature at which the austenite is transformed, with consequent change in the type of structure into which it is transformed.

Alloys, however, harden after suitable heat treatment because their atoms are indifferent to one another, and the principal, or *host*, atom can accommodate many guests on its lattice. At lower temperatures the host atom cannot accommodate all its guest atoms, and therefore rejects them. As a result, a new type of crystal (for example, low-temperature steel and high-temperature steel) forms. The time of heat treatment is important because an atom has great difficulty in moving through a mass of densely packed atoms. During carburizing, for example, the carbon atom moves through 0.06 inch, or 6 million atoms in 8 hours.

The Structure of Hardened Steel

Fully annealed eutectoid steel consists entirely of pearlite, which is the transformation product of austenite under conditions of equilibrium at the transformation point which corresponds to the Ar_1 point of the iron-carbon constitutional diagram; fully annealed hypoeutectoid steel consists of pearlite plus ferrite, and hypereutectoid steel consists of pearlite plus cementite. The hardness of a steel of stated carbon content

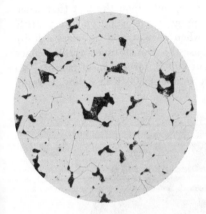

Figure 75. Ferrite and pearlite. Magnification 100×.

Courtesy United States Steel Corporation.

Figure 76. Pearlite and cementite. Magnification 1000×.

Courtesy United States Steel Corporation.

Figure 77. Martensite needles in austenite. Magnification 2400×.

Courtesy United States Steel Corporation.

Figure 78. Tempered martensite. Magnification 2000×.

Courtesy United States Steel Corporation.

depends upon the structure which takes the place of the pearlite that would exist if full annealing were carried out.

The effectiveness of heat treatment depends upon the fact that when austenite is cooled below the critical point, transformation into pearlite does not start instantaneously, and when once started requires a finite amount of time for completion. Because this transformation takes place only between the Ar_1 point and about 800°F, formation of pearlite can be prevented by cooling steel rapidly to about 800°F or below. If rapid

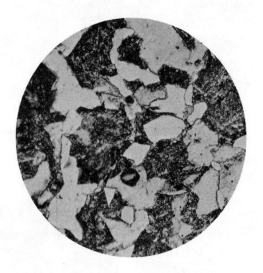

Figure 79. Quenched steel. Magnification 500 ×.
Courtesy Bausch and Lomb Optical Company.

cooling is continued to about 400°F or to room temperature, the austenite is transformed into a structure known as *martensite*. The lowest rate of cooling which results in transformation of austenite into martensite without production of any pearlite is called the *critical cooling rate;* it is largely dependent upon the carbon content of the steel, being greater for low-carbon steels than for high-carbon steels. A cooling rate of at least 54,000 F degrees per hour is necessary to obtain fully martensitic microstructures; this is therefore the critical cooling rate for full martensite, and it is usually obtained in practice by quenching the steel in an agitated liquid medium. Martensite is considered to be a supersaturated solid solution of carbon in ferrite; it is the hardest, strongest, and least ductile form of steel.

If steel is cooled rapidly to about 1000°F, and is held at any temperature between 1000°F and about 500°F for a sufficient length of time, the austenite is transformed into another material intermediate between pearlite and martensite; the structure of this material depends upon the temperature at which transformation occurs. Several names have been applied to these structures, but the current tendency is to designate them collectively as *bainite* and to distinguish between upper and lower bainite, depending upon the temperature at which transformation takes place.

Similarly, when steel is cooled rapidly to any temperature above about 800°F, and is held at that temperature for the time required for complete transformation of austenite, the resulting structure is one of those which are grouped together as *pearlite*, and designated as coarse, medium, or fine. The pearlite reaction is favored by fine-grain size.

If steel is held long enough at any transformation temperature for complete transformation to take place, the resulting structure is stable, and undergoes no further transformation when cooled to room temperature. By suitable choice of transformation temperature it is possible to produce any one of a range of structures from coarse pearlite to fine martensite.

The microstructure formed at each temperature level has definite characteristics as shown below.

TABLE 3. MICROSTRUCTURES AND CHARACTERISTIC PROPERTIES FOR APPROXIMATE TRANSFORMATION TEMPERATURE RANGES

Approximate Transformation Temperature Range	Microstructure	Characteristic Properties
1300–1000°F	Pearlite	Softest of the transformation products Lower ductility than bainite or tempered martensite at same hardness Good machinability
1000–500°F	Bainite	Substantially harder than pearlite and at the lower temperature levels approaches hardness of martensite Excellent ductility at high hardness
Below 500°F	Martensite	Hardest of the transformation products Brittle unless tempered
Reheating martensite in Temp. Range 300–1300°F	Tempered Martensite	Superior strength and toughness

It will be noted that in general the hardness increases as the transformation temperature decreases and that the microstructures corresponding to optimum properties in respect to strength and toughness are those involving transformation at the lower temperature levels—bainite or tempered martensite. These superior properties make these microstructures of particular interest to the alloy steel user.

Figure 80. Martensite. Magnification 2400 ×.
Courtesy Bausch and Lomb Optical Company.

Isothermal Transformation Diagrams

The changes that take place when steel is cooled at various rates are illustrated graphically in *isothermal transformation diagrams*, also called *time-temperature transformation* (*T.T.T.*) *curves*, and *S-curves*. The isothermal transformation diagram for a eutectoid steel is shown in Figure 81. The vertical scale is arithmetic and indicates temperatures; the logarithmic horizontal scale expresses time. The curve at the left shows the time required at any stated temperature for transformation to start, and the curve at the right shows the time required for completion of transformation. The diagram therefore provides the information required for production of a structure of the desired characteristics. It shows, for example, that transformation to coarse pearlite at 1325°F

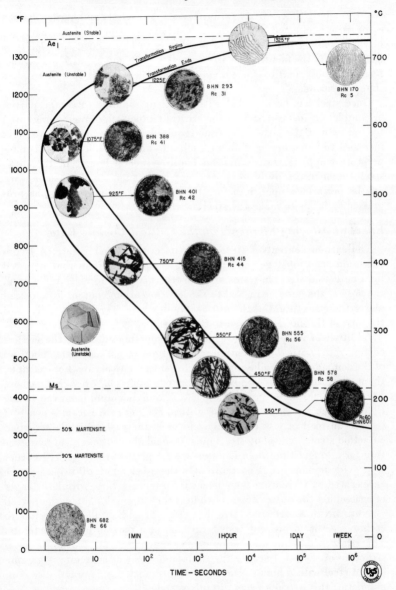

Figure 81. Isothermal transformation diagram of a eutectoid steel containing 0.89 per cent carbon and 0.29 per cent manganese; austenitized at 1625°F; grain size, 4-5.) Courtesy United States Steel Corporation.

requires a long time, whereas transformation to a finer and harder variety of pearlite is effected at 1050°F in less than 5 seconds. The importance of rapid cooling if formation of pearlite is to be avoided completely is illustrated by the fact that between about 1100°F and about 950°F, transformation into pearlite starts in one second or less, and is complete in less than ten seconds.

Once steel has been cooled with sufficient rapidity to escape partial or complete transformation in the critical temperature range represented by the *nose* of the curve, the time required for transformation to start increases to a maximum of about 8 minutes at about 425°F. At this temperature complete transformation requires several hours; if, therefore, steel is quenched rapidly to approximately that temperature, some work can be performed upon it before the formation of martensite has progressed so far as to make the steel brittle.

Effect of Alloying Elements on the Critical Cooling Rate

The carbon content of steel affects its critical cooling rate. A hypoeutectoid steel must be cooled at a more rapid rate than that required for a eutectoid steel; in terms of the isothermal transformation diagram, the nose of the curve is moved to the left. As carbon content is increased above the eutectic ratio, the critical cooling rate decreases slightly, and the nose of the curve is moved to the right.

Addition of alloying metals has a considerable effect on the critical cooling rate; with the exception of vanadium which has little effect, all of the commonly used alloying metals lower the critical rate. For example, the nose of the curve for SAE 4140 steel containing 0.37 per cent carbon, 1.0 per cent chromium, and 0.20 per cent molybdenum, is moved to the right to such an extent that the critical cooling rate becomes nearly 3 seconds instead of less than 1 second as is the case with plain carbon eutectoid steel. Some alloy steels must be held above the critical temperature for a longer time than is necessary for plain carbon steels, in order to provide homogeneous austenite. On the other hand, diffusion may be accelerated by increasing the austenitizing temperature, with less danger of coarsening the austenite grain than is the case with plain carbon steel.

RETAINED AUSTENITE   Another effect of adding alloying metals to carbon steel is to prevent transformation of a portion of the austenite. This *retained austenite* may amount to as much as 35 per cent of the total structure of a steel. Because the iron component of austenite is gamma iron, a steel with a high content of retained austenite is practically nonmagnetic; this is true of some *stainless steels*.

Hardenability

The data shown in isothermal transformation diagrams are based upon the assumption that the sample of steel subjected to treatment is of such small size that the change of temperature throughout its mass occurs at a constant rate. In any actual piece of steel of practical size, a temperature gradient necessarily exists from surface to center, and although the changes at the surface take place substantially as predicted by the isothermal transformation diagram for the rate of quenching employed, the center cools more slowly, with corresponding differences in the formation of the various microstructures. However rapid the cooling, therefore, martensite is formed only as an outside layer, while the structure of the interior of the piece may grade to coarse pearlite at the center. The depth of the outside layer or *case* of martensite depends to a considerable extent upon the size of the austenite grains at the start of cooling; the coarser the initial grain structure, the deeper is the hardening for any stated rate of cooling. Constitution of the steel is the other important factor in depth of hardening; the action of alloying metals in lowering the critical cooling rate has the effect of permitting transformation of the austenite to martensite to a greater depth than is possible with plain carbon steel. The depth to which steel can be hardened to martensite under stated conditions of cooling is called its *hardenability*. Hardenability is concerned with the rates or manner of transformation of austenite.

FACTORS AFFECTING COOLING RATE AND HARDENABILITY

Factors affecting cooling rate of object quenched

1. Quenching medium
2. Temperature of quenching medium
3. Speed of circulation of quenching medium
4. Size of object
5. Shape of object
6. Quantity of scale on object
7. Kind of scale on object
8. Temperature of object

Factors affecting hardenability of steel

1. Heating temperature; grain size
2. Prior treatment—effect on carbides
3. Heating rate—effect on carbides
4. Composition
5. Grain size

In general, alloying elements delay transformation in the pearlite and bainite regions. Their addition to steel permits formation of martensite with slower rates of cooling; this means that alloying elements increase hardenability with less distortion and cracking in light sections and permit less drastic quenches of heavy sections. Although alloying metals have a notable effect upon hardenability of steel, they have little effect upon its maximum hardness, which for any stated heat treatment is determined chiefly by the carbon content.

The Jominy Test

The Jominy test is a standardized procedure by which the hardenability of a steel is determined. A piece of the steel to be tested is machined to a solid cylinder 1 inch in diameter and $3\frac{7}{8}$ inches in length. This specimen is heated to the quenching temperature that has been found suitable for its composition, and is maintained at that temperature for 20 minutes. Then it is placed in a bracket that holds it in a vertical position, while a

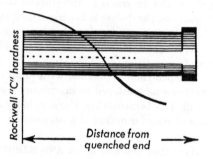

Figure 82. Standard Jominy test specimen.
Courtesy Republic Steel Corporation.

jet of water at a temperature of 75°F ± 5 degrees is played against its lower end. The orifice through which the water is admitted is $\frac{1}{2}$ inch in diameter and is placed $\frac{1}{2}$ inch below the bottom of the specimen; the water is under such pressure that the jet would reach a height of $2\frac{1}{2}$ inches if the specimen were not in place. Water is allowed to flow until the specimen is practically at room temperature, or the jet may be operated for 10 minutes, and cooling of the specimen completed by quenching fully. The maximum cooling rate produced by the Jominy test is approximately 600 (Fahrenheit) degrees per second, which represents the most rapid rate attained in industrial heat-treating operations.

Because the water strikes only the bottom face of the specimen, cooling there is at a much higher rate than at the top of the specimen; this rate varies from the maximum 600 degrees per second, which occurs at $\frac{1}{16}$ inch above the bottom, to slightly over 4 degrees per second at a distance 2 inches above the bottom. This gradient in the cooling rate produces a correspondingly wide gradient of hardness throughout the length of the sample. Hardness is measured at each $\frac{1}{16}$ inch interval from $\frac{1}{16}$ inch to 2 inches from the bottom, and the values so obtained are plotted on a chart.

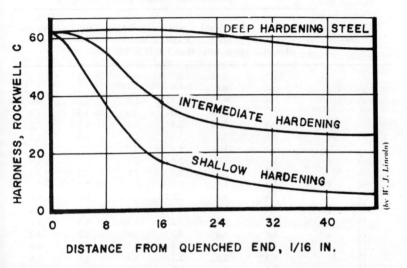

Figure 83. Typical hardenability curves obtained in (Jominy) test. (Rosenberg and Digges.)

Jominy test hardnesses have been correlated by laboratory experiments with hardnesses produced in the same steel by cooling rates of different degrees of severity, and the data so calculated are shown in the form of a chart (Table 4) which shows the diameter of the bar at the center of which the hardness obtained at each position of the Jominy test bar would be obtained.

Because performance of steel in use has been found to be dependent to a great extent upon its hardness, the distinctive pattern of hardenability indicated by the Jominy test is an important characteristic. The American Society for Testing Materials, the Society of Automotive Engineers, and the American Iron and Steel Institute have recognized

TABLE 4. TABLE OF JOMINY END-QUENCH DISTANCE vs. BAR DIAMETER

When the end-quench hardness curve of a steel has been found, this table enables one to estimate the hardnesses that would be obtained at the centers of quenched round bars of different diameters, when that same steel is quenched with various severities of quench.

For each successive $\frac{1}{16}$-inch position, the hardness obtained in the end-quench test would be found at the center of the bar size shown.

Distance from End in End-Quench Test Inches		*Equivalent Bar Diameter when Quenched*					
		Still Oil $H = 0.25$	Circulated Oil $H = 0.45$	Still Water $H = 1.0$	Circulated Water $H = 1.5$	Still Brine $H = 2.0$	Infinite or Idealized Quench $H = \infty$
$\frac{1}{16}$		0.1	0.15	0.3	0.35	0.4	0.7
$\frac{2}{16}$	$\frac{1}{8}$	0.2	0.3	0.5	0.65	0.75	1.15
$\frac{3}{16}$		0.35	0.55	0.85	1.0	1.25	1.6
$\frac{4}{16}$	$\frac{1}{4}$	0.5	0.80	1.15	1.3	1.5	1.9
$\frac{5}{16}$		0.6	0.95	1.4	1.6	1.75	2.2
$\frac{6}{16}$	$\frac{3}{8}$	0.8	1.2	1.6	1.8	2.0	2.4
$\frac{7}{16}$		1.0	1.4	1.8	2.0	2.3	2.7
$\frac{8}{16}$	$\frac{1}{2}$	1.1	1.5	2.1	2.3	2.5	2.9
$\frac{9}{16}$		1.3	1.7	2.3	2.5	2.7	3.2
$\frac{10}{16}$	$\frac{5}{8}$	1.4	1.9	2.5	2.7	2.9	3.4
$\frac{11}{16}$		1.6	2.1	2.8	3.0	3.2	3.6
$\frac{12}{16}$	$\frac{3}{4}$	1.7	2.2	3.0	3.2	3.4	3.8
$\frac{13}{16}$		1.9	2.4	3.2	3.4	3.5	4.0
$\frac{14}{16}$	$\frac{7}{8}$	2.0	2.5	3.3	3.5	3.7	4.2
$\frac{15}{16}$		2.1	2.7	3.5	3.7	3.9	4.4
$\frac{16}{16}$	1	2.3	2.8	3.7	3.9	4.1	4.6
$\frac{17}{16}$		2.4	3.0	3.9	4.1	4.2	4.7
$\frac{18}{16}$	$1\frac{1}{8}$	2.5	3.1	4.0	4.2	4.4	4.9
$\frac{19}{16}$		2.6	3.3	4.1	4.4	4.5	5.0
$\frac{20}{16}$	$1\frac{1}{4}$	2.7	3.4	4.3	4.5	4.7	5.1
$\frac{21}{16}$		2.8	3.5	4.4	4.7	4.8	5.3
$\frac{22}{16}$	$1\frac{3}{8}$	2.9	3.6	4.5	4.8	4.9	5.4
$\frac{23}{16}$		3.0	3.7	4.7	5.0	5.1	5.5
$\frac{24}{16}$	$1\frac{1}{2}$	3.1	3.8	4.8	5.1	5.2	5.6
$\frac{25}{16}$		3.2	4.0	4.9	5.2	5.3	5.8
$\frac{26}{16}$	$1\frac{5}{8}$	3.3	4.0	5.0	5.3	5.4	5.9
$\frac{27}{16}$		3.4	4.1	5.1	5.4	5.5	6.0
$\frac{28}{16}$	$1\frac{3}{4}$	3.5	4.2	5.2	5.5	5.6	6.1
$\frac{29}{16}$		3.6	4.3	5.3	5.6	5.6	6.2
$\frac{30}{16}$	$1\frac{7}{8}$	3.6	4.4	5.4	5.7	5.7	6.2
$\frac{31}{16}$		3.7	4.5	5.5	5.8	5.8	6.3
$\frac{32}{16}$	2	3.8	4.5	5.5	5.8	5.9	6.4

Courtesy United States Steel Corporation.

that fact, and have accepted the Jominy test as standard for structural alloy steels. Tentative specifications have been prepared to include hardenability tolerances; the new steels made under these specifications are known as *H steels*. Specifications previously in effect, which stated only the ranges of content of carbon and other alloying elements permissible in each steel, have not been entirely satisfactory because it has not

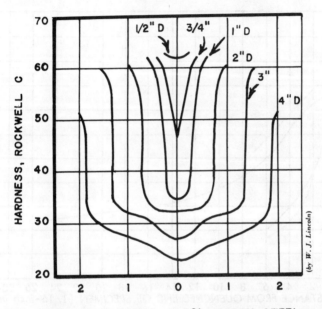

RADIUS, IN MATERIAL: 0.5% CARBON STEEL

Figure 84. Variation in hardness across section of different size rounds quenched in water from 1530°F.

been found possible to predict hardenability with sufficient accuracy from a knowledge of only the chemical composition of steel.

The new specifications fix closer tolerances of hardenability than could be met in steel specified in terms of chemical composition. At the same time it has been found that the limits of chemical composition can be widened while maintaining the specified tolerances of hardenability. The increased chemical tolerances permit a steelmaker to balance his chemical composition more readily to produce steel within the required hardenability band. If the chemical analysis made before tapping a heat of steel shows the content of one or more alloying elements to be too high, for example, he can reduce the amounts of other alloying elements to be added, and so bring the steel within the hardenability band specified. Under the old specifications, if one or more elements were

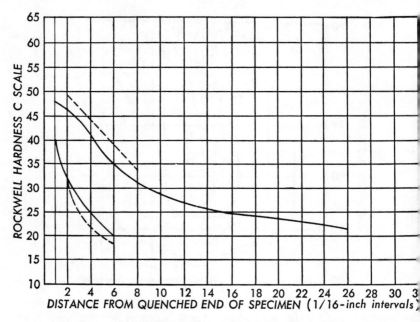

Figure 85. Tentative hardenability band for 8620 H steel. (The broken line is for standard chemical composition ; the solid line is for 8620 H.)

From *Materials and Methods*, April 1947, courtesy Reinhold Publishing Co.

in excess, the steelmaker still would be obliged to add enough of each of the others to bring the content of each up to the minimum chemical proportion, with the result that the steel would be too high in hardenability although within the chemical specifications.

Figures 85 and 86 show tentative upper and lower limits for 8620 H and 8630 H steels, respectively ; the space between the curves delineates the *hardenability band* in each case. On the same charts are shown the upper and lower hardenability curves of 8620 and 8630 steels, respectively, the specifications for which are based upon chemical content only. The charts show graphically that the variations in hardenabilities of the 8620 and 8630 steels are materially greater than the variations in the hardenabilities of the 8620 H and 8630 H steels, which are permitted a wider tolerance in their proportions of alloying elements.

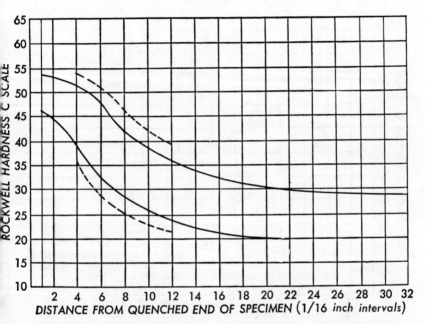

Figure 86. Tentative hardenability band for 8630 H steel. (The broken line is for standard chemical composition; the solid line is for 8630 H.)

From *Materials and Methods*, April 1947, courtesy Reinhold Publishing Co.

The hardenability of a standard H steel may be specified in any one of several ways as recommended by Charles M. Parker of the American Iron and Steel Institute, in an article which appeared in *Materials and Methods* for March 1947:

a. The minimum and maximum distances at which any desired hardness occurs, as illustrated by points *A-A* of Figure 87.
b. The minimum and maximum hardness values at any desired distance, as shown by the points *B-B*.

c. Two maximum hardness values at two desired distances, as shown by points *C-C*.

d. Two minimum hardness values at two desired distances, as shown by points *D-D*.

e. Any point on the minimum hardenability curve, plus any point on the maximum hardenability curve.

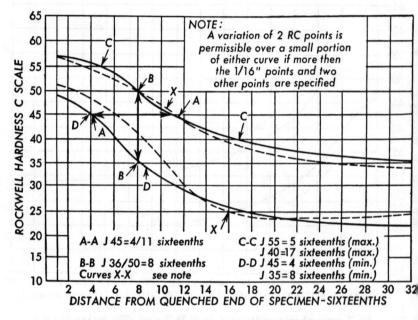

Figure 87. Methods of specifying hardenability requirements.

From *Materials and Methods*, April 1947, courtesy Reinhold Publishing Co.

Figure 87 shows the usual method of writing each of the above methods of specifying the hardenability desired.

Questions

1   Define heat treatment.
2   What is the purpose of the heating cycle in heat treating?
3   What is a *full anneal*?

4 What constitutents may be included in hardened steel?
5 Why can two pieces of steel of the same carbon content have different hardnesses?
6 How is an isothermal transformation diagram made?
7 What is meant by the *critical cooling rate*?
8 Define hardenability.
9 Describe briefly the Jominy test.
10 Why are Jominy test hardenabilities of value in the industry?

15 / Annealing, Hardening, and Tempering Steel

Annealing

Annealing may be performed by one of several methods, depending upon the results desired. The purpose of annealing may be: (1) to remove stresses that have occurred during casting or as a result of work done on steel; (2) to soften steel for greater ease in machining, or to meet stated specifications; (3) to increase ductility in order to make steel suitable for drawing operations; (4) to refine the grain structure and make the steel homogeneous; (5) to produce a desired microstructure.

Full Annealing

For full annealing, steel usually is placed in tightly closed boxes, heated to a temperature about 100°F above the critical range, and held at that temperature for a period of at least one hour for each inch of maximum section of the piece. Steel expands during heating except through the critical range, where it contracts. Because of these changes in volume, heating must be slow in order to avoid a considerable difference in temperature between outside and center of the piece; an excessive temperature differential would produce stresses which would tend to cause warpage.

After the heating period, steel ordinarily is allowed to cool in the furnace at an initial rate not exceeding 70 Fahrenheit degrees per hour, until the temperature has dropped below that which corresponds to the nose of the isothermal transformation diagram for the steel being annealed, in order to prevent formation of any bainite or martensite. When that temperature is reached at such a slow rate of cooling, transformation of the austenite into pearlite is complete, and the steel may be cooled safely in air at room temperature. Cooling may be accomplished also by removing the steel from the furnace and placing it in some insulating material to prolong the time of cooling as compared with unrestricted cooling in air.

Fully annealed steel is free of internal stress, and is soft and ductile. In the case of hypoeutectoid steel, which is preferred for full annealing, the resulting microstructure consists of ferrite and well-segregated pearlite.

Normalizing

A piece of steel in which stresses have been set up as a result of cold work, or as a result of cold work followed by local heating which occurs in welding and forging operations, may be *normalized*. This process relieves stresses and produces sufficient softness and ductility for many purposes, at the same time producing steel which is harder and which consequently has greater tensile strength than has steel which has been annealed fully. This increased hardness is often advantageous in promoting easy machinability, particularly in low carbon steels; in high carbon steels and in some alloy steels, normalizing produces too much hardness for easy machinability.

The steel is heated to the same or a slightly higher temperature than that used for full annealing, and is held at that temperature just long enough for complete transformation to austenite. After completion of heating, it is removed from the furnace and allowed to cool in still air at room temperature. Air cooling is considerably more rapid than the initial rate of cooling permitted in full annealing, but nevertheless slow enough to permit formation of some pearlite. Steel with carbon content between 0.5 and 0.9 per cent is most suitable for normalizing.

Normalizing is often the final treatment to which steel is subjected to bring it to required specifications. The process is used on large articles which, because of their size, might crack if quenched and drawn.

Subcritical Annealing

When a full anneal is not necessary to secure adequate relief of stress produced by work, *subcritical annealing*, also called *process* or *work annealing*, may be used to restore malleability and ductility. It is used to a considerable extent in continuous strip mills and wire mills.

The pieces to be annealed are placed in containers packed with protective material and heated to just below the critical range, or to about 1050°F to 1200°F. The rate of subsequent cooling that is permissible depends upon the carbon content, the rate decreasing with increasing carbon. It is somewhat more rapid than the rate used for a full anneal for steels that contain 0.5 per cent or less of carbon, and approximately the same for steel of higher carbon content. The resulting product is less soft than fully annealed steel, but is practically free of stresses.

Spheroidizing

To soften high-carbon steel sufficiently to make it readily machinable, it is *spheroidized*. The cementite is caused to assume a rounded or globular

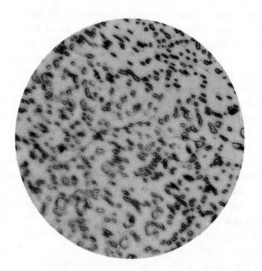

Figure 88. Spheroidized pearlite steel. Magnification 2000 ×.
Courtesy Bausch & Lomb Optical Company.

shape, leaving larger areas of ferrite free from cementite; this produces the softest steel possible for the same chemical composition.

Spheroidizing is accomplished by prolonged heating at a temperature slightly below the critical point. It is general practice to heat first to a temperature less than 100°F above the critical range; the closer this temperature is to the transformation temperature, the greater is the tendency to spheroidize. After this heating the temperature is allowed to fall to just below the critical range, and is maintained there for an extended period. Slow cooling is the final step.

Hardening

The rate of cooling, which determines hardness for any stated composition of steel, is adjusted by choice of quenching method. Because cooling takes place first at the outer surface of a quenched article, the most suitable method is chosen on the basis of hardenability as shown by the Jominy test, size of the piece being hardened, and hardness required at various points within the piece. Figure 89 shows cooling rates for the surface, center, and a point midway between surface and center of a one-inch round quenched in water, and for one quenched in oil. Figure 90 is the same as Figure 89 except that time is plotted on a logarithmic scale.

Quenching

Severity of quenching may vary over a wide range, from the extremely mild quench obtained by cooling in still air (normalizing) to the severe quench obtainable by a water spray or by a brine quench.

If the degree of hardening obtainable by normalizing is adequate for the purpose for which the steel is required, that process has the advantage of giving a product with minimum internal stress, because the transformation takes place relatively uniformly throughout the mass of steel. With some steels of high alloy content, the transformation process is so slow that cooling in still air produces sufficient hardness to make the steel suitable even for cutting tools.

Quenching in oil is relatively slow, but can be made sufficiently rapid to produce full hardening in hypereutectoid steel and in some low-alloy steels, when quenched below about 120°F. Because cooling in oil is less rapid than in water, oil quenching produces less rapid change of volume and consequently less distortion than does water quenching. The oil is often a mineral oil of flash point higher than 120°F, and of sufficiently low viscosity to permit free circulation around the piece being quenched. The quenching property of oil is increased materially by vigorous agitation.

Quenching in water at temperatures below 100°F provides rapid cooling, and is used frequently for carbon steels of a wide range of carbon content, and for medium-carbon low-alloy steels. For low-carbon steels only water quenching is sufficiently rapid to give full hardness. For maximum effect the water may be agitated violently or may be applied as a spray. When a piece of hot steel is plunged into water, bubbles of steam form around it, momentarily insulating the hot steel from the action of the cooling water. This may result in soft spots on the finished article.

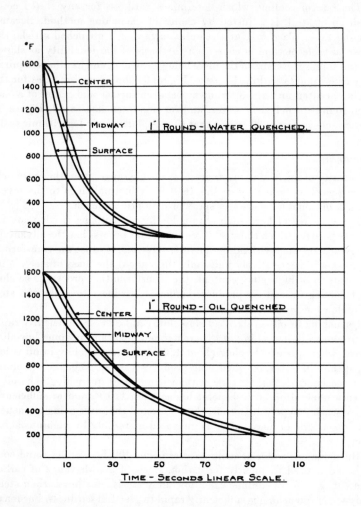

Figure 89. Cooling rates for one-inch rounds during quenching, linear scale.
Courtesy United States Steel Corporation.

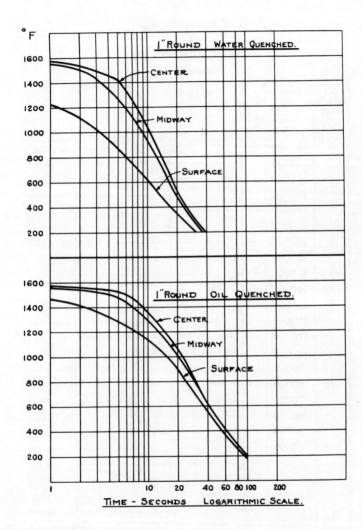

Figure 90. Cooling rates for one-inch rounds during quenching, logarithmic scale.
Courtesy United States Steel Corporation.

AISI-C 1137, Coarse Grain
(Oil Quenched)
PROPERTIES CHART
(Average Values)

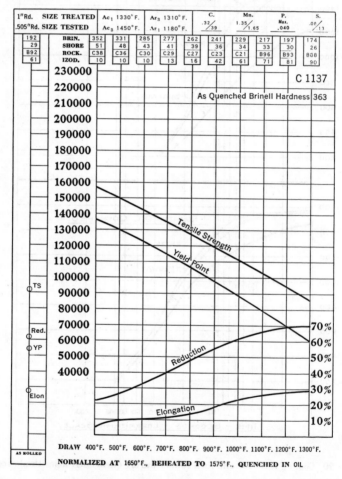

Figure 91. Physical properties resulting from oil quench.
Courtesy Bethlehem Steel Company.

AISI-C 1137, Coarse Grain
(Water Quenched)
PROPERTIES CHART
(Average Values)

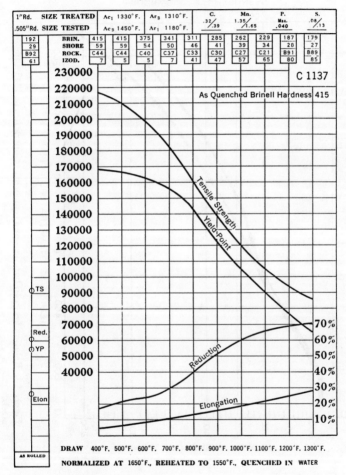

Figure 92. Physical properties resulting from water quench.
Courtesy Bethlehem Steel Company.

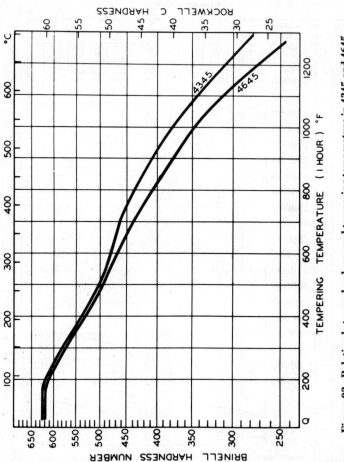

Figure 93. Relation between hardness and tempering temperature in 4345 and 4645 steels.

Courtesy United States Steel Corporation.

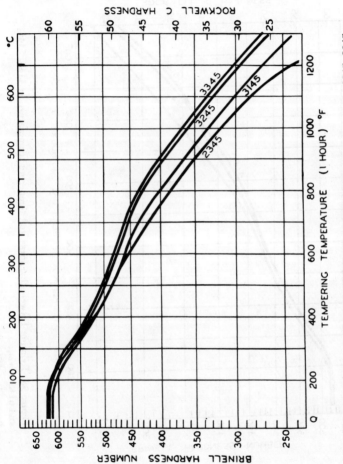

Figure 94. Relation between hardness and tempering temperature in 2345, 3245, and 3345 steels.

Courtesy United States Steel Corporation.

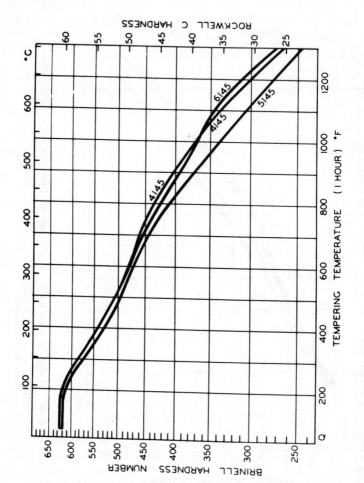

Figure 95. Relation between hardness and tempering temperature in 4145, 5145, and 6145 steels.

Courtesy United States Steel Corporation.

Quenching in a brine containing 10 per cent by weight of sodium chloride prevents this action. The salt crystals that crystallize near the surface of the steel as water is vaporized explode as they come into contact with the hot steel, and agitate the solution sufficiently to break down the bubbles of steam and break the scale from the work. This expedites cooling and makes it more uniform over the entire surface. In a water quench the outside of the piece of work is transformed into martensite much more rapidly than is the interior. Because the transformation of austenite into martensite involves an increase in volume, subsequent transformation of the austenite in the interior of a piece of steel of considerable size may cause cracking during quenching.

The best quenching medium for a stated purpose is the one that acts at the lowest rate consistent with obtaining the required hardness.

Tempering

When a piece of steel has been hardened fully, it is hard, brittle, and internally stressed to such an extent that it may fail in service. Martensite has a tendency to contract on aging, with the result that an article of fully hardened steel is dimensionally unstable. Any retained austenite tends to transform slightly to martensite on aging, and to a greater extent when the piece of steel is put into service. This tendency contributes further to dimensional instability, because the transformation results in an increase in volume as the dense austenite becomes less dense martensite. It is necessary, therefore, to apply to a piece of hardened steel some sort of after treatment to make it less brittle and therefore tougher, to relieve internal stress, and either to stabilize the retained austenite or to cause it to transform into a dimensionally stable structure.

The treatment applied is a reheating operation called *tempering*, which is carried out immediately after hardening and before the work has cooled to room temperature, but after it has cooled below 400°F. This reheating releases the carbon held in supersaturated solution (martensite), and permits it to form carbide crystals; the resulting microstructure is called *tempered martensite*. The piece of work to be tempered is heated to a temperature above 300°F and below the critical temperature, depending upon the nature of the steel and the reduction in hardness which can be permitted. It is held at that temperature for a suitable length of time and then is allowed to cool in still air. The details of time and temperature required for proper tempering vary considerably with the composition of the steel.

Hardness and Strength

The relationship between hardness and strength is definite and regular. The curve of Figure 96 expresses the generally accepted relationship between hardness on the Brinell scale and ultimate strength; the marked

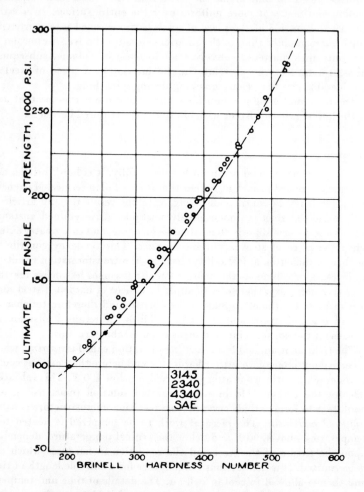

Figure 96. Relation between hardness and tensile strength.

Courtesy United States Steel Corporation.

points are those determined by precise measurements for the three steels named on the diagram. Because knowledge of the ability of steel to carry loads is more important than information as to its ultimate breaking strength, a more useful comparison is that of hardness with yield strength.

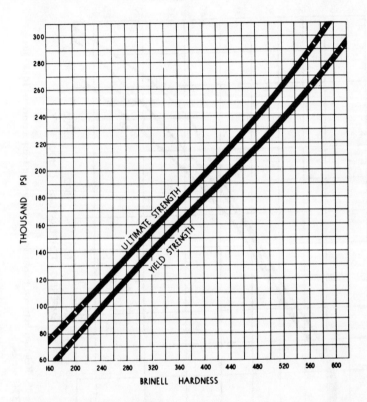

Figure 97. Ultimate strength and yield strength at different hardnesses.
Courtesy United States Steel Corporation.

Because yield strength is more difficult to measure precisely, and because different methods of measurement are used by different investigators, its correlation with hardness is less definite than is correlation of ultimate strength with hardness. Figure 97 shows values obtained statistically from a large number of tests, for both ultimate strength and yield strength,

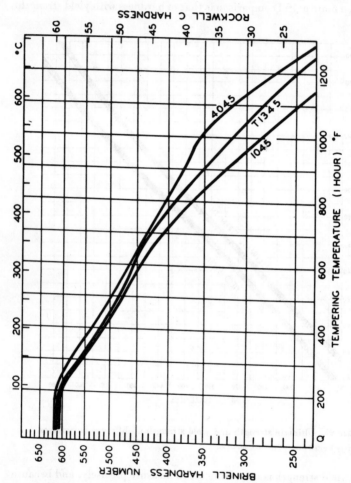

Figure 98. Relation between hardness and tempering temperature in 1045, 4045, and T1345 steels.

Courtesy United States Steel Corporation.

as related to Brinell hardness. This relationship applies for practical purposes to all steels regardless of chemical composition. It is also a fact that fully quenched steels of the same carbon content have practically the same hardness regardless of their chemical constituents. Steels of the same carbon content but with different contents of alloy metals, however, temper at different rates, and the tempering procedure must be adjusted appropriately.

Toughness, which is the ability to withstand load without breaking, is an indication that steel has sufficient plasticity to permit some plastic deformation. This is an important attribute, and the tempering method that gives the best compromise between strength and toughness is selected.

By referring to Figure 97, the hardness which corresponds to the required strength is found. For a stated hardness, tempering should be done at the highest temperature which permits that hardness to be retained, because toughness in general is higher for higher tempering temperatures. Figure 98 shows the effect of tempering at various temperatures three varieties of steel which had approximately the same initial hardness after quenching.

Interrupted Quenches

An interrupted quench is one which is not carried through to the temperature at which transformation of austenite to martensite commences, but is interrupted at some higher temperature, in order to suppress transformation of austenite into pearlite and at the same time avoid formation of martensite. The quenching bath is kept at a stated temperature appropriate to the formation of the microstructure which is desired for the steel being treated, and the work is kept in the bath for a suitable period. Hardening by this method is preceded by the usual heating cycle to convert the steel into homogeneous austenite.

Quenching and the subsequent isothermal treatment usually are carried out in a bath of molten salt of a variety which does not attack steel chemically, and which can be washed off readily after completion of the process. Another requirement is that the bath shall not be too viscous to permit vigorous agitation at the temperature at which it is to be used. Sodium nitrate and sodium nitrite frequently are used for this purpose. Molten lead sometimes is used, but is less convenient because it is more dense than steel, and the pieces being treated must be held forcibly below the surface. Highly finished parts may be hardened in baths of molten salt or lead without scaling or impairment of their smoothness; they are practically free also from surface carburization or decarburization. Figure 99 shows the rate of cooling in molten salt baths, compared with rates in water, oil, and air.

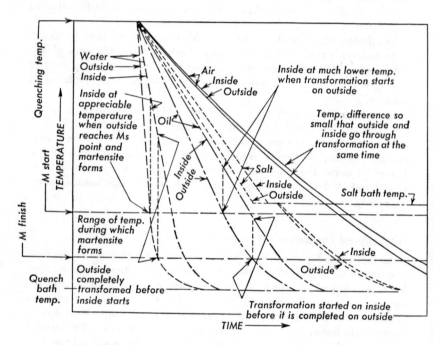

Figure 99. Rates of cooling in water, oil, molten salt, and air.

From *Metal Progress*, October 1944, courtesy American Society for Metals.

Austempering

Austempering is an interrupted quenching process which consists in quenching in a bath of molten salt at a temperature between 450°F and 900°F, depending upon the microstructure desired, and maintaining that temperature until transformation of austenite into bainite is complete. Because the steel is held at the same temperature for the entire period during which transformation is taking place, little internal stress or distortion are developed. The result is a steel which has greater toughness

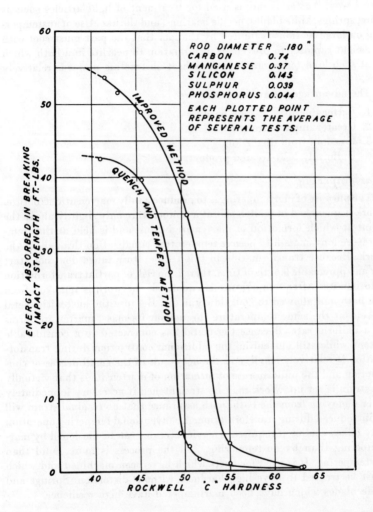

Figure 100. Comparison between austempered and quenched-and-tempered steel.
Courtesy United States Steel Corporation.

and greater ductility than one that is hardened and tempered in the usual way. Austempering is used for treatment of light articles such as wire, springs, knife blades, needle bearings, and the like. Use of austempering on large sections is limited by the fact that the part must cool with sufficient rapidity to prevent transformation to pearlite in a bath which is at such high temperature that its heat-abstracting power is relatively low.

The advantages of austempering are:

1. Better ductility at high hardness.
2. Greater impact strength.
3. Freedom from distortion.
4. Uniformly heat-treated product.

Martempering

The purpose of martempering is to produce a fully martensitic structure. Work is quenched in molten salt at a temperature only slightly above the point at which formation of martensite begins, and is held at that temperature long enough to permit temperature equalization throughout the work. Because transformation in this temperature range does not start for an appreciable length of time, there is no risk of partial transformation into bainite. After temperature is equalized, the work is removed from the bath and allowed to cool slowly in air. Because the austenitic metal is kept at the same temperature throughout its mass, martensite forms at a uniform rate. Because the metal has contracted to a considerable extent while still austenitic, the additional contraction during transformation is relatively small, and the net result of the small degree of contraction and the uniform rate of formation of martensite is that virtually no strain is set up. Work may be straightened if necessary immediately after removing from the bath, with assurance that no residual strain will be introduced during transformation. A conventional tempering operation may follow cooling if required. Heavier sections can be hardened by martempering than by austempering, and the process is more rapid than austempering. It is used for parts which have been machined and which must be treated to have high hardness without distortion. Springs and knife blades which have been martempered have high resilience.

Isothermal Quenching

Another type of isothermal quenching has not been given a distinctive name as have austempering and martempering. In this process the austenitized steel is quenched in an agitated bath of molten salt at about

450°F, and is held at that temperature for the time required for completion of isothermal transformation. Upon removal from the bath, the work is immersed immediately in another bath, usually of the same composition, which is at a higher temperature. Finally the steel is cooled in air, completing the tempering operation. This method of hardening produces

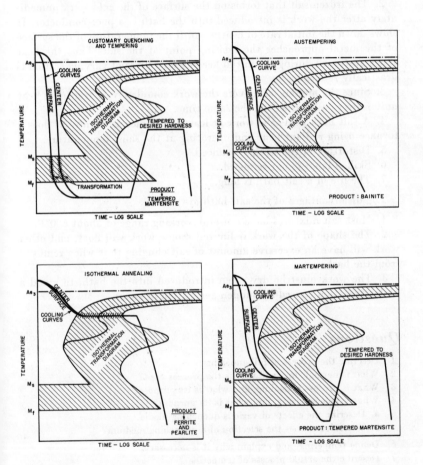

Figures 101, 102, 103, and 104. Comparison of heat treating processes.
Courtesy United States Steel Corporation.

little internal strain, and there is relatively little danger of cracking, because tempering follows quenching without intermediate cooling.

The advantages of the electrically heated salt bath are:

1. The surface is automatically protected from the air while in the bath and after removal from the bath.

2. The frozen salt that forms on the surface of the cold work immediately after the work is introduced into the bath is a poor conductor. It slows down the initial rate of heating until the temperature of the surface of the metal approaches the melting point of the salt. Thus the heat shock sustained during the initial heating on a piece of varying section is drastically reduced.

3. Since the molten salt covers the work completely, it transmits heat uniformly to all surfaces at the same time.

4. A salt bath heats the work at least four times as rapidly as does a furnace using radiant heat and operating at the same temperature.

5. Distortion of the work is reduced.

6. Skilled labor is not required.

7. The life of a salt bath is long.

The disadvantages of the salt bath are:

1. Most salt mixtures have a useful working range of about 600°F.

2. The shape of the work is limited. Some work will float, and other work will have an excessive amount of salt clinging to it when removed from the bath.

3. The work must be dry when introduced into the bath, to avoid a violent reaction caused by sudden generation of steam.

Questions

1 Describe the process of full annealing.
2 What is normalizing, and why is the process used?
3 What is process annealing, and what is its purpose?
4 What is spheroidizing, and what is its purpose?
5 a. Describe the effects of various quenching media.
   b. What determines the selection of a quenching medium?
6 Define tempering, and explain why it is necessary.
7 Describe the actual process of tempering.
8 Describe the process of austempering.
9 Describe the process of martempering.
10 State the important difference between austempering and martempering.

16 / Surface Hardening of Steel

The purpose of surface hardening is to impart to a piece of low-carbon finished or semifinished steel a hard surface which is resistant to wear and abrasion, while retaining the tough and ductile composition of low-carbon steel in the interior. The hard surface layer is called the *case*, and the interior of the piece is called the *core*.

Carburizing

The oldest method of case hardening is *carburizing*, a process which consists of treating the surface of a piece of steel in such a way that it absorbs additional carbon. Steel which is to be carburized should be free-cutting, should be capable of absorbing carbon uniformly, and should be of such composition that it does not become excessively coarse grained and therefore brittle upon prolonged heating. Plain carbon steel with carbon content between 0.12 and 0.18 per cent, and steels which contain small amounts of nickel or of nickel and chromium, are suitable. Some steels which are used for carburization contain molybdenum.

Carburization is followed by suitable heat treatment to produce the desired case and core. If the steel is sufficiently fine grained it is necessary only to reheat it to about 1660°F and quench, or it may be quenched directly from carburizing temperature; the core of low-carbon steel then has the finest possible grain. If the grain of the case is so coarse that refinement is necessary, the piece is reheated again, this time to about 1360°F, and is quenched from that temperature; a final tempering operation usually is necessary.

Pack Carburizing

Pack carburizing is essentially the same as the method by which cementation steel was made from wrought iron before modern methods of making steel were developed. The articles to be carburized are cleaned and packed loosely in a metal box with carbonaceous material, or with a commercial carburizing compound. Carbonaceous materials that have been used include coal, charcoal, charred bone, bone meal, wood, and hide scraps. Commercial carburizing compounds usually are hardwood charcoal or bone charcoal, and include carbonates of barium, calcium, and sodium, which act as energizers in hastening the reaction.

It is believed that carbon is not absorbed by the steel directly, but indirectly by formation of carbon monoxide. Carbon monoxide is formed by reaction between the oxygen of the air in the box and the carbonaceous material, and by decomposition at carburizing temperature of the carbonates added to serve as energizers. Carbon monoxide reacts with iron to form iron carbide and carbon dioxide; the carbon dioxide reacts with more carbon to form carbon monoxide, and the cycle repeats. The reactions are:

$$3\,Fe + 2\,CO \rightarrow Fe_3C + CO_2$$
$$CO_2 + C \rightarrow 2\,CO$$

In general, the higher the carburizing temperature, the greater is the carbon content of the case, and the greater is its depth. The depth of case increases also with the time during which the steel is held at carburizing temperature. It is customary, however, to use a temperature range between 1675°F and 1700°F for pack carburizing, and to adjust the time to the nature of the steel and the depth of case desired. That temperature range produces adequate depth of case without incurring danger of excessive grain growth.

Pack carburizing is the simplest method of case hardening, and requires the least skilled labor. It is slow, however, and requires a considerable amount of unskilled labor.

The average carbon atom moves 0.06 inch, or about 6 million atoms in about 8 hours during carburizing.

Gas Carburizing

Gas carburizing is carried out in an atmosphere of carburizing gases, including carbon monoxide and such hydrocarbons as butane, ethane, methane, and propane. The process is flexible and more accurately controllable than is pack carburizing; it can be used to produce almost any desired hardness, depth, or carbon content of the case. Skilled operators are required to obtain optimum results from the method, however.

Portions of the work which do not require hardening may be protected from carburization by a layer of copper plating, which is not penetrated by the carbon formed in the reaction. As a practical matter, parts often are copper plated all over, and the copper is removed by machining from those portions which require hardening.

Nitriding

Nitriding is a process by which extremely high surface hardness combined with exceptionally high wear resistance can be obtained on steel. Nitrided steel is resistant to corrosion and fatigue. Parts may be machine finished before nitriding, because virtually no distortion occurs during the process, and further heat treatment is not required. The hardness imparted by nitriding is maintained at temperatures as high as 1100°F. Large sections of work can be hardened by nitriding.

During the nitriding process a very slight, but constant increase in volume occurs as a result of nitride precipitation. Hence proper allowances must be made in the finish machining dimensions of the article, prior to nitriding. The article is usually finished to size before the nitriding treatment.

Steels that are to be nitrided are usually preheat treated before nitriding (1) so that the mechanical properties of the core will be those that are desired in the finished article; (2) so that the grain will be sufficiently refined by preheat treatment that the nitrided case will be reasonably tough.

The process consists of forming a surface layer of complex nitrides. Iron nitride, which is formed by nitriding plain carbon steel, has ample hardness, but is too brittle, and special steels therefore have been developed which give adequate hardness while retaining good core properties. Aluminum, chromium, and molybdenum are the nitride-forming alloying elements principally used in steels intended for nitriding, although other varieties of alloy steel sometimes are nitrided.

In the most commonly used process of nitriding, completely machined and heat-treated steel parts are subjected to action of ammonia gas at a temperature between 950°F and 1000°F (usually 975°F). The ammonia breaks down into nitrogen and hydrogen in accordance with the reaction,

$$2\,NH_3 \rightarrow N_2 + 3\,H_2$$

and the nascent nitrogen precipitates the nitride-forming alloy elements from the solid solution as nitrides. Because nitrogen can diffuse further into the steel only after these nitrides have been precipitated completely, increased alloy content decreases the depth of hardness obtainable for a stated time of treatment. Depth increases, but hardness decreases, with increase in temperature. At temperatures between 1000°F and 1050°F, a depth of from 0.005 to 0.010 inch is obtained on stainless steel containing approximately 18 per cent chromium and 8 per cent nickel.

If any areas of a piece of work are to be protected against hardening by this process, a layer of tin about 0.0003 to 0.0005 inch thick is plated on the work, and is ground off the area to be hardened.

Nitriding is used for surface hardening of such articles as cams, gears, gages, dies, crankshafts, spindles, shackle bolts, and the like.

Cyaniding

Cyaniding is in effect a combination of carburizing and nitriding. In this process a thin case between 0.001 and 0.015 inch in thickness is produced by immersing steel in a molten salt bath containing a cyanide, usually sodium cyanide. The concentration of cyanide varies in different shops, but the *Metals Handbook* of the American Society for Metals recommends from 30 to 75 per cent; the remainder of the bath may be a mixture of sodium chloride and calcium chloride, or a mixture of sodium chloride and sodium carbonate. The bath is used at a temperature of 1500°F or higher; it produces carbon monoxide which carburizes steel in the same manner as it does in pack carburizing, and nascent nitrogen which has the same nitriding effect as does the nascent nitrogen formed by decomposition of ammonia. The reaction involved is believed to be:

$$2\,NaCN + 2\,O_2 \rightarrow Na_2CO_3 + CO + N_2$$

Penetration is at the rate of about 0.007 to 0.009 inch in the first hour; thereafter the rate decreases rapidly, and it has been found uneconomical to form a case deeper than 0.015 inch.

Parts are quenched in water or brine immediately upon removal from the cyaniding bath; the structure is a hard martensite case and a fine-grained core of low carbon content. The surface is clean, and has a lightly mottled appearance as a result of local oxidation produced by the steam formed in the quenching operation.

Cyaniding is sometimes done by sprinkling dry sodium cyanide over the work and heating, but that method has the disadvantage of sometimes leaving soft spots in the case.

The principal uses of cyanide hardening are:

1. To secure a shallow and low-cost wear-resisting surface of high quality on such articles as screws, grease fittings, business machine parts,

roller chain components, dental burrs, sewing machine parts, shafts, and bolts.

2. To harden heavy-duty high-quality gears in the automotive field.
3. To reheat carburized work.

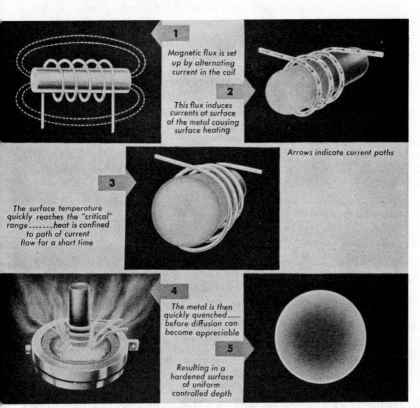

Figure 105. Induction hardening.

Courtesy Westinghouse Electric Corporation.

Induction Hardening

The induction-hardening process is based upon the heating effect produced in a piece of steel by placing it in a high-frequency electric field generated in a suitably shaped water-cooled coil of copper wire or tubing.

Heating is produced at the start by a combination of hysteresis and eddy currents. After the temperature of the steel reaches the Curie point (about 1420°F), the steel becomes nonmagnetic, and hysteresis therefore ceases to exist; from that point on, the entire heating effect is produced by eddy currents. The piece of steel may be considered analogous to a short-circuited secondary winding of a transformer, and eddy currents correspond to the current which would flow in such a secondary winding.

At low frequencies eddy currents pass throughout the entire mass of the steel being heated and therefore tend to heat it virtually uniformly. At higher frequencies, however, the current tends to stay in an increasingly thin layer at the surface; this is known as the *skin effect*. The depth of case produced therefore can be regulated with great accuracy by adjusting the frequency of the voltage used. Frequencies from 1000 cycles per second up to those in the radio-frequency spectrum are used.

The coils, or *inductors*, by means of which the electromagnetic field is produced must be shaped appropriately to handle pieces of the shape and size required. In automatic operation, the inductors are provided with openings through which the water or brine for quenching is introduced after the required temperature has been reached.

The induction-hardening process may be operated on an automatic cycle when once set up, and because of the accuracy of timing, the product has a high degree of uniformity. Because heating of the surface is extremely rapid, little heat is conducted to the core, and the core retains its original structure with consequent minimum internal stress. Another advantage of the process is that heating can be localized in any desired area.

Flame Hardening

In flame hardening, a concentrated oxyacetylene flame is passed rapidly over the portions of the piece of work which are to be hardened, producing an extremely high surface temperature. The flame is followed immediately by a jet of water or an air blast to quench the heated surface layer; thus the surface is hardened before the core can become heated appreciably by conduction. Flame hardening is similar to induction hardening in that it is rapid and may be operated with unskilled labor.

The process is used to harden cams, gears, shafts, lathe beds, and similar articles which might crack or warp if hardened by conventional furnace heating and subsequent quenching. It is used on both steel and gray cast iron.

We suggest that you write a non-committal letter to the Westinghouse Electric Corp. (or any other company that makes these machines), asking for more information related to price and available machines. State that you are interested.

Figure 106. Induction-hardening machine.
Courtesy Westinghouse Electric Corp. and The Oliver Corp.

Because neither carbon nor nitrogen is added during induction hardening or flame hardening, the carbon content of the steel must be higher than that of steels which are to be carburized; the range is usually 0.4 to 0.5 per cent.

Questions

1 What is the purpose of surface hardening?
2 What is carburizing, and what are the steps in the process?
3 What type of steel is best suited to carburizing?
4 Describe the process of pack carburizing.
5 Explain the use of an energizer in pack carburizing.
6 Describe the heat treatment which follows carburizing.
7 What is nitriding, and what are its advantages?
8 What is cyaniding, and when is it used?
9 Describe induction hardening. Why are different frequencies used?
10 Describe flame hardening. When is it used?

17 / Alloy Steels

Alloy steels in great variety have been developed to overcome limitations which make plain carbon steels unsuitable for many purposes. An alloy steel is one to which have been added one or more elements in addition to the carbon and the small amounts of sulfur, phosphorus, manganese, and silicon which are present in all plain carbon steels. Alloying elements change the properties of steel in so many ways that generalization is difficult. One important characteristic which can be imparted to steel by addition of suitable alloying elements is increased hardenability, resulting in retention of toughness while permitting use of a type of heat treatment which produces great hardness and strength. Other characteristics which can be improved by use of alloying elements are resistance to wear, resistance to corrosion, and performance at high temperatures.

Classification of Alloying Elements

Some alloying elements, including aluminum, molybdenum, tungsten, and vanadium, raise the thermal critical points of steel; others, including copper, manganese, and nickel, lower the critical points. Most alloying elements shift the eutectoid point of the iron-carbon diagram to the left, so that a eutectoid steel which contains an alloying element usually contains less carbon than does a plain carbon steel of eutectoid composition.

The elements of one group used for alloying purposes have the property of dissolving to a considerable extent in alpha iron and forming a solid solution with it. This results in stronger ferrite, imparting increased hardness to the steel, but without loss of ductility at ordinary temperatures. Steels which contain these elements usually also have high surface stability against corrosion and oxidation. The elements of this group are aluminum, cobalt, copper, nickel, and silicon.

Elements of a second group tend to combine with the carbon in steel and to form carbides, which usually are of complex nature. Like iron carbide, these carbides increase hardness and strength, but decrease ductility; their solubilities and the times and temperatures required for solution differ widely. Elements of this group are chromium, niobium (formerly termed columbium), manganese, molybdenum, titanium, tungsten, and vanadium. Niobium, titanium, and vanadium have the

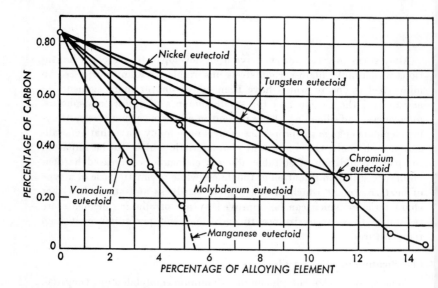

Figure 107. Effect of various elements on the eutectoid point.

Reproduced by permission from *Modern Steels*, copyright 1939 by American Society for Metals, Cleveland, Ohio.

strongest carbide-forming properties; chromium, molybdenum, and tungsten have lower carbide-forming tendencies; and manganese has only somewhat greater tendency to form carbides than has iron. Most carbide formers are also soluble in alpha iron, and when insufficient carbon is available to convert the total amount of any carbide former to its carbide, the remainder goes into solid solution. With the exception of chromium most carbide-forming elements are added to steel in relatively small quantities.

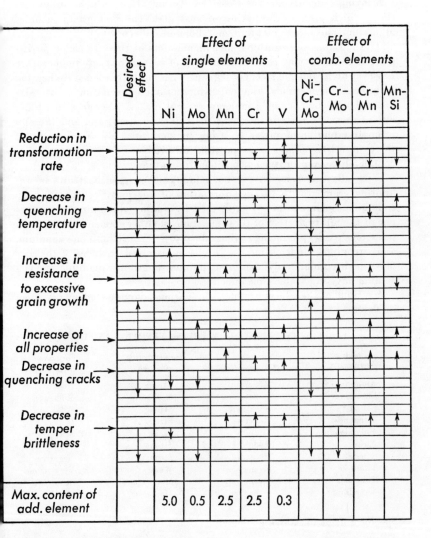

Figure 108. Effect of alloying elements on structural and engineering steels (after Sourdillon).

From *Metal Progress*, November 1944, courtesy American Society for Metals.

Alloying elements may be classified also into those which inhibit and those which promote formation of austenite. The first group includes aluminum, chromium, niobium, molybdenum, silicon, titanium, tungsten, and vanadium. Addition of sufficient amounts of these so-called *ferrite-forming* elements may prevent formation of austenite at any temperature however high, and the steel consequently cannot be hardened by heating and quenching methods. Elements which promote formation of austenite are carbon, cobalt, copper, manganese, and nickel. The effect of adding any of these elements is to retard the transformation rate, and thereby promote retention of austenite at low temperatures. By judicious use of these *austenite formers* it is possible to produce steel which remains completely austenitic even at ordinary room temperatures.

Some alloying elements exert a refining effect upon austenite by inhibiting grain growth through formation of fine particles of refractory compounds. Aluminum oxide, formed by adding aluminum to steel, is one of the most effective of these substances. Other elements which form refractory oxides producing similar effects are silicon, titanium, vanadium, and zirconium. Under some conditions undissolved particles of the carbides of chromium, molybdenum, titanium, tungsten, vanadium, and zirconium have much the same effect.

A résumé of the effects produced by various alloying elements is given in Table 5.

TABLE 5   INFLUENCE OF ALLOYING ELEMENTS IN STEEL

Alloying Element	Increased Hardenability	Strengthening of Ferrite	Formation of Carbide
Aluminum	Moderate	Strong	Negative
Chromium	Moderate	Moderate	Moderate
Cobalt	Negative	Strong	Weak
Manganese	Moderate	Strong	Moderate
Molybdenum	Strong	Weak	Strong
Nickel	Moderate	Strong	Negative
Silicon	Moderate	Strong	Negative
Titanium	Strong	Weak	Very strong
Tungsten	Strong	Moderate	Strong
Vanadium	Strong	Moderate	Very strong

Specific Effects of Alloying Elements
Chromium

Chromium is used in large quantities as an alloying element of steel. Although it is primarily a carbide-forming element, it is added to some steels in amounts greater than those required to combine with available carbon, and the excess dissolves in and strengthens the ferrite. The critical range of temperatures is increased by addition of chromium, and the eutectoid point is moved to the left to such an extent that if steel contains from about 12 to 14 per cent of chromium, a stated carbon content produces approximately three times the hardening effect which the same carbon content could produce in plain carbon steel.

The various carbides of chromium and the complex carbides of chromium and iron are extremely hard and resistant to wear, and confer these properties upon chromium alloy steel without making it as brittle as would the same degree of hardness if produced by increase in carbon content alone. Addition of chromium tends to refine the grain of steel; this effect, in combination with its ferrite-strengthening property, increases toughness. The dual effect of increasing both toughness and hardness simultaneously makes chromium one of the most useful alloying elements. The presence of chromium impedes the transformation of steel either to or from the austenitic condition, and consequently increases its capacity for deep hardening.

Alloy steels containing from 14 to 27 per cent chromium, with carbon content not in excess of about 0.35 per cent, and without other alloying element, are ferritic and cannot be hardened to any appreciable extent by heat treatment because they do not form austenitic phases at any temperature.

Nickel

Nickel is another of the alloying elements used in large quantities, and is one of those which have been used over the longest period of time. Addition of nickel to steel moves the eutectoid point to the left and increases the critical range of temperatures. Nickel forms neither a carbide nor an oxide when added to steel as an alloy. It is soluble in gamma iron in all proportions and in alpha iron to the extent of about 25 per cent regardless of carbon content, and because of this high solubility it is used in many steels as a means of increasing strength by strengthening ferrite without decreasing ductility. As much as 5 per cent of nickel may be added to increase strength and toughness without increasing hardness to any extent.

When present in moderate amounts, nickel in steel of medium-or high-carbon content lowers the critical cooling rate sufficiently to permit hardening by quenching in oil; about 3 per cent of nickel causes high carbon steel to retain some austenite when quenched in water, and 12 per cent or more produces austenitic steel which cannot be quench-hardened.

Steels containing from 25 to 35 per cent of nickel are extremely tough and dense, have high resistance to shock, are resistant to corrosion, and have low coefficients of expansion; an alloy steel containing 36 per cent of nickel is *Invar*, which shows practically no dimensional change with variation of temperature in the ordinary room temperature range. Nickel intensifies the effects produced by molybdenum and chromium.

Addition of chromium to a nickel alloy steel results in a steel which retains the toughness and ductility provided by nickel, combined with the hardness and resistance to wear which are characteristic of chromium alloy steel. Optimum results for most purposes are obtained by using about 40 per cent as much chromium as nickel.

Manganese

Manganese is present to some extent in all steel, because it always is added during the process of making steel as a deoxidizing and desulfurizing agent. Steel usually is classified as a manganese alloy steel, however,

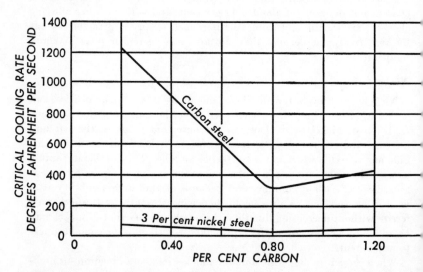

Figure 109. Effect produced on the critical cooling rate by adding 3 per cent of nickel.

Courtesy International Nickel Co., Inc.

only when the manganese content is 1 per cent or more, and when the sulfur content is low. Manganese dissolves in ferrite and hardens it to a marked degree, and also forms carbide. Addition of manganese lowers the critical range of temperatures and moves the eutectoid point to the left. It has sufficient effect upon the time required for transformation to improve hardenability to such a degree that oil quenching is practicable; water quenching of manganese alloy steel tends to make it too brittle for most purposes.

Steel which contains more than about 9 per cent of manganese together with about 1 per cent or slightly more of carbon remains austenitic upon slow cooling to room temperature. Although there is no transformation from austenite to pearlite, steel of this composition is given heat treatment for the purpose of refining grain size and to retain the carbides in solution; such steel combines toughness with strength.

Silicon

Some silicon is necessarily present in all steel because it is added in some stage of steel making to act as a deoxidizer. When it has been added only for its deoxidizing properties, the content usually does not exceed about 0.5 per cent, but as an alloying element it may be present in amounts as high as 5 per cent. The presence of silicon produces little or no effect upon the position of the eutectoid point, but raises the range of critical temperatures. The principal function of silicon as an alloying element is that of strengthening ferrite. It does not form carbide, but on the contrary, has a tendency to decompose other carbides and thus form graphite; for that reason it is necessary to add another element such as chromium or molybdenum to stabilize the carbide in steel which has relatively high contents of both carbon and silicon.

Addition of silicon to steel which contains extremely small amounts of both carbon and manganese produces steel which has high magnetic permeability and low hysteresis loss. Steels containing from 1.25 to 4 or even 5 per cent of silicon are used for the sheets used in such electric equipment as generators and transformers. Such steels have coarse grain structures and are free from internal stress; they are also hard and brittle, but that is not a disadvantage because they are not used in applications where mechanical strength is required.

Molybdenum

Molybdenum added to steel forms a complex carbide of iron and molybdenum, which stabilizes martensite; it also dissolves to some extent

in ferrite and thereby adds toughness to steel. The presence of molybdenum in steel increases the critical range of temperatures, moves the
eutectoid point slightly to the left, and increases hardenability and uniformity of hardening to a marked degree. Progressive increase in molybdenum content has a strong effect in lowering the transformation point.
Molybdenum is therefore more effective than any other common alloying
element in imparting oil-hardening properties and air-hardening properties to steel. It has the greatest hardening effect of any element except
carbon, but at the same time minimizes enlargement of grain, with the
result that toughness is retained. In steel of high molybdenum content,
molybdenum has a tendency to form an oxide which volatilizes at rolling
or forging temperatures unless protected by a flux such as borax.

Molybdenum is used alone in some steels, but usually as a supplement
to other alloying elements, particularly nickel or chromium or both, because it augments the desirable properties imparted by the other alloying
elements.

Vanadium

Vanadium is a powerful deoxidizing agent, and forms complex carbides
with iron. It forms finely divided particles of refractory vanadium oxide
which become distributed throughout the mass of steel and act as nuclei
for crystallization during solidification, thereby producing a fine grain

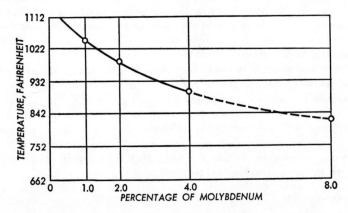

Figure 110. Lowering of transformation temperature by molybdenum.
From *American Machinist*, March 14, 1946, courtesy McGraw-Hill Publishing Co.

structure. The oxide particles also inhibit grain growth when steel is heated above the critical range for heat treatment. Some of the vanadium added to steel dissolves in the ferrite, with consequent strengthening and toughening effect.

Vanadium has a greater tendency toward formation of carbide than have any other common alloying elements except niobium and titanium; for that reason only small amounts are needed to produce steel of high hardness, and few steels except tool steels contain more than about 0.2 per cent vanadium. An outstanding property of vanadium alloy steel is its resistance to softening by tempering.

Tungsten

Tungsten as an alloying element of steel has properties similar to those of molybdenum, but considerably more tungsten than molybdenum is required to produce an equal effect. It has a strong tendency to form carbides, and also dissolves to some extent in ferrite, thus providing both toughness and hardness. Tungsten increases the ability of steel to remain hard at red heat, which makes it suitable for use as a component of some tool steels. Tungsten forms hard undissolved carbide particles which aid its cutting properties and inhibit grain growth. For the most part, tungsten is used in combination with other alloying elements.

Cobalt

Cobalt added to steel forms little carbide; its chief function is that of strengthening ferrite. It is one of the few alloying elements which move the eutectoid point to the right, and also one of the few which reduce hardenability. Used in tool steel in combination with tungsten, it helps in hardening ferrite to such an extent that cutting edges can be maintained at red heat. Magnet steel containing cobalt in combination with chromium and tungsten retains high residual magnetism.

Code Numbers of Steels

Steels are identified by a code of symbols adopted by the American Iron and Steel Institute. This code, known as the *AISI* code, is based upon the use of a number of four or five digits which describes the approximate chemical composition of each steel, preceded by a letter symbol which designates the manufacturing process by which the steel is made. The letter symbol is used because two steels with substantially the same content of carbon and alloying elements may show slightly different analyses for incidental elements; for example, a basic open-hearth steel usually contains greater amounts of phosphorus and sulfur than does an electric furnace steel of similar carbon and alloy contents. The *SAE* code of the Society of Automotive Engineers uses the same number for each steel, but the letter prefix is omitted.

LETTER PREFIXES The processes designated by the letter prefixes are:

A Open-hearth alloy steel
B Acid bessemer carbon steel
C Basic open-hearth carbon steel
D Acid open-hearth carbon steel
E Electric furnace steel of both carbon and alloy types

NUMERICAL DESIGNATIONS The first two digits of the four-digit or five-digit number which designates a steel indicate the type of alloy; the last two or three digits indicate the approximate middle of the permissible range of carbon content. The classifications of carbon and alloy steels are shown in Table 6.

NATIONAL EMERGENCY STEELS During the war, shortage of some alloying elements made it necessary to develop replacement steels, known as *National Emergency* or *NE steels*, in which the more plentiful alloying elements could be used. These steels were developed to have the same hardenability of the grades which they replaced. Some will no doubt be continued in use by industry, and permanent code designations will be assigned; at present the code number of each is preceded by the letters NE.

TABLE 6 AISI OR SAE CODE FOR STEELS

Designation	Type of Steel
10xx	Basic and acid open-hearth and acid bessemer carbon steel grades, nonsulfurized and nonphosphorized
11xx	Basic open-hearth and acid bessemer carbon steel grades, sulfurized but, nonphosphorized
12xx	Basic open-hearth carbon steel grades, phosphorized
13xx	Manganese 1.60 to 1.90 per cent
23xx	Nickel 3.50 per cent
25xx	Nickel 5.00 per cent
30xx	Nickel 0.70 per cent; chromium 0.70 per cent
31xx	Nickel 1.25 per cent; chromium 0.60 per cent
32xx	Nickel 1.75 per cent; chromium 1.00 per cent
33xx	Nickel 3.50 per cent; chromium 1.50 per cent
40xx	Molybdenum
41xx	Chromium-molybdenum
43xx	Nickel-chromium-molybdenum
46xx	Nickel 1.65 per cent; molybdenum 0.25 per cent

TABLE 6   AISI OR SAE CODE FOR STEELS—*continued*

Designation	Type of Steel
48xx	Nickel 3.25 per cent; molybdenum 0.25 per cent
50xx	Low chromium
51xx	Medium chromium
52xxx	Chromium, high carbon
61xx	Chromium-vanadium
86xx	Nickel 0.55 per cent; chromium 0.50 per cent; molybdenum 0.20 per cent
87xx	Nickel 0.55 per cent; chromium 0.50 per cent; molybdenum 0.25 per cent
92xx	Manganese 0.80 per cent; silicon 2.00 per cent
94xx	Manganese 0.95 to 1.15 per cent; silicon 0.50 per cent; nickel 0.35 per cent; chromium 0.30 per cent; molybdenum 0.12 per cent
95xx	Manganese 1.35 per cent; silicon 0.50 per cent; nickel 0.55 per cent; chromium 0.50 per cent; molybdenum 0.20 per cent

Corrosion-Resistant Steel

Corrosion of steel at ordinary temperatures is in most cases a process of gradual oxidation. This is produced by combined action of moisture and oxygen, by action of various chemical substances such as acids, or by the action of corrosive atmospheres which result from some chemical process. At high temperatures, corrosion by formation of oxides and sulfides is often so rapid that a scale forms on the surface of steel. The best protection against corrosion is the formation of a closely adherent film of oxide on the surface of the steel, but unfortunately the ferric oxide which forms initially on iron or steel is not sufficiently adherent to afford any useful protection; it is therefore necessary to add an alloying element or elements capable of forming an efficient film of oxide. Although copper and nickel have some beneficial effect in protecting steel from corrosion, the most potent alloying element has been found to be chromium; consequently all corrosion-resistant (stainless) steels have high contents of chromium. Oxidation of chromium starts more rapidly than does oxidation of iron, with the result that a thin transparent film of chromium oxide forms on the surface of chromium alloy steel, protecting it from further action of moisture and oxygen. Because tendency to corrode increases with increasing temperature, steel which is to be used in high-temperature service must have greater resistance to corrosion and therefore greater content of chromium than is required in steel used at ordinary temperatures.

For a low-carbon steel that contains about 0.15 per cent or less of carbon, approximately 11 per cent of chromium is required to provide adequate resistance to corrosion by moisture and air, or by fruit juices. As the carbon content of steel is increased, some of the chromium acts to form carbide and is therefore not available to form the protective oxide film; to compensate, about 1.0 per cent more chromium for each additional 0.05 per cent of carbon is required to maintain resistance to corrosion. Resistance to corrosion increases almost proportionally to chromium content of steel; other alloying elements are added to stainless steel chiefly for their effects upon the physical properties rather than for such assistance as they may give in prevention of corrosion. Many varieties of stainless steel have been developed to serve specific purposes.

Martensitic Stainless Steels

Steels of this type contain typically less than 18 per cent chromium, together with small amounts of aluminum, niobium, molybdenum, tungsten, nickel, or silicon. They are magnetic and may be hardened and tempered by methods similar to those used for plain carbon steels. These steels harden uniformly throughout even sections of large cross section, with oil or air cooling. With minimum chromium content of about 11.5 per cent, steel of this variety is resistant to weather, water, steam, and mildly corrosive chemicals. Rustless 16-2 contains a maximum of 0.20 per cent carbon, 15 to 17 per cent chromium, and 1.25 to 2.5 per cent nickel; it can be hardened to about 400 to 440 Brinell. Rustless 17-C-100 is a high-carbon steel containing 0.95 to 1.10 per cent carbon, 16 to 18 per cent chromium, but no nickel; it can be hardened to about 620 to 630 Brinell.

Ferritic Stainless Steels

Ferritic stainless steels contain usually more than 18 per cent of chromium, but with such low carbon content that they do not form austenite at any temperature however high; they therefore cannot be hardened by heat treatment, but can be given a hard surface by nitriding. Because the iron content has the alpha structure, these steels are magnetic. Some varieties contain small amounts of copper, molybdenum, nickel, nitrogen, silicon, or tungsten. Their resistance to corrosion is considerably greater than that of martensitic stainless steels, and they are suitable for use with nitric and other strongly oxidizing acids.

Austenitic Stainless Steels

These steels contain enough nickel in addition to the chromium to cause the steel to retain the austenitic structure at all temperatures; because

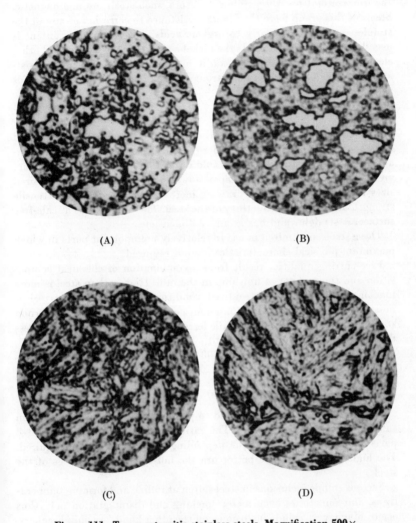

(A)   (B)

(C)   (D)

Figure 111. Two martensitic stainless steels. Magnification 500×.

(A) Structure of annealed rustless 17-C-100 Type 440C. (B) Structure of hardened rustless 17-C-100 Type 440C. (C) Structure of annealed stainless rustless 16-2 Type 431. (D) Structure of hardened stainless rustless 16-2 Type 431.

Courtesy Armco Steel Corporation, Baltimore Works.

the iron content has the gamma structure, such steels are nonmagnetic. Steels of this group have the highest resistance to corrosion of any of the stainless steels, particularly to organic acids. The chromium content is usually approximately double the nickel content, or vice versa, and other elements sometimes are added; addition of molybdenum increases resistance to corrosion by brine and strong reducing agents such as sulfurous acid.

Free-Cutting Steels

Free-cutting steels are plain carbon steels to which more sulfur and manganese have been added in order to make them more readily machinable. Most free-cutting steels are cold drawn at the mill; hot rolled bars are passed through dies that taper down in size to produce the desired sizes and shapes. This cold drawing increases the hardness and tensile strength and lowers the ductility of the steel. The product has a scale-free surface, is straight, and has accurate dimensions.

These steels are limited in use to relatively unimportant parts in which maximum physical characteristics are not required.

Free-cutting qualities result from a combination of chemical composition and the manner of finishing in the mill. High-carbon steel is more machinable if in the spheroidized condition. Medium-carbon steel is more machinable if in the laminated pearlitic condition. Low-carbon steels, which are mainly soft ductile ferrite, are difficult to machine unless they are made free-cutting by addition of sulfur or lead.

Free-cutting steels are in the 1100 series of the AISI and SAE designations.

Chemistry of Free-cutting Steel

Manganese and sulfur are added to plain carbon steel to produce iron sulfide and manganese sulfide. The ratio of manganese to sulfur is 3–8 parts manganese to 1 part sulfur, depending on the grade of steel required. The higher this ratio, the better are the hot-working properties of the steel.

Ferrous sulfide inclusions in steel soften at rolling and forming temperatures, and sometimes even melt; cracking and disintegration may thus occur. Manganese sulfide inclusions, on the other hand, are chip breakers, and are more refractory, less likely to segregate at grain boundaries, and sufficiently plastic to deform with the metal during rolling or forming.

The sulfur acts to reduce rubbing friction between the chip and the tool in machining, thus reducing the force on the tool. It also improves chip

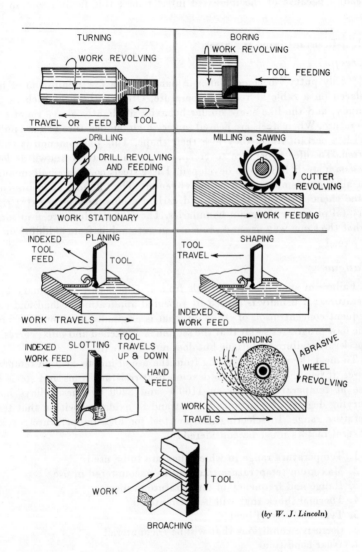

Figure 112. Types of machining.

(by *W. J. Lincoln*)

formation, causing the chips to coil more tightly and thus break more readily because of the increased interference; this facilitates chip disposal.

Heat-Resistant Irons and Steels

Creep

"Silly putty" becomes flat as a pancake in a few minutes after it is placed on a table at room temperature. Other substances such as tar, candy, and the like show similar behavior at slightly above room temperature. When steels are heated to very high temperature and are pulled with a certain force, they lose their shape. This phenomenon is called *creep*. To offer resistance to creep, special alloy steels known as *heat-resistant* steels have been developed. These steels contain small amounts of alloying elements such as molybdenum. The alloying elements combine chemically with the iron and carbon atoms, and form extra rigid crystal plates in the grain boundaries. The plates reduce creep in somewhat the same way as sand prevents an automobile from skidding on an icy road.

Fatigue

Failure of steel or other metals from *fatigue* occurs at ordinary temperatures; it usually is caused by repeated application of load and consequent concentration of stress on what is originally a minor flaw, until the flaw progresses so far that the section is weakened and fails; corrosion accelerates failure by fatigue but does not initiate it.

To meet the problems arising from the use of metals at high temperatures, it became necessary to develop heat-resistant irons and steels for use at temperatures between 1100°F and 2200°F. These alloys have varying degrees of resistance to heat and to oxidation within that temperature range. In selecting iron or steel for high-temperature service, several factors must be considered:

1. Temperature range in which they are to be used
2. Maximum temperature that will be encountered in use
3. Range and frequency of temperature cycling
4. Thermal shock that will be encountered
5. Type and size of load
6. Corrosive conditions that will be encountered
7. Wear conditions
8. Ease of fabricating and replacing

The strength and load-carrying abilities of ferrous alloys decrease with increase of temperature above 600°F. When ferrous alloys are stressed, they first suffer elastic deformation; when the stress becomes sufficiently high, they deform in a plastic manner. An increase in temperature lowers the stress point at which plastic deformation begins, so that it is important to determine the stress that produces the maximum allowable plastic deformation at a specified temperature, for long periods of time.

Cast Irons

Cast irons show little creep up to 600°F. Above 800°F their value is measured by their ability to resist growth and oxidation. Above 850°F they deteriorate rapidly, and are unsafe for use.

Malleable Irons

Malleable iron castings should not be used at temperatures above 1200°F, because their carbon dissolves above 1400°F. These castings have excellent stability at elevated temperatures.

Steels

Although plain carbon steels have comparatively low strength, they are widely used between 300°F and 1000°F because of their resistance to oxidation. These steels are relatively inexpensive and easy to weld and form.

Molybdenum and chromium-molybdenum steels are of pearlitic or martensitic type, and are used at temperatures below 1200°F. They are of low carbon content and are used in the annealed or normalized condition. They have good creep strength, and at intermediate temperatures their resistance to oxidation is relatively low. Because of their excellent ductility, they are easy to work and to weld.

Chromium-nickel austenitic stainless steels are widely used at elevated temperatures because of their high strength and excellent resistance to scaling. They have, however, higher coefficients of thermal linear expansion than the ferritic steels. Unlike plain carbon steels, the thermal conductivity of austenitic stainless steels increases significantly with increase in temperature in the range 300–1500°F. In general, austenitic stainless steels are less resistant to carburization than ferritic stainless steels.

Some heat-resistant steels used at high temperatures are substantially ferritic, and are not hardenable by heat treatment. Others are substantially martensitic and are hardenable by heat treatment. They have a good combination of oxidation resistance and strength, with low coefficient of thermal expansion. Analysis shows 11–30 per cent chromium, 0.1–0.3 per cent carbon, plus minor amounts of molybdenum, vanadium, tungsten, and cobalt, which, since they are carbide formers, cause complex carbides to form. However, since they diffuse slowly, the growth of stable carbides is slow. As a result, softening of these steels is greatly retarded, and they maintain good hardness and strength after relatively high-temperature tempering. All steels of this group are air hardening.

Tool Steel

Originally all tool steel was plain carbon steel of relatively high carbon content, and plain carbon tool steel with carbon content ranging from 0.70 to 1.30 per cent still is used in large quantities. Steel of this type is more easily worked than are many alloy steels, may be hardened to almost any desired degree, and has considerable resistance to shock. For tools of simple design which would not be susceptible to breakage on hardening, and for work in which the temperature of the cutting edge of the tool is not raised above about 500°F, plain carbon steel is satisfactory. Some low-alloy tool steels which contain small amounts of chromium, manganese, silicon, or vanadium have properties similar to those of plain carbon steel, and usually are classified with it. As rapid production methods were introduced in industry, it became necessary to operate machine tools at higher speeds, with consequent production of higher temperatures which often keep a tool at red heat. At such high temperatures, plain carbon steel softens rapidly, and it became necessary to develop alloy steels which retain hardness at red heat. The development of tools of intricate design which could not be made of plain carbon steel and hardened without risk of breakage necessitated development of suitable grades of alloy steel. Alloy tool steels are made to give special characteristics for special purposes; the important requirement may be resistance to deformation on hardening, ability to operate continuously at red heat without softening, resistance to shock, or resistance to abrasion.

In an article in the March 14, 1946 issue of *American Machinist*, S. C. Spalding presented a logical classification and discussion of the several types of tool steel; Table 7, which shows the chemical compositions of typical steels of each class, is reproduced from that article by permission.

The steels listed in the second group in Table 7 are those of medium total alloy content. Steels of composition similar to that of steel No. 4 are oil hardening and practically nondeforming during heat treatment because of the small change in volume which occurs, but they cannot be sharpened to give a keen cutting edge. Steel No. 5 contains enough tungsten to impart additional hardness and permit its use at somewhat higher cutting speeds than would be practical with plain carbon steel; No. 6 and No. 7 are used for cutting and swedging tools and for special purposes. Steels No. 8 to No. 11 inclusive are shock-resistant types used for such tools as punches, chisels, rivet sets, heavy-duty shears, and tools used in boilermaking.

In the third group are listed four air-hardening steels which have great resistance to wear, are not subject to deformation during hardening, and can be oil hardened or air hardened throughout large cross sections. They have greater resistance to heat than have the steels of the preceding groups. No. 14 and No. 15 are lower in cost than are the other two.

The hot-work steels in the fourth group are listed in order of decreasing resistance to heat. They are used in shaping processes which are carried out at elevated temperatures, such as hot drawing, die casting, forging, extrusion, punching, and the like.

High-speed steels, of which typical examples are shown in the fifth group of Table 7, remain hard and retain cutting edges even when used at red heat. High-speed steel is also abrasion resistant at ordinary temperatures because it contains hard carbide particles embedded in its matrix. High-speed steels therefore are useful in hand tools because they hold an edge longer than do plain carbon steels or low alloy steels. After high-speed steel has been quenched, it contains an appreciable amount of residual austenite in addition to martensite. When it is tempered at increasingly high temperatures, the hardness increases because some of the residual austenite becomes converted into martensite; this same phenomenon, called *secondary hardening*, occurs when tools made of high-speed steel become heated in use.

Stellites

Stellites are corrosion-resistant nonferrous alloys which have a high degree of red hardness, and which therefore are useful for cutting purposes. They sometimes are used also for hard facings and for surgical instruments and mirrors. Stellites are made in different compositions, but in general they contain large proportions of cobalt, moderate proportions of chromium, and some tungsten, together with small amounts of carbon and traces of other elements such as iron and silicon.

Questions

1 State the advantages and disadvantages of plain carbon steel.
2 Name the elements usually found in plain carbon steel, and state the effect of each.
3 What is an alloy steel? Why are alloy steels important?
4 Give three reasons for using alloy steel.
5 What is the AISI number of a steel?
6 State the effect of each alloying element used in steel.
7 Define corrosion and explain how it can be prevented.
8 What is a free-cutting steel?
9 Explain the chemistry of a free-cutting steel.
10 Define *creep* and *fatigue* and their effects.
11 What purposes are served by the use of heat-resistant irons and steels?
12 What is an air-hardening steel?
13 Why are tool steels important to industry?
14 What is the most important characteristic of high-speed steel?

TABLE 7 CLASSIFICATION OF TOOL STEELS

Type	C	Mn	Si	Cr	W	Mo	Ni	Co	V
I—CARBON TOOL STEELS									
1. Carbon Tool	0.70/1.30	0.25	0.25						0.20
2. Carbon Vanadium	0.70/1.30	0.25	0.25						
3. Carbon Chrome	0.80/1.30	0.25	0.25	0.50					
II—ALLOY TOOL STEELS									
4. Oil-Hardening	0.80/1.0	1.0/1.5	0.25	0.60*	0.60*	0.60*			
5. Tungsten Finishing	1.2/1.5	0.30	0.30	1.25*	3.0/4.5				0.25*
6. Tungsten Tap	1.0/1.3	0.30	0.10	0.25/0.80	1.0/1.5	0.30*			0.25*
7. Chrome Roll	0.85/1.25	0.15/06.5	0.35	1.1/1.85	2.50	0.50*			0.25
8. Chrome-Tungsten	0.40/0.60	0.15/0.70	0.15/1.15	1.50					0.25
9. Chrome-Vanadium	0.45/0.60	0.20/0.90	0.35	0.80/1.20		0.40*			0.50*
10. Chrome-Nickel	0.50/0.80	0.25/0.65	0.25	0.60/1.20		1.50*	1.25/1.75		
11. Silicon-Manganese	0.45/0.70	0.30/1.20	0.70/2.25	0.50*					0.50*
III—HIGH-CARBON, HIGH-CHROMIUM, AND AIR-HARDENING DIE STEELS									
12. High-Chromium Oil-Hardening	1.7/2.4	0.35	0.35	10.0/14.0		1.0*		0.60*	1.0*
13. High-Chromium Air-Hardening	1.2/1.7	0.35	0.35	11.0/14.0		0.50/1.0		3.5*	1.0*
14. Air-Hardening Die	0.90/1.10	0.55	0.35	0.5/6.0		0.75/1.25			0.70*
15. Air-Hardening Manganese	0.95/1.05	2.0	0.30	0.80/2.0		0.80/1.10			1.0*
IV—HOT-WORK STEELS									
16. Tungsten Hot-Work	0.25/0.55	0.30	0.30	3.5	7.0/18.0				0.40/1.20
17. Chrome Tungsten Hot-Work	0.30/0.45	0.30	1.0	5.0	4.5				
18. Chrome Tungsten Molybdenum Hot-Work	0.30/0.45	0.30	1.0	5.0	1.25	1.50			0.40*
19. Chrome-Molybdenum Hot-Work	0.30/0.45	0.30	1.0	5.0		1.50			0.40
20. Chrome Hot-Work	0.80/1.10	0.30	0.30	4.0		0.50*			
V—HIGH-SPEED STEELS									
21. Tungsten 18-4-1	0.70	0.30	0.30	4.0	18.0	1.0			1.0
22. Tungsten High-Vanadium	0.75/1.25	0.30	0.30	4.0	14.0/20.0	1.0			2.0/4.0
23. Tungsten-Cobalt	0.70	0.30	0.30	4.0	14.0/20.0	1.0		3.5/12.0	1.0/2.5
24. Tungsten-Molybdenum	0.80/1.3	0.30	0.30	4.0	6.0	5.0			2.0/4.0
25. Molybdenum-Tungsten	0.80	0.30	0.30	4.0	1.5	8.5			1.2
26. Molybdenum	0.80	0.30	0.30	4.0		8.0			2.0
27. Molybdenum-Cobalt	Either 4, 5, or 6% Molybdenum plus 3.5—8% Cobalt								

* Optimum percentage.

18 / Corrosion or Rusting

General

Corrosion takes a toll of millions of dollars annually, and if the proper preventatives and the proper coatings were not employed the toll would run many times higher. Because corrosion problems affect all consumers of metals, a knowledge of the causes of corrosion or rusting and its prevention is important to every user or fabricator of metals.

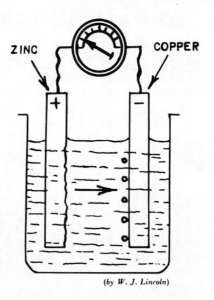

(by W. J. Lincoln)

Figure 113. Simple galvanic cell.

Corrosion is due to the inherent tendency of most metals, particularly ferrous metals, to change to more stable compounds, generally the same compounds found in the ore from which the metal was extracted. Corrosion may be described as the destructive chemical attack of a metal by agents with which it comes into contact such as rain, air, and sea water. The most familiar type of corrosion is the attack on metal articles by the atmosphere and by moisture. Corrosion is generally accompanied by the setting up of small electric currents caused by the interchange of electrons (current flow) between the corroding medium and the base metals.

New corrosion problems have been created by the higher temperatures and pressures involved in modern chemical reactions, which, in turn, have greatly increased the corrosion rate of equipment used in the chemical industry. Also, the higher steam pressure and hot gases used in modern gas turbines have created new corrosion problems for the power equipment industry.

1. Atmospheric
2. Bacterial
3. Chemical
4. High-temperature oxidation
5. Soil
6. Stray current electrolysis
7. Underwater

Electrochemical Corrosion

Research has discovered that moisture is present in almost every environment in which metals operate, and as a result, virtually all corrosion is primarily an electrochemical process. The actual corrosion results from the presence of many small local galvanic cells which are similar to small batteries.

In a battery the electric current is produced by suspending two dissimilar metals (electrodes) in an electrolyte. When the circuit is completed by a wire, one metal dissolves in the solution, thereby setting up a difference in potential. This causes an electric current to flow through the solution from the dissolving metal to the second metal.

By applying the same principle to a piece of metal in contact with moisture, it can be seen that the metal will behave like a small battery, since the presence of particles of an impurity or contact with some other metal or difference in the environment from point to point sets up a slight difference of voltage thereby causing a very small current to flow. The

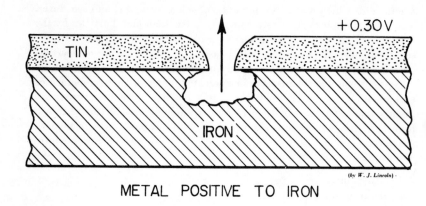

METAL POSITIVE TO IRON

METAL NEGATIVE TO IRON

Figure 114. Corrosion of metal positive to iron ; galvanic protection by zinc coating (negative to iron).

moisture-containing air or some dissolved chemical will conduct the electricity and cause local attack to begin. During the rusting of iron, for example, iron replaces hydrogen or a metal in a weak electrolyte. Since the air contains carbon dioxide, oxides of sulfur, small amounts of salt, and moisture, it serves as an electrolyte.

The ability of every metal or alloy to replace hydrogen or other metals in electrolytes is known as its electropotential. The difference in potential results from the fact that different metal surfaces possess different solution or electrode potentials. A metal that dissolves readily in solution possesses a high electrode potential; a metal that does not dissolve readily in solution has a low electrode potential. Magnesium and aluminum belong to the former group; copper and silver, to the latter group.

The difference in potential responsible for electrochemical corrosion can be established by:

1. Use of dissimilar metals—galvanic corrosion
2. Inhomogeneity in a single metal—direct electrochemical corrosion
3. Inhomogeneity in the contacting solution—concentration cell corrosion

Galvanic corrosion is the most rapid and destructive type.

Two reactions occur simultaneously during electrochemical corrosion: (1) metal ions dissolve at areas of higher potential (anodes); and (2) hydrogen ions pass out of solution at areas of lower potential (cathodes). As a result, the current will flow from the metal of high potential to the metal of lower potential. It is this flow of current which increases the amount of corrosion over and above that which would occur if the metals were not coupled galvanically.

The extent of corrosion caused by any galvanic couple is determined by the following five factors:

1. The electrode potential of the two metals
2. The polarizing and film-forming characteristics of the metal in the corroding solution
3. The relative areas of the anode and cathode
4. The internal and external resistances of the galvanic circuit
5. The factors that normally influence general corrosion processes

Electromotive Series

In the electromotive series the metals have been arranged in increasing order of activity when immersed in solutions of a definite strength, with their electropotential expressed in volts. For convenience the zero point has been placed at iron. Two metals close to each other in the series cause minimum corrosion to take place, whereas two metals at opposite ends of the series cause maximum corrosion to occur, everything else being equal.

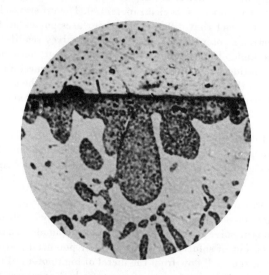

Figure 115. Photomicrograph of a joint showing complete wetting of the aluminum (*bottom*) by the zinc-base solder (*top*). Magnification 500 ×.

Courtesy *Bell Laboratories Record*.

TABLE 8   ELECTROMOTIVE SERIES

Metal	Electro-potential	Metal	Electro-potential
Gold	+1.86	Nickel	+0.19
Platinum	+1.64	Iron	0.00
Silver	+1.24	Cadmium	+0.04
Copper	+0.78	Chromium	−0.27
Hydrogen	+0.44	Zinc	−0.32
Lead	+0.31	Aluminum	−1.23
Tin	+0.30		

Based on data from W. W. Latimer, *The Oxidation States of the Elements and Their Potentials in Aqueous Solutions*. New York: Prentice-Hall, Inc., 1938.

A study of the table shows that tin has less tendency to go into solution and therefore less tendency to rust than iron. Zinc, though it is a less durable material, is preferred as a protective coating for iron and steel, because it provides galvanic protection of any iron that may be exposed

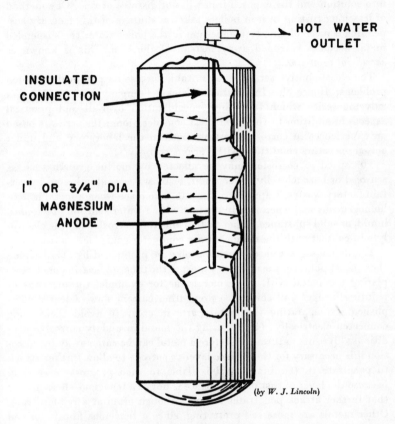

HOT WATER OUTLET

INSULATED CONNECTION

1" OR 3/4" DIA. MAGNESIUM ANODE

(*by W. J. Lincoln*)

Figure 116. Cathodic protection in a water-heater tank.
Courtesy *Materials and Methods*, July, 1949.

at breaks in the coating. The coating should be thin for economical and mechanical reasons and therefore will probably be imperfect and have scratches, pits, and other breaks in the surface. These openings permit the corrosive electrolyte to come in contact with the iron and coating metal. In the case of tin and iron, the iron is the more active metal and will therefore dissolve.

In the case of zinc and iron, the zinc is the more active metal and as long as any zinc remains near the break in the surface the zinc will go into solution and the exposed iron will not dissolve or rust. One method of lessening rusting in iron boilers takes advantage of this fact. By suspending a piece of zinc or magnesium in the boiler water the suspended metal becomes corroded away instead of the iron. This is known as *sacrificial protection*.

The electromotive series is not suitable for use on practical corrosion problems. Hence the International Nickel Company has developed a galvanic series which takes into consideration over-all and practical aspects in addition to the theoretical considerations; the series is based on experiences in corrosion testing, both in the laboratory and under actual operating conditions with various corrosives.

The speed of corrosion may be affected by marine growth such as seaweed or barnacles, by sewer water, or by water contaminated by certain factory wastes. Rapid corrosion can take place when a metal is non-homogeneous and when it is in contact with a water solution of gaseous, liquid, or solid substance. Other factors affecting corrosion are the electrolyte in contact with the metal; a poor conductor causes less corrosion.

The damaging action of corrosion may be minimized by: (1) keeping the contact points of the metal dry most of the time; (2) coating or electroplating the metal with other metals, as for example, tinning cans or painting bridges and ships; (3) preventing current flow, a feat accomplished by separating the metal parts in order to avoid their being connected electrically; (4) not using the metal in unduly corrosive conditions; (5) using nature—for before a metal can be eaten away by corrosion it is necessary for the surface layer to corrode to allow further attack to penetrate it. (Ordinary steel is liable to such progressive rusting, because the layer of rust on the surface is not of a tenacious character, so that further attack penetrates inwards, though often at a reduced rate. Other metals are more self-protecting, since a tenacious film is formed which seals the metal from further attack. Aluminum provides an illustration of this, for it forms a thin but strong film of aluminum oxide on its surface which, when the metal is pure, prevents further corrosion from taking place. Stainless steels containing chromium are similar, their corrosion resistance being due to the passivity caused by the formation of a protective oxide film. Stainless steels are corroded appreciably only by reagents, such as chlorides, that break down their passivity.)

TABLE 9   GALVANIC SERIES OF METALS AND ALLOYS

Corroded End (anodic, or least noble)

Magnesium
Magnesium alloys

Zinc

Aluminum 2 S

Cadmium

Aluminum 17 ST

Steel or Iron
Cast Iron

Chromium-iron (active)

Ni-Resist

18–8 Stainless (active)
18–8–3 Stainless (active)

Lead-tin solders
Lead
Tin

Nickel (active)
Inconel (active)

Brasses
Copper
Bronzes
Copper-nickel alloys
Monel

Silver solder

Nickel (passive)
Inconel (passive)

Chromium-iron (passive)
18–8 Stainless (passive)
18–8–3 Stainless (passive)

Silver

Graphite
Gold
Platinum

Protected End (cathodic, or most noble)

From *Aspects of Galvanic Corrosion*, International Nickel Co., Inc.

(6) isolating the metal from the corrosive media by some physical barrier such as an organic, metallic, or ceramic coating; (7) cathodic protection, a method which prevents or virtually eliminates corrosion by passing an electric current opposite in direction to the small current formed in the corroding medium.

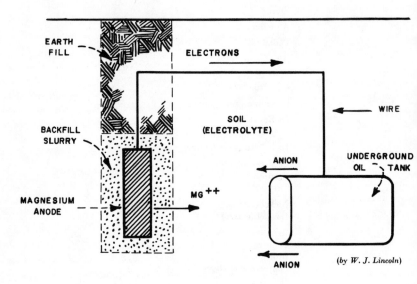

Figure 117. Cathodic protection, with soil acting as electrolyte.

Cathodic Protection

It has been stated that corrosion is primarily electrochemical in nature. It may result from the flow of current from one point on a buried metal (the anode) through the earth and back to the metal at another point (the cathode).

As the current flows, it pulls minute pieces of metal into the soil with it. Actually the surface of a corroding metal is like a series of tiny batteries in which some areas act as anodes, others as cathodes, the earth or water surrounding the metal serving as the electrolyte. The current required by the cathodic areas is supplied by the anodic areas.

Cathodic protection seeks to stop corrosion by means of an external anode. The latter supplies the current required by the local cathodes, thereby saving the local anodes from corrosion. Cathodic protection is sometimes applied to large steel domestic hot and cold water tanks, super-heaters, heat exchangers, and to sewer pipelines, conduits, domestic oil tanks, etc., buried in corrosive soils. Magnesium, zinc, and aluminum are widely used for this purpose. They serve as the anodes of a battery which is set up between these anodes in the soil, and the object to be protected is connected to them by means of an insulated wire, the soil acting as the electrolyte. The current thus generated flows from the anode, through the soil, and onto the structure (the cathode). As a result, the destructive current is reversed. In the absence of cathodic protection the current flow is from the structure, thereby causing corrosion. The anodes are called *sacrificial* or *expendable anodes* because they are consumed in the process of protecting the buried structure and must be replaced from time to time. In some instances it is more desirable to use an external source of current as a rectifier, with the current being supplied through ordinary iron or graphite electrodes buried in the soil.

In the case of domestic tanks, cathodic protection is accomplished by suspending sacrificial anodes in as nearly a central location in the tank as possible so as to obtain equitable distribution of current.

The metal used for cathodic protection must not only be anodic to the metal to be protected, but the potential difference developed must be sufficient to polarize properly the local cathodes of the protected metal.

Questions

1   Explain the difference between corrosion and rusting.   How is the meaning of *corrosion* broader than the meaning of *rusting?*
2   How is the speed of corrosion affected by temperature?
3   Explain why modern household tanks are often lined with glass.
4   Steel coated with zinc is commonly termed *galvanized steel.* What advantage is gained by galvanizing steel?
5   What is the cause of corrosion?
6   How can corrosion be minimized?
7   Why is corrosion primarily an electrochemical process?
8   Explain *electropotential.*
9   What is the electromotive series?
10   What is sacrificial protection?
11   Explain how cathodic protection works.
12   How does soil act as an electrolyte?

19 / Copper and Copper Alloys

The outstanding characteristics of copper are its high electric conductivity, its resistance to many kinds of corrosion, and the ease with which it can be worked. Because it has the highest specific conductivity of any metal except silver, copper is used almost exclusively in every application pertaining to transmission or conduction of electricity. Its ductility permits copper to be drawn readily into wire of any degree of fineness, and to be rolled, forged, pressed, or spun into desired shapes without danger of cracking. Copper in the pure state is not readily machinable because its softness and toughness cause it to leave a cutting tool in long curls instead of breaking away cleanly; addition of small amounts of lead or tellurium improves machinability. As cast, the tensile strength of copper is between 17,000 and 20,000 pounds per square inch, but it can be increased to over 65,000 pounds per square inch by cold working. Copper cannot be hardened by any type of heat treatment, but copper that has been hardened by cold work can be softened by annealing. Since copper is strong, it is easy to finish by plating or lacquering. When added to iron or steel, copper increases resistance to atmospheric corrosion. Copper can be welded, brazed, or soldered satisfactorily.

Commercial Grades of Copper

The purest commercial grades are *electrolytic* and *low resistivity lake* copper, both of which are more than 99.9 per cent pure. Such copper ordinarily is allowed to retain from 0.02 to 0.03 per cent of oxygen; this *tough pitch* copper can be cast with a level surface, whereas oxygen-free copper solidifies with a deep shrinkage pipe that makes casting more expensive. Oxygen-free copper is electrolytic copper free of copper oxide; it is obtained by deoxidizing, generally by use of phosphorus.

Copper of highest purity is required for electrical work, because most impurities increase resistivity; arsenic and phosphorus are particularly detrimental in that respect, and even silver, which is the only metal which in the pure state has higher conductivity than that of copper, decreases the conductivity of pure copper when present in small amounts as an impurity.

A grade of copper known as *fire refined other than lake* is refined directly from native copper or roasted ore by smelting. It contains about 99.7 per cent of copper, but there are also present small amounts of impurities which reduce its conductivity below that of electrolytic copper; the impurities may include as much as 0.10 per cent of arsenic. Copper of this grade is used for rolling into sheets and for other mechanical purposes.

A third grade, *arsenical* or *high resistance lake* copper, may contain as much as 0.50 per cent of arsenic. Its greater hardness and tensile strength in the cold-worked condition make it more suitable for some purposes than purer grades. It is better suited to high temperature applications because its recrystallization temperature is higher than that of other grades.

About twenty grades of copper are furnished for special uses, including oxygen-free, deoxidized, silver-bearing, tellurium-bearing, and others.

Copper Ore

Copper ores of the United States average somewhat less than 1 per cent copper. The largest portion of copper is obtained from low-grade ores which contain copper usually as a sulfide. The most abundant of the sulfide ores is one which contains *chalcopyrite*, a double sulfide of copper and iron; *bornite*, another double sulfide of copper and iron, and *chalcocite*, a copper sulfide, are the other important copper-containing minerals found in sulfide ores.

Concentration

Copper ore as received from a mine contains only a small proportion of copper, and therefore must be concentrated greatly before smelting if the smelting process is to be economically practicable. The first operation is reduction of the size of the pieces of ore to about that of fine sand.

Froth or Chemical Flotation Process

In 1911 the froth flotation process was developed for separating the useful material from the waste matter. This process consists in causing the mineral particles to adhere to gas bubbles in an aqueous suspension of the particles of ore, and to be carried to the top of the tank, where the useful material is collected and removed. The effectiveness of the process is increased by treating the ore with a substance called a *collector* or *promoter*, which makes the sulfide material nonwettable without affecting the wettability of the gangue. The mineral surface which thus has been made nonwettable has an increased tendency to adhere to a bubble. Some promoters also have the property of acting selectively on some minerals,

ORE-FROTH FLOTATION

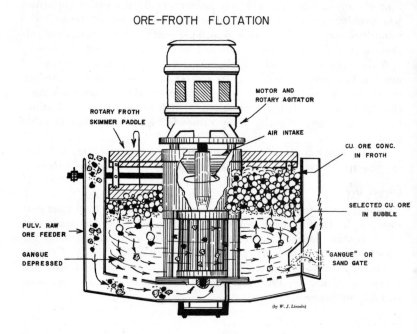

Figure 118. Cutaway diagram of froth-flotation cell.

so that by suitable choice of a promoter and another chemical called a *depressant*, selective flotation and separation of any desired metallic sulfide can be accomplished. The depressant counteracts any effect which the collector may have on the unwanted sulfides, and in the case of copper sulfide flotation prevents them from being floated with those sulfides which contain copper; lime is added, for example, to prevent flotation of iron sulfide.

When air is blown into water, bubbles form and rise to the surface; there they break because of the high surface tension. Obviously a flotation process could not be operated if the bubbles were to break upon reaching the surface of the water, because the floated mineral matter would drop back to the bottom of the flotation tank. To lower the surface tension and thus to permit the bubbles to retain their form after reaching the surface of the water, a *frothing agent* is added. It is usually an organic compound of high molecular weight and of relatively low solubility in water; pine oil and cresylic acid are two substances frequently used for the purpose. The concentration of the frothing agent determines the type of bubbles which form, and therefore is an important factor in the efficiency of the process. If insufficient frothing agent is present, the bubbles are too large and do not carry as much mineral as do smaller bubbles with correspondingly more aggregate surface area. At the optimum concentration the bubbles form a froth which appears dry because it carries relatively little water between the small bubbles; such a froth is most effective in floating the desired mineral particles, and promotes most rapid operation of the process.

A flotation machine consists of a series of cells so arranged that the suspension of mineral particles in water flows successively through them. In each cell there is a rotating rubber agitator which keeps the suspension uniformly mixed and forces the collecting agent into contact with the particles of ore. The bubbles are formed by a stream of air forced into the bottom of each cell. While the bubbles remain below the surface they are separated from each other by water, so that the entire surface of each bubble is available to collect the particles which it is to carry to the surface with it. When the bubbles reach the surface there is no longer water between them, and the bubbles, with the particles of mineral adhering to them, coalesce into a froth. The froth is scraped from the top of each cell and subsequently is thickened, filtered, and dried, preparatory to roasting.

This first step in the concentration of ore results in the elimination of so much unwanted matter that the copper content is about 25 per cent, as compared with about 1 per cent or even less in the original ore.

Roasting

The dried concentrate from the flotation tank, containing about 10 to 12 per cent of water, is roasted in a multiple-hearth furnace to burn off excess sulfur and such elements as arsenic and antimony which form volatile oxides. Much of the iron sulfide is converted to iron oxide during the roasting process, but the reaction is so adjusted that enough sulfur

remains in the calcined ore to combine with all of the copper and the
greater part of the iron during the smelting operation.

Smelting

Smelting increases still further the concentration of copper by melting
the calcined ore and thus permitting separation of much of the unwanted
material by changing it to slag. Smelting is conducted in a reverberatory
furnace heated by pressure-fed pulverized bituminous coal, oil, or gas to
a temperature between 2500°F and 2700°F. At that temperature all
copper compounds are converted to cuprous sulfide, and a considerable
portion of the iron is converted to ferrous sulfide; these two sulfides
dissolve in each other in the molten state to form *matte*, which contains
about 38 to 45 per cent of copper. That portion of the iron which does not
form sulfide remains as iron oxide, and enters the slag with the alumina,
lime, and silica. The matte, which is considerably heavier than the slag,
settles to the bottom of the furnace and is tapped off as required. The slag
either is wasted or is chilled rapidly in water to form a granular product
sometimes used for roadbeds. *Rock wool* is made by blasting molten slag
by a steam jet in special equipment.

Converting

The chemistry of extraction of copper from sulfide ores consists principally in the conversion of ferrous sulfide to ferrous oxide and then to
ferrous silicate, and the subsequent conversion of cuprous sulfide to
metallic copper by oxidation of the sulfur. The net reactions involved are:

$$2\,FeS + 3\,O_2 \rightarrow 2\,FeO\ + 2\,SO_2$$
$$FeO\ + SiO_2 \rightarrow FeSiO_3$$
$$Cu_2S + O_2\ \ \rightarrow 2\,Cu\ \ + SO_2$$

Reduction of the copper sulfide of the matte to metallic copper is
effected in a horizontal cylinder with basic lining, somewhat similar to
the bessemer converter used in making steel. The charge in a converter of
usual size is about 65 tons of matte, together with enough raw ore to
supply silica and alumina for slagging purposes. The process of conversion
takes place in two distinct stages.

During the first stage the iron sulfide is decomposed; the iron is oxidized
to iron oxide by the air forced through the converter, and the sulfur burns
to sulfur dioxide, which escapes from the mouth of the converter. Raw gold
and silver ores are added at intervals to supply the alumina and silica required to combine with the ferrous oxide to form a slag. During this

SIMPLIFIED FLOW SHEET OF COPPER SMELTING AND REFINING
(SULFIDE MINERALS)

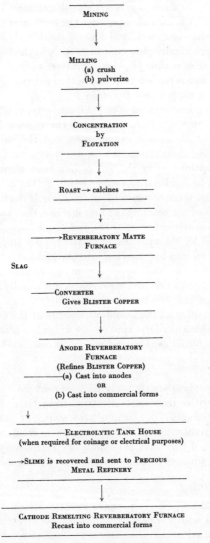

MINING

↓

MILLING
(a) crush
(b) pulverize

↓

CONCENTRATION
by
FLOTATION

↓

ROAST → calcines

↓

→REVERBERATORY MATTE
FURNACE

SLAG

↓

CONVERTER
Gives BLISTER COPPER

↓

ANODE REVERBERATORY
FURNACE
(Refines BLISTER COPPER)
(a) Cast into anodes
OR
(b) Cast into commercial forms

↓

ELECTROLYTIC TANK HOUSE
(when required for coinage or electrical purposes)

→SLIME is recovered and sent to PRECIOUS
METAL REFINERY

↓

CATHODE REMELTING REVERBERATORY FURNACE
Recast into commercial forms

NOTE: Gases from all furnaces are led to COTTRELL PRECIPITATOR.

period the operator judges the progress of the reaction by watching the color of the flame issuing from the mouth of the converter. When the reaction starts, the flame is green; this color changes through blue to deep violet. The violet color indicates that the converter is *high*, which means that the iron has been oxidized almost completely and has entered the slag, leaving practically pure cuprous sulfide in the bottom of the furnace. The air blast then is shut off, the converter is tilted, and the molten slag is poured off and returned to the smelting furnace for recovery of the 2 to 3 per cent of copper that it retains.

After the slag has been removed the converter is tilted up, the air blast is turned on, and the second stage of the process begins. During this period the cuprous sulfide decomposes, and the sulfur burns to sulfur dioxide, leaving metallic copper in the converter; the reaction is not allowed to continue after the sulfur has been burned off, because the copper would be oxidized to copper oxide by the action of the air blast. As with the first stage of conversion, the color of the flame is an indication of the progress of the reaction; at first it is red, then passes through orange and yellow to nearly colorless at the end of the blow. The end point, however, is judged by the appearance of the copper shell which solidifies on the surface of an iron bar inserted into the melt.

At this time the molten copper is poured into ladles, and may be sent in the molten state to a fire-refining furnace, but usually is cast into cakes for shipment to a refinery. When the copper solidifies, a considerable quantity of gas is given off, and the bubbles breaking at the surface of the copper leave marks which have the appearance of broken blisters; for that reason it is known as *blister copper*, and the cakes are called *blister cakes*.

Refining

Blister copper is approximately 98.8 per cent pure, and contains enough metallic impurities to keep its electric conductivity low and to interfere with its malleability. It contains also sufficient sulfur to prevent the making of a usable casting. To remove these residual impurities, blister copper is refined first in a reverberatory furnace by chemical methods and finally in aqueous solution by electrolytic methods.

Furnace Refining

Furnace refining is a preliminary process designed to remove some of the impurities which would contaminate the electrolytic bath if blister copper were used as anodes. Another effect of removal of impurities is production of anodes of more nearly homogeneous structure; use of non-homogeneous anodes would result in severe fluctuations in the electric current, which would cause trouble during electrolytic refining.

A charge of about 175 tons of blister copper is placed in a reverberatory furnace similar to the smelting furnace, and is melted under an oxidizing atmosphere. Air at a pressure of from 16 to 100 pounds per square inch is forced through the copper bath by means of iron pipes inserted through side doors; this air oxidizes the remaining sulfur and some other impurities. Sulfur dioxide gas boils up through the melt and escapes at the top. Some oxygen combines with part of the copper to form cuprous oxide, which dissolves in the bath and must be reduced again to the metallic state. To effect this reduction, the slag containing iron added by the blowpipes is skimmed off, and timbers or green trunks of hardwood trees are inserted into the furnace and forced beneath the surface of the molten copper, which is covered simultaneously with coke. This operation is called *poling*. Gases produced by the moisture content and by decomposition of hydrocarbons which form the structure of wood agitate the bath violently and force any cuprous oxide in the bath into intimate contact with the coke at the surface, causing its reduction to metallic copper. The copper, now about 99.3 per cent pure, is tapped into copper molds coated with bone ash to prevent the castings from burning or sticking. These castings, which are slabs weighing from 500 to 750 pounds each, serve as anodes in the electrolytic refining process. The furnace slag is relatively rich in copper, and is returned to the converter for reclamation of the copper content.

Electrolytic Refining

Electrolytic refining is carried out in a tank containing a solution of copper sulfate acidified with sulfuric acid. The copper is removed electrolytically from the anodes, which are connected to the positive side of a direct-current line, and is deposited on the cathodes of pure sheet copper which are connected to the negative side of the electric circuit.

The solution tanks are made of Douglas fir or concrete, and are lined with antimonial lead. They are in cascaded groups, each with a separate system for circulating the electrolyte at the rate of 6 gallons per minute. The electrolyte is heated initially to about 140°F, and when it leaves the last tank of the cascade, flows back to the heating tank for reheating.

The electrolyte when made up consists of a solution containing about 3.2 per cent of copper as copper sulfate and about 16 per cent of sulfuric acid. During the refining process the copper content increases, and the solution becomes contaminated by some of the impurities which were in the anodes; this makes it necessary to withdraw a portion regularly and to replace it with fresh solution to maintain its chemical composition within proper limits. The contaminated solution is concentrated to crystallize out the greater part of the copper sulfate, and the antimony and bismuth are removed chemically; the remaining strongly acid solution is returned for use in mixing fresh electrolyte. If the electrolyte contains much nickel, the copper is removed by electrolysis and the nickel is crystallized out as nickel sulfate.

With the exception of those which dissolve in the electrolyte, practically all impurities which were contained in the original anodes are precipitated and enter the slime at the bottom of the tank. The slime is removed at intervals and is refined to recover the gold, silver, selenium, and tellurium that are included among the impurities.

The process is operated until the anodes have been reduced to about 8 to 10 per cent of their original weights, when they are removed from the electrolyte and are sent to be melted down into new anodes.

Melting and Refining

The pure electrolytic copper cathodes must be converted into ingots and bars of various shapes and sizes for convenient use by industrial fabricators; for this purpose they are melted down in a furnace similar to that used in preliminary furnace refining. During the melting the copper is oxidized to some extent because of the slightly oxidizing atmosphere of the furnace, with the result that the molten copper contains between 0.15 and 0.30 per cent of oxygen combined with copper as cuprous oxide. It then is blown to about 0.80 per cent of oxygen to eliminate any sulfur picked up from the fuel or any present as copper sulfate from the electrolyte. To reduce the greater part of this cuprous oxide, the procedure is identical with that of the preliminary furnace refining except that the process is controlled more carefully to produce copper in which the oxygen content is within the limits of tolerance. When the oxygen content has been reduced to between 0.02 and 0.03 per cent the molten copper is tapped and poured into molds of the desired shapes.

Alloys of Copper

Copper forms useful alloys with aluminum, lead, nickel, silicon, tin, and zinc. Special alloys of high strength are made with beryllium. In some alloys small amounts of iron and manganese are included. In general, alloys of copper and zinc are called *brasses*, and alloys with aluminum, silicon, tin, and some other metals are called *bronzes*. Other important alloys are copper-nickel alloys such as monel metal, cupronickel, and nickel silver containing also zinc.

Brasses

Commercial brasses contain zinc in amounts varying from about 5 per cent to about 45 per cent. Brasses are readily machinable, and the compositions can be varied to give a wide range of physical properties.

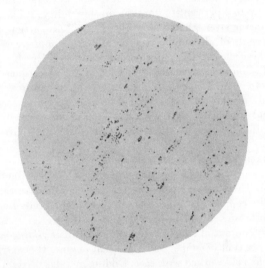

Figure 119. Tough pitch copper showing particles of cuprous oxide. Unetched. Magnification 100 ×.
Courtesy The American Brass Company.

Although brass is resistant to many types of corrosion, a brass which is in an internally strained condition has a tendency to develop cracks along the grain boundaries when subjected to attack by corrosive agents, even those in air. This condition is called *season cracking* or *stress corrosion cracking*. Liability of a brass to this type of failure can be detected by immersing a sample for 4 hours in a solution containing 1 per cent of mercurous nitrate and 1 per cent of nitric acid; if cracks appear during the test, the brass is likely to fail in service. Low-temperature annealing relieves the stresses sufficiently to remove danger of season cracking without reducing the strength of the brass. Season cracking seldom occurs in brasses containing less than 15 per cent of zinc.

Another type of corrosion in brass is *dezincification*. The alloy first dissolves, and the copper content immediately deposits galvanically on the surface in spongy condition. The net effect is removal of the zinc. Brasses of high zinc content are more liable to dezincification than are those low in zinc.

Small amounts of other metals sometimes are added to brass to secure physical properties desired for specific purposes. Lead sometimes is added in amounts up to 3 per cent to improve machinability; it has the adverse effect of reducing strength and ductility. A small proportion, never more than 3 per cent, of aluminum may be added to improve resistance to corrosion; these alloys tend to give porous castings and are hard to handle. Addition of aluminum increases strength but decreases ductility. Tin in amounts up to about 1.5 per cent is added chiefly to improve resistance to corrosion; it increases hardness and decreases ductility. Iron increases hardness and tensile strength to some extent, but is rigidly controlled in brass mills because its presence interferes with control of grain size.

ALPHA BRASSES   Brasses that contain about 38 per cent or less of zinc consist of a solid solution designated *alpha*, which is very ductile but also fairly strong. These *alpha brasses* are the varieties most suitable for cold working. Like copper, alpha brasses can be hardened only by cold work, but when so hardened can be annealed by heating.

The lowest content of zinc is in the alloy known as *gilding metal*, which contains 95 per cent of copper and 5 per cent of zinc. It is extremely ductile and can be cold worked with ease. Gilding metal is used for the manufacture of detonator caps, costume jewelry, and various stamped or forged products.

Red brass, also called *architectural bronze*, contains 85 per cent of copper and 15 per cent of zinc. It has excellent ductility and cold-working properties which permit it to be spun, drawn, stamped, and cold forged. It is stronger than pure copper and when free from lead has hot-working qualities equal to those of copper. Red brass has good corrosion resistance to salt water and is resistant to dezincification and season cracking. It is used for water piping and tubing and for other articles which are subjected to similar corrosive conditions.

Admiralty metal consists of 71 per cent of copper, 28 per cent of zinc, and 1 per cent of tin. It has good mechanical properties and is exceptionally resistant to the corrosive action of salt water. It is used particularly for condenser tubes and plates, and for other parts of heat exchangers used under mildly corrosive conditions.

Cartridge brass, also called *70-30 brass*, contains 70 per cent of copper and 30 per cent of zinc. It has the most favorable combination of ductility and strength of all brasses, and is used widely for drawing and spinning purposes.

Common brass, or *high brass*, contains about 65 per cent of copper and 35 per cent of zinc. It is used for the manufacture of brass pipe and for many drawing and stamping operations. This is the highest in zinc content of the alpha brasses, and therefore the least expensive.

ALPHA-BETA BRASSES Brasses that have zinc contents of between about 39 per cent and about 45.5 per cent contain the alpha solid solution and another known as *beta;* they are therefore known as *alpha-beta brasses*. Brasses of this variety have greater strength but less ductility than have alpha brasses, and their properties can be modified by heat treatment.

Muntz metal contains 60 per cent of copper and 40 per cent of zinc. It has relatively low ductility but considerable strength, and resists corrosion by salt water fairly well. It is used in inexpensive condenser tubes, ship sheathing, and for similar purposes, and sometimes is substituted for the more expensive admiralty metal when low cost is important. Muntz metal containing from 2 to 3 per cent of lead is a free-cutting alloy.

Naval brass is substantially Muntz metal to which tin has been added to increase resistance to corrosion. A typical composition is 60 per cent of copper, 39.25 per cent of zinc, and 0.75 per cent of tin.

Manganese bronze, despite its misleading name, is actually an alpha-beta brass to which manganese and other alloying metals have been added. A typical composition is copper, 56 per cent; zinc, 41 per cent; tin, 0.5 per cent; iron, 1.0 per cent; manganese, 0.5 per cent; aluminum, 1.0 per cent. Other compositions also are used, with copper contents varying from about 55 to about 62 per cent. These alloys have high tensile strength and are suitable for casting. Because they resist dizincification and corrosion by salt water, manganese bronzes are used for propellers; other uses are for pumps, screens in various industries, valves, gears, and the like.

Bronzes

Bronzes are more costly than brasses, but have compensating advantages. Bronzes are not subject to a type of corrosion analogous to dezincification, by which one constituent is removed. They also as a rule are stronger than brasses.

Copper and Copper Alloys

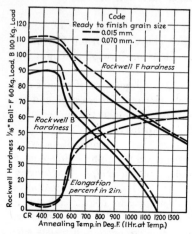

The effect of annealing on the Rockwell hardness and percentage elongation in 2 in. of 70-30 (cartridge brass) strip, previously cold-rolled 6 B. & S. Nos. (50 per cent reduction of area) from two different grain sizes, 0.015 and 0.070 mm. (69.83 % copper) (0.040-in. stock).

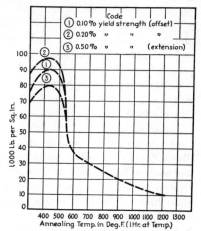

The effect of annealing on the yield strength of 70-30 (cartridge brass) strip, previously cold-rolled 6 B. & S. Nos. (50 per cent reduction of area) from a grain size of 0.015 mm. (69.83 % copper) (0.040-in. stock).

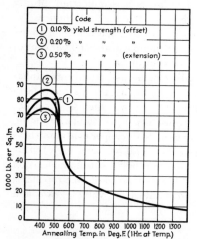

The effect of annealing on the yield strength of 70-30 (cartridge brass) strip, previously cold-rolled 6 B. & S. Nos. (50 per cent reduction of area) from a grain size of 0.070 mm. (69.83 % copper) (0.040-in. stock).

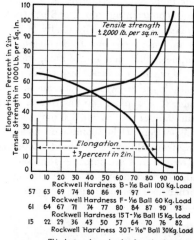

This chart can be employed to determine the approximate tensile strength and percentage elongation of 70-30 (cartridge brass) strip (69.83 % copper) when only Rockwell hardness is known. It is accurate for all thicknesses between 0.020 and 0.080 in. within the given limits.

Figure 120. Annealing temperature curves for 70-30 cartridge brass.

Reproduced by permission from Wilkins and Bunn, *Copper and Copper-base Alloys*, copyright by McGraw-Hill Book Co., Inc., New York.

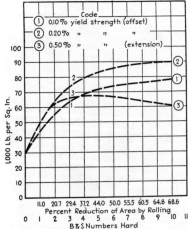

The effect of cold rolling on the yield strengths of 70-30 (cartridge brass) strip, previously annealed to a grain size of 0.015 mm. (69.83 % copper) (0.040-in. stock).

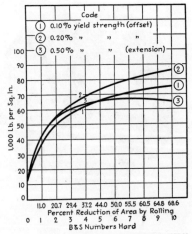

The effect of cold rolling on the yield strengths of 70-30 (cartridge brass) strip, previously annealed to a grain size of 0.070 mm. (69.83 % copper) (0.040-in. stock).

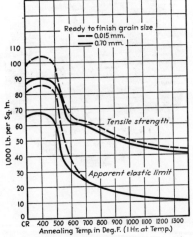

The effect of annealing on the tensile strength and apparent elastic limit of 70-30 (cartridge brass) strip, previously cold-rolled 6 B. & S. Nos. (50 per cent reduction of area) from two different grain sizes, 0.015 mm. and 0.070 mm. (69.83 % copper) (0.040-in. stock).

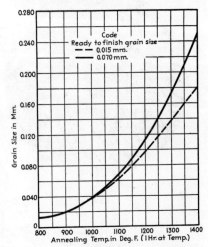

The effect of annealing on the grain-growing characteristics of 70-30 (cartridge brass) strip, previously cold-rolled 6 B. & S. Nos. (50 per cent reduction of area) from two different grain sizes, 0.015 and 0.070 mm. (69.83 % copper) (0.040-in. stock).

Figure 121. Annealing temperature curves for 70-30 cartridge brass.

Reproduced by permission from Wilkins and Bunn, *Copper and Copper-base Alloys*, copyright by McGraw-Hill Book Co., Inc., New York.

122

123

124

125

Microstructures of typical brasses. Magnification 75 ×.
Courtesy The American Brass Company.

Figure 122. Cartridge brass sheet
annealed

Figure 123. Cartridge brass sheet
cold rolled (50 per cent reduction)

Figure 124. Naval brass rod
extruded

Figure 125. Naval brass rod
hot rolled

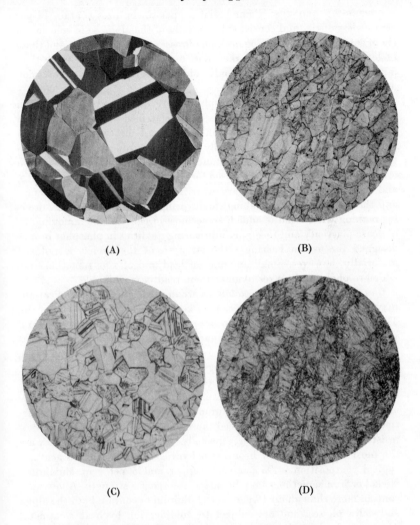

(A)

(B)

(C)

(D)

Figure 126. Microstructures of typical bronzes. Magnification 75 ×.
Courtesy The American Brass Company.

(A) Phosphor bronze sheet annealed

(B) Phosphor bronze sheet cold rolled (37 per cent reduction)

(C) Silicon bronze sheet annealed

(D) Silicon bronze sheet cold rolled (37 per cent reduction)

COPPER-TIN BRONZES   Commercial copper-tin bronzes contain 12 per cent or less of tin, and consist substantially of the alpha solid solution of tin in copper. Like the analogous alpha brasses they are ductile, and those of low tin content can be worked hot, but those containing more than 5 per cent of tin can be worked only cold. They are resistant to corrosion and have high tensile strength.

Because at least traces of phosphorus are added to most copper-tin bronzes for deoxidation, most of them are classified as *phosphor bronzes.* Some phosphor bronzes have phosphorus contents between 0.02 and 0.35 per cent added to impart materially increased hardness and wear resistance. They are also highly resistant to corrosion, fatigue, and season cracking. Phosphor bronzes are used for gears, worm wheels, clutch disks, and similar parts subject to hard wear. Phosphor bronze has resilience, and therefore is used for making small springs; another important use is for making wire. Lead is added, sometimes in considerable amounts, to increase plasticity and to supply lubricating qualities in phosphor bronze designed for use in bearings; the strength of the bronze is reduced materially, however. Small amounts of lead and of zinc sometimes are added to phosphor bronzes to make them readily machinable.

Gun metal or *government metal* in accordance with present specifications may contain as much as 4 per cent of zinc with 8 to 10 per cent of tin. The zinc serves as deoxidizer and improves casting qualities. Gun metal was used originally for making ordnance, but now is used for heavy castings for equipment which must be resistant to corrosion, such as gears, pumps, and the like, and for bearings that carry heavy loads.

ALUMINUM BRONZE   Aluminum bronze is an alloy of copper and aluminum. With 10 per cent or less of aluminum, the alloy consists substantially of the alpha phase. Mechanical properties of aluminum bronzes, particularly resistance to wear, exceed those of copper-tin bronzes, and their resistance to corrosion is better, especially at high temperatures. They are the most resistant of any bronzes to hydrogen sulfide and to acids. A typical composition is 95 per cent copper and 5 per cent aluminum. Nickel or iron sometimes may be added to increase strength. Alloys that contain more than about 10 per cent of aluminum contain both the alpha and delta phases, and are subject to modification by heat treatment. Aluminum bronzes that contain approximately 13 per cent of aluminum are exceptionally hard and are used for dies.

SILICON BRONZES   Silicon bronzes usually contain not more than 4 per cent of silicon, and in addition either iron, manganese, tin, or zinc, or sometimes more than one of these metals. If easy machinability is required, a small amount of lead is added. Some typical silicon bronzes are those sold under the trade name *Everdur*. For example, Type A Everdur has the composition: copper, 96 per cent; silicon, 3 per cent; manganese, 1 per cent. Silicon bronzes as a class have good mechanical properties, are readily weldable by electric arc or oxyacetylene torch, and are resistant to corrosive compounds, particularly hydrochloric and sulfuric acids, alkalies, and certain organic compounds.

Figure 127. Manganese bronze. Magnification 50 ×.
Courtesy Bausch & Lomb Optical Company.

BERYLLIUM BRONZE Alloys of beryllium and copper contain between about 1 per cent and 2.75 per cent beryllium, and usually small amounts of nickel for grain refinement. A typical beryllium bronze consists of 97.6 per cent copper, 2.05 per cent beryllium, and 0.35 per cent nickel. Beryllium bronzes are the strongest of all bronzes and can be hardened by heat treatment to such an extent that they are used to make cutting tools to replace steel tools for uses where the lower tendency to spark is important. They have high resistance to corrosion, high tensile strength, and great resistance to fatigue. The resilience and resistance to fatigue make beryllium bronze a useful alloy of which to make springs. This is an exceptionally versatile and valuable alloy, and its use is restricted only by its high cost.

Copper-Nickel Alloys

Copper and nickel are mutually soluble in all proportions, and many valuable alloys of different compositions are available.

Alloys of copper content between 75 per cent and 80 per cent are used for making coins, for condenser tubes, and for capping rifle bullets.

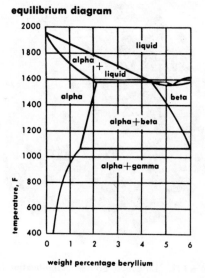

equilibrium diagram

Figure 128. Beryllium-copper constitutional diagram.
Courtesy The Beryllium Corporation.

An alloy of 70 per cent copper and 30 per cent nickel, known as *cupro-nickel*, has high strength, high ductility, and excellent resistance to corrosion by most compounds except acids. It is used for high-pressure steam fittings, condenser tubes in marine boilers, and in many types of fittings which are required to have high strength and to be resistant to corrosion. This alloy is readily worked, but is not well suited for casting because of its high shrinkage.

An alloy consisting of 60 per cent copper and 40 per cent nickel is called *constantan*; its low temperature coefficient and high electric resistance make it useful in electric equipment. Constantan wire is used frequently in thermocouples in connection with a wire of a dissimilar metal such as iron or copper.

Alloys known as *nickel silver* or *German silver* are solid solution alloys containing copper, nickel, and zinc; a typical composition is copper, 66 per cent; nickel, 17 per cent; zinc, 17 per cent. Nickel silver has a white color similar to that of silver, and a dense, fine-grained structure with excellent mechanical properties. Resistance to corrosion is high. Nickel

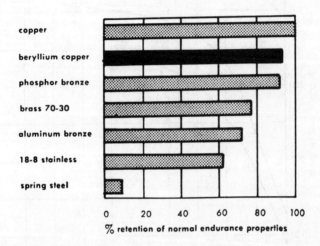

Figure 129. Chart showing the relative corrosion resistances of several materials in salt spray solution.

Courtesy The Beryllium Corporation.

silver is used for making drawing instruments, ornamental hardware of various sorts, and as a base metal for most silver-plated ware.

Several alloys of high nickel content are known as *monel*. The content of the original monel was approximately 67 per cent nickel and 28 per cent copper, with small amounts of iron and manganese; this was the composition that resulted from smelting one of the Canadian nickel ores. A typical monel is exceptionally resistant to corrosion by practically all alkalies or by acids except nitric and sulfuric, and retains its high strength at elevated temperatures to a greater degree than does any other copper alloy. This combination of qualities makes it of great value to industries in which chemical inertness is important. It is used for sinks, piping, and similar equipment.

K MONEL METAL   This alloy, a modification of the original monel, is a nickel-copper-aluminum alloy that contains approximately 2.75 per cent

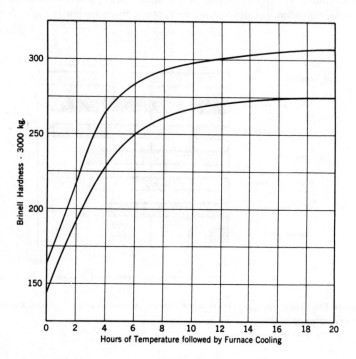

Figure 130. Age hardening of K monel and KR monel. Effect of time at 1080–1100°F.

of aluminum. In large sections, its strength and hardness are comparable with those of heat-treated alloy steels. Its resistance to corrosion is similar to that of other monels. K monel can be hardened to high levels of strength by aging. (Aging denotes simultaneous increase of hardness and of the strength factors in tension, torsion, and compression, as the result of thermal treatment.) The response of K monel to heat treatment is almost the reverse of the response of steel: K monel becomes soft when quenched and relatively hard when cooled slowly.

K monel in cold-worked condition is annealed for softening by holding

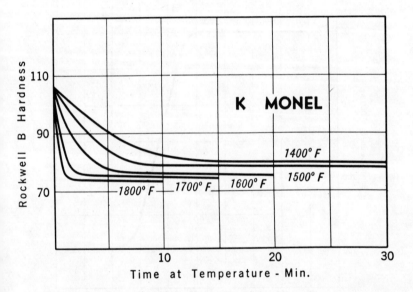

Figure 131. Approximate time required at various temperatures to produce different degrees of softness in K monel and KR monel by open annealing followed by quenching.

Courtesy International Nickel Co., Inc.

at 1600°F for 5–10 minutes, or at 1800°F for 2–5 minutes, followed by quenching.

Hardening of K monel is accomplished by a single heat treatment consisting of aging at the required temperature, followed by controlled cooling. If it is heated above the specified temperature, some softening results. The correct temperature and times are determined by the initial temper of the metal.

K monel in soft condition is age hardened by heating to 1080–1100°F for 16 hours, followed by cooling in the furnace at the rate of 15–25 degrees per hour, to a temperature of 900°F. When K monel is in fully cold-worked

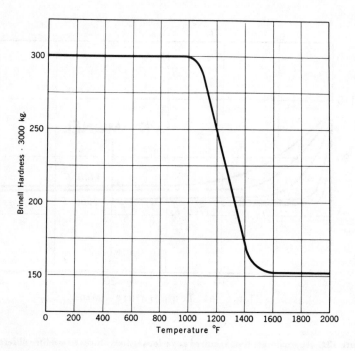

Figure 132. The softening effect of water-quenching age-hardened K monel and KR monel from various temperatures.
Courtesy International Nickel Co., Inc.

condition, it is age hardened by holding at 980–1000°F for 6–10 hours, the longer time for the softer material. Cooling is the same as for soft metal.

The increase in strength produced by heat treatment results from precipitation throughout the lattice of small cubic crystals of the composition

Ni$_3$(Al–Ti). These crystals, about 50 angstroms along the cube edge, strain the matrix lattice and thus produce hardening. The increment of hardening so obtained can be superimposed upon that obtained by prior cold working. Thus cold-worked and aged K monel is stronger than

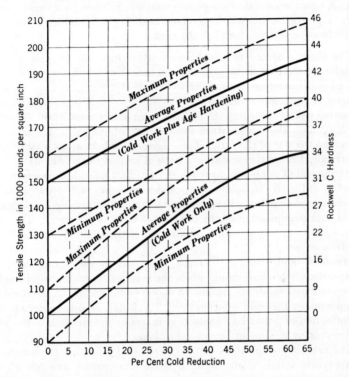

Figure 133. Effect of cold work and age hardening on K monel and KR monel.
Courtesy International Nickel Co., Inc.

annealed and aged K monel, since the annealing treatment eliminates the effect of prior cold work.

Heat Treatment of Nonferrous Alloys

Unlike iron, which exists in three allotropic forms that change from one to another at critical temperatures, a nonferrous alloy exists in only one crystalline form. For that reason, heat treatment depends solely upon using an alloying constituent the solubility of which in solid solution with the nonferrous metal increases materially as the temperature is raised. This is called solution heat treatment. It includes two essential steps:

1. A solution anneal at a high temperature, followed by a rapid quench
2. A precipitation, which involves heating the metal at a comparatively low temperature for a definite time

The alloy is heated to a high temperature until the alloying element dissolves in the base metal; the temperature at which this occurs is maintained until solution is complete. This heat treatment requires close control of the temperature, which should be as high as possible without incurring risk of exceeding the melting point of any constituent. The time during which the temperature must be maintained depends upon the alloy and the size of the article being heat treated. Heating may be in an electrically heated furnace or in a bath of molten mixed nitrates.

At the end of the heating period, the article is quenched promptly and rapidly, ordinarily in cold water; milder quenches in hot water, oil, or air blast tend to reduce the corrosion resistance of the alloy. Cooling is so rapid that precipitation of the alloying element is prevented. The alloy then consists of a supersaturated solution of the alloying element in the nonferrous metal.

Hardness and strength increase when the article stands at room temperature after quenching, while the solute is precipitating from the supersaturated solution. The solute may be an element, an intermediate compound, or the solid solution itself. Precipitation of the solute results in formation of a platelike phase that increases the strength of the alloy by preventing movement between the planes.

The rate of precipitation can be increased by heating the alloy to a temperature intermediate between room temperature and 650°F; the proper temperature depends upon the alloy. This not only accelerates the process of aging, but produces hardness and strength materially higher than those which result from natural aging. This operation is called *artificial aging*, *age hardening*, or *precipitation heat treatment*.

Normally the solution-annealing treatment is carried out at the mill where the sheet, wire, rod, or the like is manufactured, and the ultimate user applies the precipitation-hardening treatment after his forming and machining operations are completed, to obtain the physical properties required of the finished part. In the case of forgings and castings, the final user may also do his own solution annealing.

Certain alloys that are susceptible to heat treatment can be given desirable properties also by cold work alone. Other alloys that retain some alloying element in solid solution as the result of manufacturing operations show improved tensile strength when artificially aged.

Heat Treatment of Beryllium Bronze

Beryllium bronze is an alloy of the precipitation-hardening type. It does not age at room temperature, but requires artificial aging. The solution-annealing temperature lies within the range 775–825°C for a period of 1–3 hours, and it is important that the ensuing quench be as rapid as possible. The precipitation-hardening range is 250–350°C. During treatment, hardness and strength increase continuously to maxima and then decrease; it is therefore important that the time at temperature be controlled accurately. ASTM Specification B 194-55 calls for a precipitation-hardening treatment of 3 hours at 315°C to be used as a check treatment. Industrial heat treatments for finished parts are similar in nature, although there is a tendency to use slightly higher temperatures and shorter times to produce the same results.

At high temperatures, salt baths corrode beryllium bronze, but they can be used safely for precipitation hardening. Because of a difference in heating rates between muffle furnaces and salt baths, shorter precipitation times can be used to develop the same hardness in a salt bath as would be obtained in a muffle furnace for a longer period.

Questions

1 What are the outstanding physical characteristics of copper, and what is its principal use?
2 What processes have been developed to make it possible to mine profitably the low-grade copper ores of this country?
3 Describe the steps involved in dressing copper ore.
4 Describe the flotation process.
5 a. What is the charge of the converter?
   b. What is the product of the converter?
   c. What disposal is made of the slag?
6 Describe briefly the process of electrolytic refining.
7 What is poling, and why is copper poled?
8 Define alpha brasses and state their properties.
9 Define alpha-beta brasses and state how their properties differ from those of alpha brasses.

10   State the differences between brass and bronze.
11   Explain the importance of K monel metal to the engineer.
12   How is K monel hardened?
13   What does the heat treatment of nonferrous alloys depend upon?
14   Describe solution heat treatment.

I. MINING

1. Vein or underground: All metals
2. Open cut, Surface, Bench: Fe, Cu, Al
3. Dredging: Au, Sn

II. CONCENTRATION OR BENEFICIATION

1. Hand picking
2. Gravity
   a. Air
   b. Water (Jigs, Vanners, Tables)
3. Flotation
4. Combination (2) and (3)
5. Magnetic
6. Electrostatic

III. ORE MILLING

1. Crushing
   a. Primary
   b. Secondary
2. Grinding
   a. Primary
   b. Secondary
3. Screening and Sizing
   a. Grizzly
   b. Screens
4. Dewatering
   a. Sedimentation or thickening (Dorr)
   b. Filtration

IV. ORE TREATMENT PRIOR TO SMELTING

1. Roasting
   a. Oxidizing–removal of volatile elements:
      principally S, Cu, Zn, Pb, Ni, monel
   b. Blast roasting
      1. Up draft (Huntington and Heberlein) Pb
      2. Down draft (Dwight-Lloyd) Pb, Zn
   c. Sulfating roast

IV. ORE TREATMENT PRIOR TO SMELTING (CONT'D.)

1. c. Sulfating roast
   d. Flash roasting
   e. Sintering (Dwight-Lloyd): Zn, Pb
   f. Chloridizing roast
   g. Reducing roast: H, Co
2. Calcining—Decomposition of carbonates
3. Chemical treatment: Al, Mg

V. REDUCTION-SMELTING

1. Shaft or blast furnaces: Pb, Sn, Ni, Cu
2. Reverberatory furnaces: Cu, Pb, Ni, Sn, monel
3. Electric furnaces
   a. Reduction of ores same as (1) and (2)
   b. Electrolysis of fused electrolyte
      Al, Mg, all alkali metals
4. Thermit reactions
5. Distillation: Zn, Cd, Mg (Sn as chloride)
   a. Fractional distillation
6. Converting: Bessemer (Fe)
               Pierce Smith (Cu)

VI. REFINING

1. Pyrometric
   a. Reverberatory: Pb, Sn, Cu
   b. Kettles: Pb, Sn, low-melting metals
   c. Electric furnaces: Cu
2. Electrolytic
   a. Aqueous solutions: Cu, Pb, Sn, Ag, Au
   b. Fused baths: Al
3. Distillation: Zn, Hg, Mg, Cd
4. Liquation: Zn, Sn, low-melting alloys

VII. HYDROMETALLURGY

1. Chemical solution treatment: Cu, Zn, Cd, Ag, Au
2. Recovery from solutions
   a. Electrolytic: Cu, Zn, Cd
   b. Cementation: Cu, Ag, Au, Cd

20 | Aluminum and Aluminum Alloys

Although aluminum in chemical combination, usually as an oxide or as a silicate, exists in the crust of the earth in greater amounts than does any other metal, metallic aluminum is one of the newest industrial metals, having been produced in commercially useful amounts for only a little more than fifty years. This is in sharp contrast with copper, tin, and iron, all of which were known to and used by the ancients.

The Light Metals

Aluminum and magnesium are the most important commercial light metals. They differ fundamentally from heavy metals in that they have greater affinity for oxygen and nonmetallic elements. Because of this affinity, aluminum and magnesium require rather complete separation of the impurities in their ores during production. The first step in the extraction of aluminum is the production of virtually pure aluminum oxide; in the case of magnesium, virtually pure magnesium chloride must first be produced. Since aluminum and magnesium cannot be smelted directly by carbon, they must be extracted by electrolysis—a process that requires large amounts of electric power.

Compared with iron and steel, the light metals have high thermal and electric conductivities, and in general offer excellent resistance to corrosion. They lose strength at moderately elevated temperatures but increase in strength and ductility as the temperature is reduced. This loss of strength at moderately elevated temperatures is not always a disadvantage, since it permits manufacture, by extrusion, of complicated shapes of even the strongest alloys.

Physical Properties of Aluminum

Probably the most notable physical property of aluminum is its light weight, which is only about one third that of iron, copper, and their

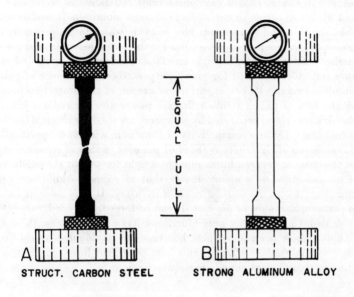

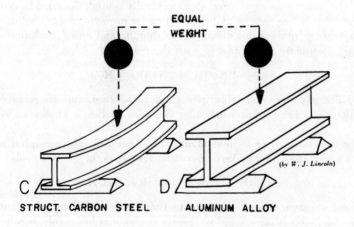

(by W. J. Lincoln)

Figure 134. Comparative strength of steel and aluminum.

respective alloys. It also has considerable resistance to corrosion by air and by many corrosive chemicals; although aluminum is readily oxidizable, the thin and transparent film of oxide which forms promptly upon exposure to air serves as protection for the metal beneath, and prevents further oxidation. Aluminum is an efficient reflector of light, and aluminum reflectors are used for many purposes because a piece of polished aluminum retains its bright surface as a result of the protection provided by the film of oxide. It has reflective power also for radiant heat, and aluminum surfaces separated by air spaces are used for thermal insulation of buildings. Electric conductivity of aluminum wire made specifically for transmission of electricity is about 61 per cent of that of standard copper on the basis of volume, but compared weight for weight the conductivity of aluminum wire is about double that of copper. Aluminum can be fabricated by every method used in industry. It may be cast in sand, in permanent molds, or die cast. It can be forged, extruded, rolled, spun, or drawn. Commercially pure aluminum has a tensile strength of about 13,000 pounds per square inch, but this can be doubled by cold work; some wrought alloys after suitable heat treatment have tensile strengths of more than 80,000 pounds per square inch.

Refining Bauxite

In the Bayer refining process, which generally is used, the dried bauxite is ground to a fine powder and digested under pressure in a hot solution of sodium hydroxide; this dissolves the alumina and forms a solution of sodium aluminate in accordance with the reaction:

$$Al_2O_3 + 2\,NaOH \rightarrow 2\,NaAlO_2 + H_2O$$

The silica and the other impurities are not dissolved, and are removed by filtration; the residue containing these impurities is known as *red mud*.

The hot solution of sodium aluminate is pumped into precipitating tanks where aluminum hydroxide crystallizes out as the solution cools:

$$NaAlO_2 + 2\,H_2O \rightarrow Al(OH)_3 + NaOH$$

The sodium hydroxide is returned to the digesters and the aluminum hydroxide, after being washed to remove the sodium hydroxide, is raised to a high temperature in rotary kilns to decompose it into aluminum oxide and water:

$$2\,Al(OH)_3 \rightarrow Al_2O_3 + 3\,H_2O$$

For every pound of silica in the ore, a pound of alumina and a pound of soda are lost in the red mud, so that treatment of bauxite high in silica is uneconomical by the Bayer process. When it became necessary to use low-grade bauxite during World War II, a new process was developed by a chemist of the Aluminum Company of America for treating ores high in silica, and the red mud residue which formerly was wasted. In this *Alcoa combination process*, the red mud is sintered with limestone and soda ash, the sinter is leached with water, and the dissolved alumina and soda are returned to the process to be digested with more bauxite and sodium hydroxide. The effect of the development has been to make available substantial amounts of bauxite not hitherto employed as sources of alumina.

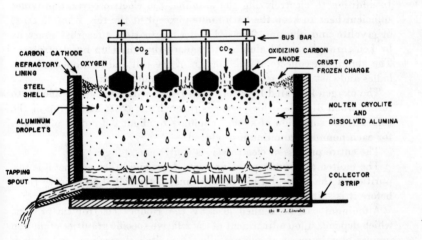

Figure 135. Cutaway diagram of Hall-Héroult electrolytic reduction cell.

Electrolytic Reduction of Aluminum Oxide

The final stage in the preparation of aluminum involves the electrolytic reduction of purified alumina by the Hall-Héroult process. The Hall-Héroult electrolytic cell is a rectangular steel shell, with a lining of carbon that forms the cathode. Carbon anodes are suspended above the cell on bus bars. Each cell is capable of producing about 500 pounds of aluminum per day.

The most suitable material for use as a solvent for alumina has been found to be *cryolite*, a double fluoride of sodium and aluminum; cryolite also has satisfactory electric conductivity for the electrothermic operation of the cell. Natural cryolite is found in useful quantities only in Greenland, but it has been found practicable to make synthetic cryolite which is equally as effective as the natural product.

When operation is started, cryolite is charged into the cell and fused by the heat generated by passage of the electric current. As soon as the cryolite is fused completely, the alumina is added, and reduction starts:

$$2\,Al_2O_3 \rightarrow 4\,Al + 3\,O_2$$

In addition to electrolyzing the alumina, the electric current provides sufficient heat to keep the bath molten except at the top, where a crust of cryolite and alumina forms. From time to time the crust must be broken through to add alumina to replace that used up by electrolysis. The aluminum collects at the bottom of the cell and is tapped off into ladles and cast into pigs weighing about 50 pounds each.

The oxygen liberated by electrolysis reacts with the carbon anodes to form carbon dioxide, which escapes through the crust at the top of the cell. The consumption of anodes amounts to about 0.8 pound of carbon for each pound of aluminum produced.

The entire process of electrolysis is continuous.

The molten aluminum picks up some dross and other nonmetallic impurities in the electrolytic cell, and must be remelted to remove them before it is suitable for industrial use.

Aluminum can be refined also by the Hoopes electrolytic process, which depends upon adjustment of the relative specific gravities of molten impure aluminum or aluminum alloy, a molten electrolyte, and molten pure aluminum. The molten impure aluminum remains at the bottom of the cell, in contact with the carbon lining that serves as anode contact. Above it is the molten electrolyte which has a specific gravity less than that of the impure aluminum; it consists of a mixture of aluminum fluoride, barium fluoride, and cryolite, saturated with alumina. The pure aluminum, transferred by electrolytic action through the layer of electrolyte, has a specific gravity less than that of the electrolyte, and therefore molten pure aluminum floats on the electrolyte. The electric circuit is completed by graphite electrodes that dip into the molten pure aluminum cathode.

Alloys of Aluminum

Alloying elements are added to aluminum primarily for the purpose of producing an alloy of greater strength than that of pure aluminum. The added strength is obtained, however, at the cost of some reduction in the resistance to corrosion to many chemical substances which is characteristic of pure aluminum. The alloying elements principally used are copper, iron, magnesium, manganese, nickel, silicon, and more recently, chromium. Aluminum alloys are classified as *casting alloys* and *wrought alloys*.

Casting Alloys

The most commonly used casting alloys contain copper or silicon or both. The original casting alloy contained about 8 per cent of copper, and was known to foundrymen as *Number 12*. At present, use of this alloy has been largely discontinued in favor of other alloys which contain, in addition to copper, other elements that improve casting characteristics and machinability. Another alloy widely used for casting contains only about 5 per cent of silicon as an alloying element; it has less strength but greater ductility and better resistance to corrosion by salt water than have the aluminum-copper alloys. Alloys containing magnesium have better physical properties than have either aluminum-copper or aluminum-silicon alloys.

Some casting alloys are susceptible to heat treatment. The one in most common use contains about 4.5 per cent of copper and 0.8 per cent of silicon. Another which has better resistance to corrosion by salt water contains about 7 per cent of silicon and 0.3 per cent of magnesium. Maximum hardness and good strength are found in an alloy containing 10 per cent of magnesium, but special foundry practices are required in casting, and the castings are not well suited for use at high temperatures.

Alloys containing copper, magnesium, and silicon commonly are used for casting pistons for internal-combustion engines; these are made in permanent molds. Aluminum-magnesium alloys are readily machinable.

Special alloys, usually containing silicon to produce high fluidity, are used for die casting.

Wrought Alloys

Wrought alloys are supplied in industrial forms such as plates, sheets, bars, rods, pipes, tubes, wire, rivets, and a variety of rolled and extruded shapes. Some wrought alloys of high strength are available in the form

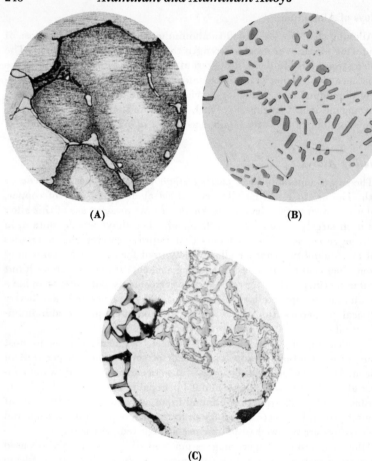

Figure 136. Microstructures of aluminum alloys, as cast. Magnification 500 ×.
(A) Al-Cu alloy.   (B) Al-Si alloy.   (C) Al-Cu-Si alloy.
Courtesy Aluminum Company of America.

of sheets coated with either commercially pure aluminum or a different
alloy which is strongly resistant to corrosion by the atmosphere and by
salt water, and thus protects the stronger alloy from corrosion. These are
known as *clad* sheets.

The aluminum to be used for producing these various forms is cast first into ingots and subsequently subjected to hot working. Finishing is by hot or cold working, depending upon the alloy used and upon the degree of strength and hardness required.

A few wrought alloys of aluminum cannot be hardened by any form of heat treatment, but can be hardened to some extent by cold working.

WROUGHT ALLOYS SUSCEPTIBLE TO HEAT TREATMENT   The first of the so-called *strong* alloys of aluminum which can be strengthened materially by heat treatment is *duralumin,* originated in 1911 by a German metallurgist, Dr. Alfred Wilm. In Dr. Wilm's experiments he discovered that when heated to a temperature of about 925°F and quenched, the tensile strength of duralumin increased to a value of between 35,000 and 40,000 pounds per square inch, and that after aging for four days, had shown further increase to 55,000 pounds per square inch. During the same period the hardness also had increased materially. Duralumin is equivalent to a domestic alloy which contains 4 per cent of copper, 0.5 per cent of magnesium, and 0.5 per cent of manganese.

For applications in which ductility is important, heat treatment is to be preferred to hardening by cold work; cold work always produces material decrease in ductility, whereas heat treatment produces much less decrease, or under some conditions may even increase ductility.

Annealing

Hardness which results from cold working pure aluminum and low alloys not susceptible to heat treatment can be removed and full softness restored almost instantaneously by heating to suitable temperatures above 650°F, to permit recrystallization; the rate of subsequent cooling is unimportant. The same softening can be produced by heating for a longer time at somewhat lower temperatures.

All work hardening and most of the hardening produced by heat treatment can be removed in the same manner from alloys which are susceptible to heat treatment, but certain precautions must be observed to avoid possible heat treatment effects. The annealing temperature should not greatly exceed 650°F, and should not be below 630°F at any part of the material to be annealed; the cooling rate is not important if the temperature has not exceeded 650°F. If it is essential that the alloy be annealed to full softness, it is heated to a temperature between 750°F and 800°F and maintained at that temperature for about two hours, followed by cooling in the furnace at a rate not exceeding 50 degrees per hour until the temperature has receded to 500°F.

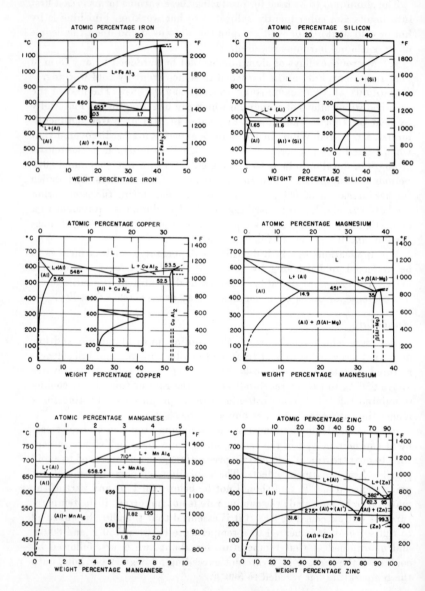

Figure 137. Aluminum end of principal constitutional diagrams for aluminum alloys.

Courtesy Aluminum Company of America.

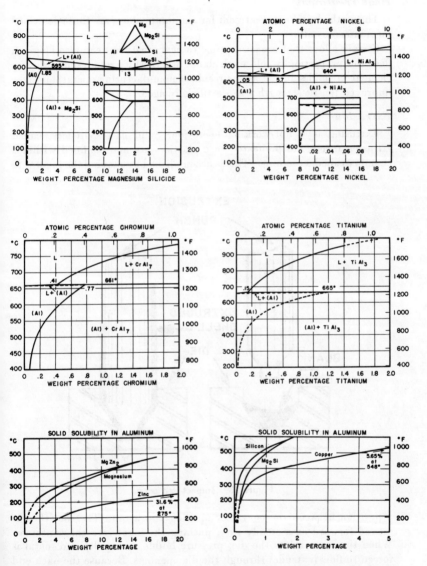

Figure 138. Aluminum end of principal constitutional diagrams for aluminum alloys.

Courtesy Aluminum Company of America.

Heat Treatment

The methods of heat treatment for aluminum alloys are those described for nonferrous alloys.

Certain alloys that are susceptible to heat treatment can also be given desirable properties by cold work alone; others which retain some alloying element in solid solution as the result of manufacturing operations show improved tensile strength when artificially aged.

Extrusion

Extrusion is a hot-working operation. Solid cast cylinders, called *extrusion ingots*, are preheated to a plastic state and confined within a chamber called a *container*. This is accomplished by moving the ingot into

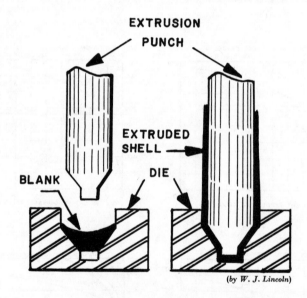

(by *W. J. Lincoln*)

Figure 139. Impact extrusion.

a position between the container opening and a ram. The ram then is moved forward at a steady rate, pushing the ingot into the container. When the ingot meets the die, pressure builds up rapidly and metal is forced to flow (extrude) through the die openings. Because the back end of the ingot may contain defects, it is discarded before being extruded.

The resulting product is of constant cross section that conforms to the shape and dimensions of the die openings.

The severe metal flow that results produces a fine-grained wrought product, which is held within close dimensional tolerances and which has uniformly high mechanical properties.

The long ingots used for extrusion are called *logs*. They must be free of such imperfections as pores, cracks, or inclusions, and must have uniformly distributed alloying ingredients, proper grain structure, and correct physical dimensions. The logs are prepared by the chill method, as follows: A molten aluminum alloy, completely degassed and cleansed of impurities, is fed into the top of a water-cooled open-end mold at a controlled rate. The molds are connected directly to the furnace. Large ingots are cast in lengths of 25 feet and up to 20 inches diameter; smaller ingots are cast in lengths of 11 feet and up to 9 inches diameter. Each log is checked ultrasonically by means of a reflectoscope, which automatically records on a cathode-ray tube the sound waves passing through the ingot. Spectrochemical analysis also is carried out. To afford better distribution of the alloying ingredients, the ingots are subjected to a *soaking* or *homogenizing* heat treatment, which equalizes the composition and relieves any stresses that may remain after casting.

HOLLOW EXTRUSIONS   To produce large tubes the ingot must be cast hollow and scalped; to produce smaller tubes the ingot is drilled. When tubes with tapered wall thickness are being extruded, a tapered mandrel can be used. As a rule the mandrel method is used only when symmetrical tubular shapes are to be extruded.

PORTHOLE EXTRUSIONS   In porthole extrusions, the metal separates to flow around the core supports, and welds together in the welding chamber before passing through the die. This method is adaptable to many intricate hollow shapes, and is generally limited to the softer alloys.

STEPPED EXTRUSIONS   In this method, the press is stopped when the ingot is only partially extruded. The original dies are removed and replaced with dies of larger opening; the push is then resumed. As a result, the product has an abrupt change in cross section.

Anodizing

Excellent protection from many types of corrosion can be given to any aluminum alloy by a process known as *anodizing*, which consists in using the aluminum as an anode during electrolysis in an aqueous electrolyte; the effect is to form a thicker film of oxide than would form naturally by

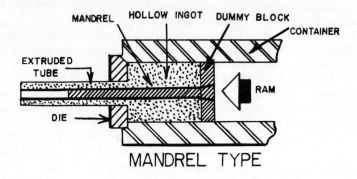

MANDREL TYPE

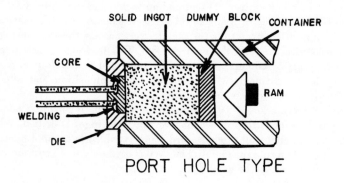

PORT HOLE TYPE

AIRCRAFT CARRIER DECK COOLING PANEL
PORT HOLE EXTRUSION

(*by W. J. Lincoln*)

Figure 140. Hollow aluminum extrusions.

Courtesy Kaiser Aluminum and Chemical Corporation.

exposure to air. The electrolyte is a solution of a strongly oxidizing compound such as chromic acid. By suitable choice of electrolyte, and by incorporating dyes in the coating. a wide variety of colors can be obtained. The anodic coating is slightly porous and forms an excellent base for paint.

Joining Aluminum and Aluminum Alloys

Aluminum alloys can be welded by oxyacetylene or oxyhydrogen torch, by electric arc, or by resistance welding methods. The methods are

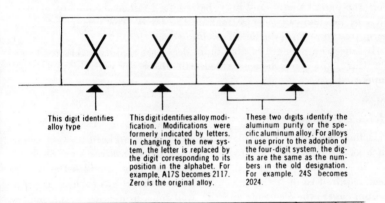

This digit identifies alloy type

This digit identifies alloy modification. Modifications were formerly indicated by letters. In changing to the new system, the letter is replaced by the digit corresponding to its position in the alphabet. For example. A17S becomes 2117. Zero is the original alloy.

These two digits identify the aluminum purity or the specific aluminum alloy. For alloys in use prior to the adoption of the four-digit system, the digits are the same as the numbers in the old designation. For example. 24S becomes 2024.

Type of Aluminum Alloy	Number Group		
Aluminum—99.00% minimum and greater.......... 1	X	X	X
Copper.. 2	X	X	X
Manganese...................................... 3	X	X	X
Silicon... 4	X	X	X
Magnesium...................................... 5	X	X	X
Magnesium and silicon.......................... 6	X	X	X
Zinc... 7	X	X	X
Other element.................................. 8	X	X	X
Unused series.................................. 9	X	X	X

Figure 141. Wrought Aluminum Alloy Four-digit Designation System.

From *Aluminum Handbook*, courtesy Aluminum Company of America.

similar to those used for steel, with the exceptions necessitated by the individual characteristics of aluminum:

1. Aluminum has a low melting point (1220°F), so that care must be taken not to melt away the parts to be welded. Since aluminum shows no red color at welding temperature, its temperature must be measured by its physical condition rather than by its appearance.

2. A very tough corrosion-resistant film of Al_2O_3 is always present on an aluminum surface. This film melts at a temperature much higher than does the parent metal, and must be removed either chemically, electrically, or mechanically before making the bond. The film must not be permitted to re-form during the welding operation.

3. Since the strength of aluminum decreases rapidly at elevated temperatures, the parts being welded will distort or collapse unless well supported.

Heliarc Welding

Heliarc welding is particularly suited for work on hard-to-weld metals such as the light metals. An arc is struck between a tungsten electrode within the heliarc torch and the workpiece; this provides the intense heat required to melt the metal. A stream of inert gas such as argon or helium covers the electrode and the weld zone, protecting both from the atmosphere. Since the gas prevents the oxygen and nitrogen of the atmosphere from entering the puddle, it prevents formation of weak and porous welds. Welds produced by heliarc welding have the same chemical and metallurgical properties as the base metal. Since the electrode is made of tungsten (which has a high melting point) and is protected from oxidation by the gas stream, it is virtually nonconsumable. Argon is recommended as the best shielding gas, because it produces smooth and uniform welds.

Heliarc welding offers advantages in welding of pipe:

1. It produces a smooth weld.

2. Welds are stronger, more ductile, and more resistant to corrosion than are welds made by any other process.

3. A clear view of the weld zone is permitted at all times, because there is no smoke or spatter.

4. No fluxing agent is required; hence there is no possibility of trapping slag within the weld.

5. It is economical.

Sigma Welding

During the welding operation an arc is maintained in a shield of argon gas between the consumable electrode of filler metal and the workpiece. No flux is required, so that pitting, flux corrosion, and entrapment are eliminated, and the welds are thus sound, clean, and smooth. Sigma welding is gaining great favor, particularly for welding pieces over $1\frac{1}{8}$ inch thick manually, and down to 0.05 inch thick automatically. It is especially useful for aluminum, but excellent for nearly every other commercial metal.

In the case of alloys which have been subjected to heat treatment before welding, the heat of welding partially removes the hardness; for that reason such parts should be welded before heat treatment, or should be given a second heat treatment after welding.

Brazing is used to join aluminum alloys without appreciable melting of the parts being joined; special fluxes and brazing compounds of aluminum alloys have been developed. Brazing has the advantage over soldering that because the brazing metal and the parts being brazed are of similar material, resistance of a brazed joint to corrosion is good.

For joining alloys after heat treatment, riveting is the most frequently used method. If steel rivets are used, the surface must be painted as a protection against corrosion.

Aluminum Wrought Alloy Designation System

Wrought aluminum and wrought aluminum alloys are designated by a 4-digit system.

In the general grouping of alloys, all the first digits, except in the 1000 series, indicate the major alloying element. The 1000 series is for aluminum of 99.00 per cent or higher purity. The second digit indicates modifications in impurity limits. The last two digits indicate the minimum aluminum percentage in terms of 0.01 per cent. Thus 1030 indicates aluminum of 99.30 per cent purity.

The first digit in the 2000 to 8000 series indicates the major alloying element of the alloy according to the accompanying table. The second digit indicates a modification of the alloy, and replaces the alphabetical designation used in the old system. Thus 17S becomes 2017, and A17S becomes 2117. The last two digits have no special significance except to differentiate the various alloys in the groups.

Temper Designation for Wrought Aluminum

This is the same as for magnesium, as set forth in the following chapter.

Questions

1 Explain why aluminum cannot be produced from bauxite without first refining the bauxite.
2 Draw a diagram of the Hall-Héroult electrolytic cell and describe its operation.
3 Describe the Bayer process.
4 What is the function of cryolite; what is its source?
5 Name four outstanding characteristics of aluminum.
6 Why is aluminum used in thermal insulation of buildings?
7 Define solution heat treatment and describe the process.
8 What is the effect on aluminum alloys of aging after solution heat treatment?
9 Describe the extrusion process.
10 What methods are used for welding aluminum, and why is aluminum more difficult to weld than is steel?
11 Give two reasons why aluminum is a good material for cooking utensils.
12 Describe heliarc welding.
13 Describe sigma welding.
14 Explain the aluminum wrought alloy designation system.

21 / Magnesium and Magnesium Alloys

Magnesium is the newest of the industrial metals. Although it had been produced in Germany for some years in the form of a powder for use in pyrotechnics and for similar purposes, it was not produced there in form suitable for structural use until 1910.

Physical Properties of Magnesium

Magnesium is the lightest of all metals used industrially in large quantities; its weight is slightly less than two thirds that of the same volume of aluminum, and only about one fifth that of an equal volume of copper. Commercially pure magnesium is approximately 99.9 per cent pure, with small amounts of aluminum, copper, iron, manganese, nickel, and silicon accounting for the 0.1 per cent of impurities. Electric conductivity on a volume basis is about 38 per cent that of standard copper. Like aluminum, magnesium can be fabricated by any of the industrial methods, although its strong tendency to oxidize rapidly makes precautions necessary in foundry and welding practices. Magnesium and its alloys are the most readily machinable of all industrial metals, and deep cuts can be made at high speeds. The amount of cold work that can be done on magnesium alloys is limited, but they are readily hot worked. Commercially pure magnesium has so little tensile strength that it is of practically no value as a material of construction, but the tensile strength of some of its alloys is in excess of 50,000 pounds per square inch.

Corrosion

Magnesium and its alloys have satisfactory resistance to corrosion by solutions of most alkalies and by many organic chemicals. They are attacked by almost all common acids except chromic and strong hydrofluoric acids, and by even weak solutions of salt. Protective coatings are

RAW MATERIALS

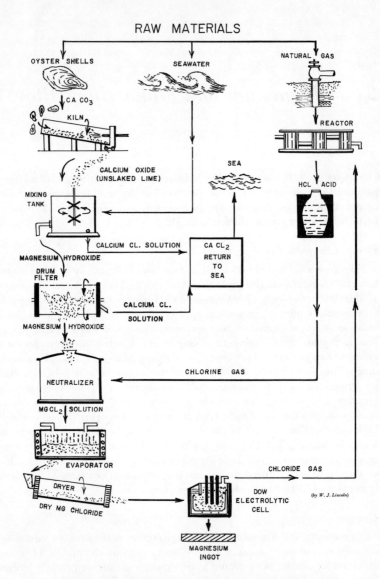

Figure 142. Flow diagram of Dow sea water process.

essential for any magnesium alloys used in the vicinity of salt water, and are recommended in all cases for protection against atmospheric corrosion. Small amounts of iron or nickel in magnesium-aluminum alloys increase susceptibility to corrosion, and these impurities are kept to the lowest possible values in the extraction processes. Because magnesium is higher in the electromotive series of the metals than are copper and iron, magnesium in contact with either of those metals or their alloys, especially in the presence of moisture, is subject to severe galvanic corrosion.

Sources of Magnesium

The sources from which magnesium is produced in commercially useful amounts are natural brines, sea water, magnesite, and dolomite.

Extraction of Magnesium by the Dow Process

The source of the magnesium chloride from which magnesium is extracted by the Dow process may be either the relatively concentrated Michigan brines, or sea water, which contains only about 0.13 per cent of magnesium. The chemistry of extraction is essentially the same, but extraction from sea water requires preliminary steps.

The first operation in the process of extracting magnesium from sea water consists in precipitating magnesium hydroxide by adding to the sea water a suspension of calcium hydroxide, which is made from oyster shells. Calcium hydroxide must be added in excess of the amount theoretically required, in order to provide excess alkalinity and so to prevent contamination of the magnesium hydroxide by the small amount of boron present in sea water.

The suspension of magnesium hydroxide so formed is allowed to settle and then is filtered and dissolved in sufficient hydrochloric acid to make a 15 per cent solution of magnesium chloride. This solution is treated with sulfuric acid to precipitate the remaining calcium as calcium sulfate, and is concentrated to a 35 per cent solution; during the process the content of sodium chloride crystallizes out. After the calcium sulfate and the sodium chloride have been filtered out, the magnesium chloride is concentrated further to a 48 per cent solution, and finally is dried to a granular solid mass containing about 74 per cent of magnesium chloride.

The bath is made up of molten magnesium chloride, sodium chloride, and calcium chloride; magnesium chloride is added continuously to replace that which is removed from the bath by electrolytic action. The metallic magnesium floats at the top of the cell and is removed by dipping with hand ladles.

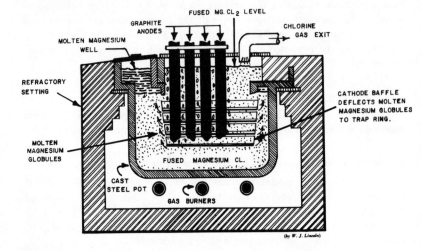

Figure 143. Side view of Dow electrolytic cell.

Extraction of Magnesium by the Elektron Process

The Elektron process, originally developed in Germany, is also based upon electrolysis of magnesium chloride in the molten state. The raw material, however, is a natural magnesium carbonate known as *magnesite*.

The magnesite is calcined to convert it into a fine powdery form of magnesium oxide. This is mixed with coal dust or pulverized coal, moistened with magnesium chloride solution, and the mixture is briquetted in a rotary kiln. The briquettes are placed in a furnace where they are converted into anhydrous magnesium chloride by reaction with a stream of chlorine gas obtained from the electrolytic cell. Metallic magnesium is extracted by subsequent electrolysis of the magnesium chloride in a closed cell which differs in design from the Dow cell.

Extraction of Magnesium by the Ferrosilicon Process

The raw material used in the ferrosilicon process is *dolomite*, a double carbonate of magnesium and calcium. After being converted by calcining to the double oxide of the two metals, it is pulverized, mixed with pulverized ferrosilicon, and the mixture is pressed into small pellets.

The pellets are packed into nickel-chromium steel tubes of between 8 and 10 inches diameter, and heated to a temperature of between 2000°F and 2200°F, under a high vacuum of between 0.05 and 0.15 millimeters of mercury. The iron remains inert during the process, and the reaction is:

$$Si + 2MgO \cdot CaO \rightarrow 2Mg + 2CaO \cdot SiO_2$$

The magnesium is vaporized under the conditions of high temperature and high vacuum, and subsequently crystallizes in one end of the tube.

After the magnesium is removed from the tubes it is melted down in pots, and its impurities, consisting principally of magnesium oxide with smaller amounts of the silicates of magnesium, calcium, and iron, are removed by use of suitable fluxes.

Magnesium Alloys

The principal alloying elements used with magnesium are aluminum, manganese, and zinc. Aluminum content between 3 and 10 per cent increases strength and hardness; aluminum in excess of about 6 per cent produces alloys susceptible to heat treatment. Manganese and zinc are added principally to increase resistance to corrosion, although they also improve the physical properties to some extent. Both cast and wrought alloys are available. Heat treatment is by solution and precipitation methods similar to those used with aluminum.

Alloy Nomenclature and Temper Designation of Magnesium Base Alloys*

A common system of alloy designation for many of the metals has been adopted. It can be used on prints and by purchasing agents in buying from any producer. The advantages of a single nomenclature system over a variety of trade names and company designations was considered by the Technical Committee of The Magnesium Association. The ASTM system of alloy nomenclature and temper designation for light metals and alloys was considered to be the one most used and least likely to cause confusion with other materials. This system has been officially adopted by. The Magnesium Association which recommends its use for all magnesium alloys.

* From ASTM designations B275 and B296, by permission of American Society for Testing Materials.

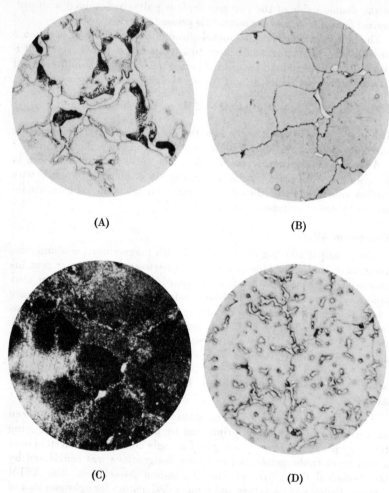

(A)

(B)

(C)

(D)

Figure 144. Microstructures of Dowmetal AZ92A. Magnification 250 × .

(A) Typical sand-cast structure, as cast.
(B) Sand cast and solution heat treated.
(C) Sand cast, solution heat treated, and artificially aged.
(D) Cast in permanent mold.

Alloy nomenclature is covered by ASTM Specification B 275-55, Codification of Light Metals and Alloys, Cast and Wrought, and temper designation by ASTM Specification B 296-56, Temper Designation of Light Metals and Alloys, Cast and Wrought. Alloy Nomenclature, which is based on chemical compositions, is determined by the following rules in ASTM B 275-55.

Figure 145. Magnesium aircraft-wheel casting with gates and risers attached.
Courtesy Dow Chemical Company.

Alloys

1. Designations for alloys consist of not more than two letters representing the alloying elements (Note 1) specified in the greatest amount, arranged in order of decreasing percentages, or in alphabetical order if of equal percentages, followed by the respective percentages rounded off to whole numbers and a serial letter (Notes 2 and 3). The full name of the

base metal precedes the designation, but it is omitted for brevity when the base metal being referred to is obvious.

2. The letters used to represent alloying elements should be those listed below.

3. In rounding off percentages, the procedure described in the Recommended Practices for Designating Significant Places in Specified Limiting Values (ASTM Designation: E 29) should be used.

4. When a range is specified for the alloying element, the rounded-off mean should be used in the designation.

5. When only a minimum percentage is specified for the alloying element, the rounded-off minimum percentage should be used in the designation.

LETTERS REPRESENTING ALLOY ELEMENTS

A	Aluminum	G	Magnesium	P	Lead
B	Bismuth	H	Thorium	Q	Silver
C	Copper	K	Zirconium	R	Chromium
D	Cadmium	L	Beryllium	S	Silicon
E	Rare Earths	M	Manganese	T	Tin
F	Iron	N	Nickel	Y	Antimony
				Z	Zinc

NOTE 1. For codification, an alloying element is defined as an element (other than the base metal) having a minimum content greater than zero either directly specified or computed in accordance with the percentages specified for other elements. The amount present is the mean of the range (or the minimum percentage if only that is specified) before rounding off.

NOTE 2. The serial letter is arbitrarily assigned in alphabetical sequence starting with A (omitting I and O) and serves to differentiate otherwise identical designations. A serial letter is necessary to complete each designation.

NOTE 3. The designation of a casting alloy in ingot form is derived from the composition specified for the corresponding alloy in the form of castings. Thus, a casting ingot designation may consist of an alloy designation having one or more serial letters, one for each product composition, or it may consist of one or more alloy designations.

Unalloyed Metals

Designations for unalloyed metals consist of the specified minimum purity, all digits retained but dropping the decimal point, followed by a serial letter (Note 2). The full name of the base metal precedes the designation, but it is omitted for brevity when the base metal being referred to is obvious.

Temper Designation

The temper designation is separated from the alloy by a dash and is determined by the following set of rules from ASTM B 296-56.

Basis of Codification

1. The designations for temper are based on the sequence of basic treatments used to produce the temper.

2. The temper designation, which is used for all metal forms except ingot, follows the alloy designation and is separated therefrom by a dash.

3. Basic temper designations consist of letters. Subdivisions of the basic tempers, where required, are indicated by one or more digits following the letter. These digits designate a specific sequence of basic treatments, but only those operations which are recognized as significantly influencing the characteristics of the product are indicated. Should some other variation of the same sequence of basic operations be applied to the same alloy, resulting in different characteristics, then additional digits are added to the designation.

The temper designations and the subdivisions are fully defined and explained in the following sections, followed by a summary table for quick reference.

Basic Tempers

—F AS FABRICATED This designation applies to products which acquire some temper qualities in the shaping processes but are not subsequently thermally treated or intentionally strain-hardened.

—O ANNEALED RECRYSTALLIZED (wrought products only) This designation applies to the softest temper of wrought alloy products.

—H STRAIN-HARDENED This designation applies to those products which have their strength increased by strain-hardening with or without supplementary thermal treatments to produce partial softening. The —H is always followed by two or more digits. The first digit indicates the specific combination of basic operations, and the following digit or digits the final degree of strain-hardening.

—W SOLUTION HEAT-TREATED This is an unstable temper. It is applicable only to those alloys which spontaneously age at room temperature after solution heat-treatment. This designation is specific only when the period of natural aging is indicated.

—T TREATED TO PRODUCE STABLE TEMPERS OTHER THAN —F, —O, OR —H This designation applies to products thermally treated to produce stable tempers with or without supplementary strain-hardening. If strain-hardening supplements the thermal treatment, it is considered in the temper designation only where it is recognized as materially influencing the characteristics of the product. The —T is always followed by one or more digits. The numerals 2 through 10 indicate types of treatment, each numeral designating a specific sequence of basic operations. When required, the numerals 11 through 19 are available for designating other types of treatment. The details of the treatment will usually be different for each alloy to produce certain desired results. The treatment usually considered standard is designated by —T followed by the numeral indicating that type of treatment. Deliberate variations of the conditions, resulting in significantly different characteristics for the product, are indicated by adding one or more digits to the treatment designation. No attempt is made to have these indicate any specific set of conditions. It should be understood that a period of natural aging at room temperature may occur between or after the operations listed. Control of this period is exercised when it is metallurgically important.

Subdivisions of the "—H" Temper

—H1 STRAIN-HARDENED ONLY The number following this designation indicates the degree of strain-hardening.

—H2 STRAIN-HARDENED AND THEN PARTIAL ANNEALED The number following this designation indicates the degree of strain-hardening remaining after the product has been partial annealed. For alloys that age-soften at room temperature, the —H2 tempers have approximately the same tensile strength as the corresponding —H3 tempers and slightly higher elongations. For other alloys, the —H2 tempers have approximately the same tensile strengths as the corresponding —H1 tempers and slightly higher elongations.

—H3 STRAIN-HARDENED AND THEN STABILIZED The number following this designation indicates the degree of strain-hardening remaining after the product has been strain-hardened a specific amount and then stabilized. This designation applies only to those alloys which, unless stabilized, age-soften at room temperature.

TABLE 10   BASIC TEMPER DESIGNATIONS AND SUBDIVISIONS

Temper designations	Definition
Basic temper	
–F	As fabricated
–O	Annealed, recrystallized (wrought products only)
–H	Strain hardened
–W	Solution heat treated Unstable temper
–T	Treated to produce stable tempers other than —F, —O, or —H
Subdivisions of the "—H" temper	
–H1, plus one or more digits	Strain hardened only
–H2, plus one or more digits	Strain hardened and then partially annealed
–H3, plus one or more digits	Strain hardened and then stabilized
Subdivisions of the "—T" temper	
–T2	Annealed (cast products only)
–T3	Solution heat treated and then cold worked
–T4	Solution heat treated
–T5	Artificially aged only Artificially aged
–T6	Solution heat treated and then stabilized
–T7	Solution heat treated and then stabilized
–T8	Solution heat treated, cold worked, and artificially aged
–T9	Solution heat treated, artificially aged, and cold worked
–T10	Artificially aged and then cold worked

Subdivisions of the "—H1", "—H2", and "—H3" Tempers

The number following these designations indicates the final degree of strain-hardening. The numeral 8 designates the temper normally regarded and arbitrarily selected as "full hard." Material having a tensile strength about midway between that of fully annealed material (—0 temper) and that of the 8 temper is indicated by the numeral 4. The numeral 2 designates material having a tensile strength about midway between that of —0 temper and 4 temper, and 6 temper material is midway in strength between 4 temper and 8 temper. Although 8 temper is considered the

TABLE 11   MAGNESIUM AND MAGNESIUM ALLOYS

Form	Alloy	ASTM Temper	Spec. No.	Federal or Military Alloy	Federal or Military Temper	Federal or Military Spec. No.	AMS No.	SAE No.	Company Nomenclature Used Prior to Adoption of Standard Nomenclature
MAGNESIUM INGOT		—	—	—	—	MIL-M-20161	—	—	AM2S
	AM80A	—	B 92-56	—	—		—	—	AM241, A
	AM100A	—	B 93-56T	—	—		—	—	AM240, G
	AZ63A	—	B 93-56T	—	—		—	—	AM265, 630, H
	AZ91A	—	B 93-56T	—	—		—	—	AM263, 910, R
	AZ91B	—	B 93-56T	—	—		—	—	910
	AZ91C	—	B 93-56T	—	—		—	—	AMA263, 910
	AZ92A	—	B 93-56T	—	—		—	—	AM260, 920, C
	M1B	—	B 93-56T	—	—		—	—	AM403, M
	AZ81A	—	—	—	—		—	—	
SAND CASTINGS	AM100A	-T6	B 80-56T	—		—	4420	—	AM240-T61
	AZ63A	-F	B 80-56T	AZ63	AC	QQ-M-56	—	50	AM265-F(1), 630, H-F, 15A-T5
	AZ63A	-T5	B 80-56T	AZ63	ACS	QQ-M-56	—	—	AM265-T51, H-T5
	AZ63A	-T4	B 80-56T	AZ63	HT	QQ-M-56	4422	—	AM265-T4, H-T4, 15A-ST
	AZ63A	-T6	B 80-56T	AZ63	HTA	QQ-M-56	4424	—	AM265-T6, H-T6, 15A-STA
	AZ63A	-T7	—	AZ63	HTS	QQ-M-56	—	—	AM265-T7
	AZ91C	-F	B 80-56T	AZ91	AC	QQ-M-56	—	504	AMA263-F, 910
	AZ91C	-T4	B 80-56T	AZ91	HT	QQ-M-56	—	—	AMA263-T4
	AZ91C	-T6	B 80-56T	AZ91	HTA	QQ-M-56	—	—	AMA263-T6
	AZ92A	-F	B 80-56T	AZ92	AC	QQ-M-56	—	500	AM260-F(2), 920, C-F, 56
	AZ92A	-T5	B 80-56T	AZ92	ACS	QQ-M-56	—	—	AM260-T51, C-T5
	AZ92A	-T4	B 80-56T	AZ92	HT	QQ-M-56	4434	—	AM260-T4, C-T4, 56-ST
	AZ92A	-T6	B 80-56T	AZ92	HTA	QQ-M-56	—	—	AM260-T6, C-T6, 56-STA
	AZ92A	-T7	—	AZ92	HTS	QQ-M-56	—	—	AM260-T7
	EK30A	-T6	B 80-56T	—		—	—	—	AMA130-T6
	EK41A	-T5	B 80-56T	—		—	4440	—	517, AMA130-T5

EK41A	-T6	B 80-56T	E233A	-T5	MIL-M-9433	4441	—	517, AMA130-T6
E233A	-T5	B 80-56T	M1	AC	QQ-M-56	4442	—	AMA131-T5, ZRE1
M1B	-F	B 80-56T	ZK51	-F	MIL-M-8213	—	—	AM403-F, M
ZK51A	-F	—	ZK51	-T5	MIL-M-8213	4443	—	Z5Z
ZK51A	-T5	B 80-56T						520, Z5Z
ZK61A	-T6	B 80-56T						
AM40*	-F							AM244-F
AZ81A	-T4	B 80-56T						
HK31A	-T6	B 80-56T				4445		
HZ32A	-T5	B 80-56T						ZT1
ZE41A	-T6	B 80-56T						RZ5
ZH62A	-T5	B 80-56T						TZ6
AM120*	-F	B 60-56T						AM246-F
AM120*	-T51	—						AM246-T51
PERMANENT MOLD CASTINGS								
AM100A	-F	B 199-56T	A10	AC	QQ-M-55	—	502	AM240-F, G-F
AM100A	-T5	B 199-56T	—					AM240-T51
AM100A	-T4	B 199-56T	A10	HT	QQ-M-55			AM240-T4, G-T4
AM100A	-T61	B 199-56T	A10	HTA2	QQ-M-55			AM240-T61, G-T61
AM100A	-T6	B 199-56T	A10	HTA1	QQ-M-55			AM240-T6, G-T6
AM100A	-T7	—						AM240-T7
AZ92A	-F	B 199-56T	AZ92	AC	QQ-M-55		503	AM260-F(2), 920, C-F, 15B
AZ92A	-T5	B 199-56T	AZ92	ACS	QQ-M-55			AM260-T51, C-T5
AZ92A	-T4	B 199-56T	AZ92	HT	QQ-M-55			AM260-T4, C-T4, 15B-ST
AZ92A	-T6	B 199-56T	AZ92	HTA	QQ-M-55	4484		AM260-T6, C-T6, 15B-STA
AZ92A	-T7	—						AM260-T7
DIE CASTINGS								
AM100B		—					500	AM263-F, 910, R, 15C, AZ91
AZ91A		B 94-52			QQ-M-38	4490	501A	910, RC
AZ91B		B 94-52			—			130
AM10*		B 94-52						—

* Alloys whose composition is not shown in ASTM specifications. Designation follows ASTM rules but no final letter is given.

(1) AM266 has similar nominal composition but wider impurity limits.

(2) AM262 has similar nominal composition but wider impurity limits.

TABLE 11 MAGNESIUM AND MAGNESIUM ALLOYS—continued

Form	ASTM Alloy	ASTM Temper	Spec. No.	Federal or Military Alloy	Temper	Spec. No.	AMS No.	S.A.E. No.	Company Nomenclature Used Prior to Adoption of Standard Nomenclature
EXTRUSIONS	AZ31B	-F	B 107-56T	AZ31B	—	QQ-M-31a	—	52	AMC52S, FS1
	AZ61A	-F	B 107-56T	AZ61A	—	QQ-M-31a	4350	520	AMC57S, J1
	AZ80A	-F	B 107-56T	AZ80A	-F	QQ-M-31a	—	523	AMC58S-F, O1-F
	AZ80A	-T5	B 107-56T	AZ80A	-T5	QQ-M-31a	—		AMC58S-T51, O1-T5
	M1A	-F	B 107-56T	M1A		QQ-M-31a	—	522	AM3S, M
	ZK60A	-F	B 107-56T	ZK60A	-F	QQ-M-31a	4352	524	AMA76S-F
	ZK60A	-T5	B 107-56T	ZK60A	-T5	QQ-M-31a	4352	524	AMA76S-T5
	MZ10*								MF
	AZ51*								JS1
	HM31XA	-F							
	Mg 99.8*								AM2S
TUBING	AZ31B		B 217-56T	AZ31X	—	W-W-T-825	—	52	AMC52S-F, FS1
	AZ61A		B 217-56T	AZ61X	—	W-W-T-825	4350	520	AMC57S-F, J1
	M1A		B 217-56T	M1	—	W-W-T-825	—	522	AM3S, M
	ZK60A	-F	B 217-56T				4352	524	AMA76S-F
	ZK60A	-T5	B 217-56T				4352	524	AMA76S-T5
	Mg 99.8*								AM2S
	MZ10*								MF
	AZ51*								JS1
FORGINGS	AZ31B		B 91-56T				—		FS1
	AZ61A	-F	B 91-56T	AZ61A	—	QQ-M-40A	4358	531	AMC57S, J1
	AZ80A		B 91-56T	AZ80A	-F	QQ-M-40A	4360	532	AMC58S-F, O1-F
	AZ80A	-T5	B 91-56T	AZ80A	-T5	QQ-M-40A	—		AMC58S-T5, O1-T5
	TA54A			TA54A	—	QQ-M-40A	—	53	AM65S
	ZK60A						—		AMA76S-F
	ZK60A	-T5	B 91-56T	ZK60A	-T5	QQ-M-40A	4362		AMA76S-T5
	M1A			M1A	—	QQ-M-40A	—	533	AM3S, M

SHEET

Alloy	Temper							Designation
AZ31A	-O	B 90-56T	AZ31A	-O	QQ-M-44	4375	510	FS1-O, AMC52S-O
AZ31B	-F							
AZ31B	-H10							
AZ31B	-H11							
AZ31B	-H23							
AZ31A	-H34	B 90-56T	AZ31A	H24	QQ-M-44A	4376		FS1-H24, AMC52S-H24
AZ31A	-H26		AZ31A	H26	QQ-M-44A			FS1-H26
AZ31B	-O		AZ31B	-O	QQ-M-44A			FS1W-O
AZ31B	-H24		AZ31B	H24	QQ-M-44A			FS1W-H24
M1A	-O	B 90-56T		-A	QQ-M-54	4370	51	M-O, AM3S-O
M1A	-H24	B 90-56T		-H	QQ-M-54			M-H24, AM3S-H24
AZ51*	—							JS1-O
								JS1-F
								JS1-H24
HK31A	-O							
HK31A	-H24							MF-O
HM21XA	-T8							MF-F
MZ10*	—							MF-H24

* Alloys whose composition is not shown in ASTM specifications. Designation follows ASTM rules but no final letter is given.

"full hard" temper, a slightly harder temper, designated by the numeral 9, is produced for special applications.

Subdivisions of the "—T" Temper

—T2 ANNEALED (cast products only) This designation is applied to castings only, to indicate a type of annealing operation used to improve ductility and increase dimensional stability.

—T3 SOLUTION HEAT-TREATED AND THEN COLD-WORKED This designation applies to those products where cold work is performed for the primary purpose of improving the strength, and also applies to those products in which the effect of cold-work (such as flattening or straightening) is recognized in applicable specifications.

—T4 SOLUTION HEAT-TREATED AND NATURALLY AGED TO A SUB-STANTIALLY STABLE CONDITION This designation applies when the product is not cold-worked after heat-treatment, and also when applicable specifications do not recognize the effect of cold work in flattening and straightening operations.

—T5 ARTIFICIALLY AGED ONLY This designation applies to products which are artificially aged without prior solution heat-treatment. The artificial aging of these products may improve mechanical properties and/or dimensional stability.

—T6 SOLUTION HEAT-TREATED AND THEN ARTIFICIALLY AGED This designation applies to products which are not cold-worked after solution heat-treatment, and in which the effect, if any, of flattening or straightening is not recognized in applicable specifications.

—T7 SOLUTION HEAT-TREATED AND THEN STABILIZED This designation applies to products in which the temperature and time conditions for stabilizing are such that the alloy is carried beyond the point of maximum hardness, providing control of growth and/or residual stress.

—T8 SOLUTION HEAT-TREATED, COLD-WORKED, AND THEN ARTIFICIALLY AGED This designation applies when the cold-working is done for the purpose of improving strength, and also when the cold-working effect of flattening or straightening is recognized in applicable specifications.

—T9 SOLUTION HEAT-TREATED, ARTIFICIALLY AGED, AND THEN COLD-WORKED This designation applies when the cold-working is done for the purpose of improving strength.

—T10 ARTIFICIALLY AGED AND THEN COLD-WORKED This designation applies to products which are artificially aged without prior solution heat-treatment, and then cold-worked for the purpose of improving strength.

Protective Treatments

Because magnesium is so susceptible to corrosion, several types of protective treatment have been developed for use on the various alloys.

In the *chrome-pickle* treatment the parts are immersed for one minute in an aqueous solution containing sodium dichromate and nitric acid; they then are rinsed in cold water, dipped in hot water, and dried. This treatment produces a surface color similar to that of yellow brass. The treatment removes as much as 0.0006 inch, and therefore is not suitable for parts that have been machined to close tolerances. The treatment is suitable for nearly all alloys except M1 for protection during shipment, storage, and machining.

For better protection and adhesion, the *dichromate* process is used. The parts are immersed for 5 minutes in hydrofluoric acid of 15 to 20 per cent strength, at room temperature, followed by boiling for 45 minutes in a 10 to 15 per cent solution of sodium dichromate to which a fluoride sometimes is added to improve corrosion resistance. The parts finally are rinsed in hot water. This treatment produces a dark color but no dimensional change.

As an alternative to the dichromate treatment, the *sealed chrome-pickle* treatment sometimes is used. The parts are given the regular chrome-pickle treatment, followed by boiling for 30 minutes in a 10 to 15 per cent solution of sodium dichromate.

All alloys may be given protection by *galvanic anodizing*, which produces a dark surface color but no dimensional change. Either this treatment or the sealed chrome-pickle treatment is required for the M1 alloy.

Questions

1 What are the principal sources of magnesium?
2 What processes are used for industrial production of magnesium?
3 Describe one of the processes.
4 Name three valuable properties of magnesium.
5 Name two harmful impurities in magnesium.
6 Why are aluminum and zinc used as alloying metals in magnesium alloys?
7 Why does alloying magnesium reduce its melting point?
8 What type of heat treatment is used on magnesium alloys?
9 Under what conditions are magnesium alloys corroded?
10 Name the protective treatments commonly used.

22 / Zinc

Evidently zinc was known in ancient times, because it was used in the brasses and bronzes produced by primitive man. Metallic zinc, known then as *spelter*, was produced and used in Europe as early as the sixteenth century. Zinc smelting started in this country in 1850, and since 1860 the United States has been the leading producer and consumer of this metal.

Zinc has considerable resistance to atmospheric corrosion because, as is the case with aluminum, the oxidation formed initially on its surface protects it from further attack. For this reason it is used as a protective coating for iron sheets. Zinc is rolled into sheets to be used for making electric dry battery cells and for some construction purposes. Increasingly large amounts of zinc are being used for making die castings. Estimates for 1959 by the American Zinc Institute indicate that the largest amounts of zinc will be consumed for die castings, 368,000 tons; galvanizing (formerly the largest consumer), 355,000 tons; and brass, 131,500 tons.

Zinc Ore

The most common source of zinc is the zinc sulfide ore known as *sphalerite* or *zinc blende*, mined principally in the Joplin mining district in Missouri, Kansas, and Oklahoma. Another important ore found in the same district is *smithsonite*, a zinc carbonate ore. Other ores, found for the most part in New Jersey, are: *zincite*, a zinc oxide ore; *franklinite*, an ore consisting of mixed oxides of iron, manganese, and zinc; *calamine* and *willemite*, two silicate ores.

Concentration and Roasting

Zinc ore is crushed, concentrated by gravity concentration methods or by flotation, and roasted, all by much the same procedures as those used for copper ores. Roasting eliminates excess sulfur and converts the ore

276

to a calcine consisting of oxides and sulfates; ore from which the zinc is to be extracted electrolytically is roasted at a relatively low temperature in order that the calcine may consist largely of sulfates. Some small particles of the ore are carried away as dust by the gases of combustion, and are recovered by passing the gases through a Cottrell electrostatic precipitator.

Extraction of Metallic Zinc

After roasting, metallic zinc is extracted from the calcine by electrolysis or by distillation either in retorts or by an electrothermal process. Zinc extracted by electrolysis is of the highest purity of any commercial grade, and is used for die casting.

Electrolytic Extraction

The calcine is leached with dilute sulfuric acid, the amount of which is so adjusted that its acidity is neutralized completely by reaction with the calcine. Zinc and other metals which form soluble sulfates, particularly cadmium and copper, are dissolved; compounds of insoluble metals such as gold, iron, lead, and silver, remain in the residue, which is sent to a lead smelter for treatment and recovery of values.

Zinc dust is added to the solution to precipitate the metals that are electropositive to zinc, and which if allowed to remain in the solution would be deposited on the cathodes simultaneously with zinc by the process of electrolysis. The metals so precipitated are principally cadmium and copper.

After having been filtered to remove the precipitate, the neutral solution of zinc sulfate is electrolyzed in a tank containing 25 lead anodes and 26 cathodes of rolled sheet aluminum connected in parallel. Zinc is stripped from the cathodes at intervals, and is melted in a reverberatory furnace and cast into slabs of 99.99 per cent pure zinc. During electrolysis the zinc removed from the electrolyte is replaced by hydrogen, making the solution rich in sulfuric acid; when exhausted as a plating bath, the solution is used for leaching calcine.

Electrothermic Distillation

Electrothermic distillation was developed by the St. Joseph Lead Company, and is used by them exclusively.

To prepare the ore for electrothermic reduction, it is sintered to convert it into pieces of suitable size and strength to hold together in the electric furnace and to be sufficiently conductive to conduct electricity during the process. Sintering also effects further elimination of sulfur, and the elimination of other impurities such as lead and cadmium. The powdery calcine is mixed with various fines, reclaimed dust, and other residues containing zinc. Crushed coke and sand are added, and the whole is mixed in a rotary mixer with about 11 per cent of water. The material then is ignited by gas. After ignition, combustion is maintained by the coke, the carbon in some of the residues, and the sulfur content of the remaining sulfide material, and proceeds at a temperature of about 2900°F. The

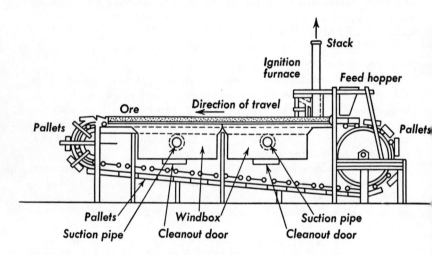

Figure 146. Dwight-Lloyd sintering machine.

gases evolved from the sintering operation contain dust consisting of zinc oxide, lead, and cadmium, and are precipitated in a Cottrell precipitator for subsequent recovery of values.

Distillation is performed in an electric furnace of the resistance type, in which the charge is the conducting material that is heated by the passage of electric current.

The charge consists of approximately equal volumes of coke and sinter. It is preheated to about 1470°F and is distributed automatically in such a manner that the larger particles, principally coke, roll toward the axis of the furnace. The result is that electric conductivity, and therefore operating temperature, is greater near the center of the furnace than at the wall. The range of temperature is from about 1650°F near the wall to 2350°F to 2550°F at the center.

The ore is reduced by the carbon of the coke, and the reaction products are zinc vapor and carbon monoxide gas. These products leave the furnace through the vapor ports, and the zinc vapor is condensed to the molten state in a vacuum condenser designed especially for the process. Molten

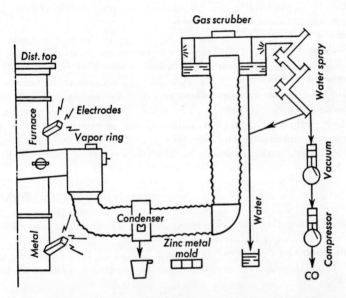

Figure 147. Section of electrothermic zinc furnace, showing condenser.
Courtesy American Institute of Mining and Metallurgical Engineers.

zinc is tapped from the condenser and cast into slabs weighing about 47 pounds each.

A similar process is used to produce zinc oxide. In the oxide process, the zinc vapor and carbon monoxide gas are removed from the furnace through refractory-lined flues known as *tewels*, where they are mixed with. enough air to cause them to oxidize immediately to zinc oxide and carbon dioxide, respectively. In general, the lower the temperature of combustion, the smaller are the particles produced. The gas in which the zinc oxide is suspended is passed through a centrifugal separating device known as a *cyclone* to remove particles of unsuitably large size; finally the zinc oxide is collected by blowing the suspension through cloth filter bags which retain the fine particles.

Protective Coatings of Zinc

Iron and steel may be protected from corrosion by application of a surface coating of zinc; the coating forms a thin layer of iron-zinc alloy, covered by a thicker layer of pure zinc. Even when a break occurs in the continuity of the coating, the presence of zinc provides electrolytic protection against corrosion of the iron.

Such a coating may be applied by one of the processes known as *galvanizing.* In the hot dip process, the ferrous metal is cleaned by pickling in dilute acid, and then is immersed in molten zinc. This process produces on wrought iron a coating that is from 25 to 40 per cent thicker than that obtainable on other varieties of iron and steel. For objects too large to be dipped, molten zinc is sprayed on by a gun; this is known as *metallizing.*

Other methods of applying a coating of zinc are electrodeposition, also called *electrogalvanizing*, and *sherardizing*, accomplished by heating small articles packed with powdered zinc at a temperature of about 650°F.

Die Casting Alloys

A large proportion of all die castings are made from alloys of which the principal component is zinc. Such alloys are made from the purest zinc obtainable, together with aluminum and often magnesium; some alloys also contain copper for increased tensile strength. The amounts of cadmium, iron, and tin present as impurities must be kept extremely low. Zinc die castings are unsuited for use at temperatures in excess of about 200°F.

Questions

1   Why is zinc resistant to atmospheric corrosion?
2   Name three zinc ores.
3   What is the purpose of roasting zinc ore?
4   Why are no fluxes required in the reduction of zinc ore?
5   Describe the process of electrolytic extraction.
6   Describe the electrothermic process for extraction of zinc.
7   Why must an excess of carbonaceous material be used in distilling zinc?
8   Why does a coating of zinc protect ferrous metals from corrosion?
9   What methods are used for coating iron and steel with zinc?

23 / Control and Testing

Metallurgical processes demand close control, and modern practices require use of instruments so designed that the needed information can be obtained from them quickly and accurately by operators who have no knowledge of the physical principles upon which they are based. The same is true of many of the instruments designed for testing metals for flaws, for chemical composition, or for physical structure.

Measurement of Temperature

Means for measuring and controlling temperature are, of course, essential to the operations of smelting ores to obtain metal, and to the operation of refining the metal so obtained. In making steel in any of the furnaces used for the purpose, and in the making of other alloys of which nonferrous metals are the components, adequate means for temperature measurement and control are required. Heat treatment of alloys requires control of temperature, sometimes within relatively narrow limits.

Any instrument used for measurement of temperature is called a *pyrometer*. Technically, the term includes the instruments based on the principle of thermal expansion of metals, liquids, or gases, and popularly called *thermometers*; these instruments have relatively little application in control of metallurgical processes. Another type of pyrometer which is suitable only for measurement of relatively low temperatures is the *resistance thermometer*; its operation is based upon the change of electric resistance of a wire with change of temperature. In addition to their applications in measurement of temperature, pyrometers can be adapted to actuate control devices automatically.

Thermocouples

A thermocouple consists of a pair of wires of dissimilar metals welded together at one end. When this junction is heated to a temperature higher

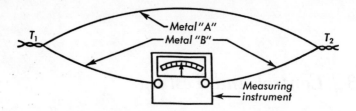

Figure 148. The fundamental thermocouple circuit.
Courtesy The Bristol Company.

than that of a similar junction in the same electric circuit, a difference of thermoelectric potential is set up between the two junctions, and if the circuit is closed a current flows. This current can be measured by a suitable electric measuring instrument inserted into the circuit. The potential difference, and therefore the current, vary directly, but not necessarily proportionately, with the difference in temperature between the hot and the cold junction. In industrial practice an actual cold junction is not used, but is replaced by the measuring instrument, usually a

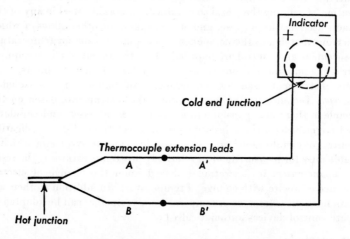

Figure 149. Typical thermocouple circuit.

Reproduced by permission from W. C. Stimpson, B. L. Gray, and J. Grennan, *Foundry Work*, Chicago, American Technical Society, 1944.

millivoltmeter, and the associated external wiring. The meter then measures the difference between the temperature of the hot junction and the temperature of the meter. If the meter is kept at a constant temperature its scale may be calibrated to read the high temperature directly.

For maximum accuracy in measurement, the voltage of an electric dry cell of known voltage is balanced by means of a potentiometer against the voltage set up by the thermocouple. By this arrangement, because no current flows in the circuit, any changes in resistance which may occur in the connecting wires or elsewhere in the external circuit do not produce inaccurate readings by the meter. The dry cell is checked frequently by comparison with a standard cell.

The hot junction and its wires are enclosed in a protective housing which may be of metal or of a refractory ceramic material depending upon the maximum temperature to be measured. Inside the protective housing the wires are insulated from each other above their junction point by enclosing one or both in electric insulation formed of a material suitable for the temperatures to be measured.

The principal metals and alloys used for thermocouple wire are:

Alumel (nickel, 94 per cent; manganese, 3 per cent; aluminum, 2 per cent; silicon, 1 per cent)
Chromel (nickel, 90 per cent; chromium, 10 per cent)
Constantan (nickel, 60 per cent; copper, 40 per cent)
Copper
Iron
Platinum.
Platinum-rhodium alloy (platinum, 90 per cent; rhodium, 10 per cent)

A thermocouple of alumel and chromel wires is most widely used for measurement of temperatures not exceeding about 2000°F. It has the advantage of producing a potential gradient almost proportional to temperature, and is extremely resistant to oxidation because of the high nickel content of both alloys.

For measurement of temperatures above the range of the alumel-chromel thermocouple, one of platinum and platinum-rhodium alloy wires is used. It is suitable for use as high as 2700°F or even 3000°F, and with proper care and protection has unlimited life and stability of calibration.

Iron-constantan thermocouples are suitable for measurements as high as about 1800°F under reducing conditions, or about 1400°F under oxidizing conditions. Copper-constantan thermocouples are limited in range to a maximum of between 500°F and 600°F.

Optical Pyrometers

The principle of the optical pyrometer is that of comparing the brightness of an incandescent filament, contained in the instrument and operating at a known temperature, with the brightness of a glowing object the temperature of which is to be measured. The filament is contained in a small telescope, and an image of the glowing object is focused upon the plane of the filament, with the result that the filament appears to the observer against the glowing object as a background. The operator adjusts the current supplied to the filament until its brightness and that of the object match, as indicated by disappearance of the image of the filament against its background. This indicates that the temperature of the filament has been made equal to that of the object to be measured. By means of a circuit incorporating a potentiometer, the indication of filament current on a meter is balanced, and the temperature is read on the potentiometer dial, which is calibrated to read directly in temperatures.

For measurement of temperatures between $1400°F$ and $2250°F$, readings are compared directly; for higher temperature measurements an absorption screen is interposed between filament and object to reduce the apparent brightness of the object, and temperature is read on another scale.

Radiation Pyrometers

A radiation pyrometer is a tube containing a thermocouple, and a lens that focuses upon the thermocouple the heat rays radiated by the object, the temperature of which is to be measured. The temperature is measured by any of the methods ordinarily used with thermocouples.

Radiation pyrometers are suited particularly for reading surface temperatures of roofs, walls, ducts, and linings of a furnace, or of the work in the furnace.

Photoelectric Measurement and Control

Photoelectric measurements are made by use of a light-sensitive cell of chemical nature or by a special type of vacuum tube; in either case, flow of electric current through the cell varies directly, but not necessarily proportionately, with the intensity of light falling upon the cell.

An example of the application of the photoelectric cell to metallurgical processes is the determination of the end point of the blow in the Bessemer process of making steel. The process is one that is largely automatic until the end point of the blow is reached, when the converter must be tilted down promptly. The progress of the reactions can be observed with considerable accuracy by measuring the intensity of light emitted by the flame. For this purpose a photoelectric cell is fitted with a filter that passes only a narrow band of radiation in the infrared portion of the spectrum. The indications of the cell are recorded on a chart by a suitable recording instrument. The chart shows a low reading during oxidation of silicon, but increases as the silicon is burned off and oxidation of carbon takes

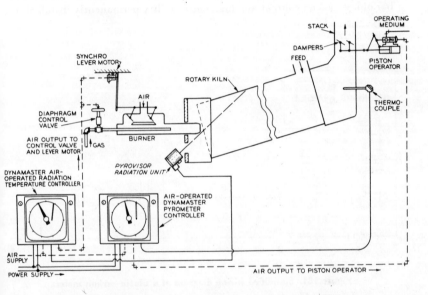

Figure 150. Schematic diagram of radiation temperature recorder.
Courtesy The Bristol Company.

place. When the carbon content has dropped to about 0.15 per cent, the photoelectric current rises for a brief time and then drops sharply, indicating a marked decrease in the energy radiated by the flame, and the end of the blow.

Some optical pyrometers incorporate a photoelectric cell that provides readings on the dial of a meter. Such instruments are more accurate than the comparison type using an incandescent filament, and can be used by inexperienced operators.

Magnetic Test for Carbon Content of Steel

The magnetic permeability of steel decreases with increased carbon content, provided the alloy content of the steel and its heat treatment remain the same. This fact provides a basis for determination of the carbon content of a sample of steel drawn from the furnace, within a period of about $2\frac{1}{2}$ minutes, and with an accuracy of about ± 0.02 per cent.

Determinations are made with a Leitz-Blosjo carbon meter, a simplified wiring diagram of which is shown in Figure 151. The standard is a steel bar of low carbon content and high permeability permanently installed in

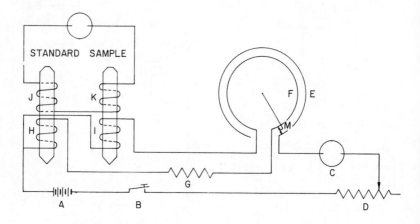

Figure 151. Simplified wiring diagram of a Blosjo carbon meter.
Courtesy E. Leitz, Inc.

the meter. When an electric current is passed through coils H and I of Figure 151, both the standard and the sample are magnetized; when the current is interrupted, collapse of the lines of magnetic force in the bars produces momentary induced voltages in coils J and K. If the currents through H and I are the same, and if the permeabilities of the standard and the sample are the same, these induced voltages cancel each other, and no deflection of the galvanometer results. If the permeability of the sample is less than that of the standard, more current must be allowed to pass through coil I than through coil H in order to obtain equal induced voltages in J and K upon interruption of the current. This is accomplished by setting the slide-wire potentiometer; the carbon content corresponding to the reading obtained from the potentiometer dial is found upon a calibration chart.

In making a test, a sample of the steel is drawn from the furnace, is de-oxidized by adding aluminum, is cast into a mold of the required size, and is quenched in water after solidification, all in accordance with a detailed procedure stated in the directions accompanying the carbon meter. The

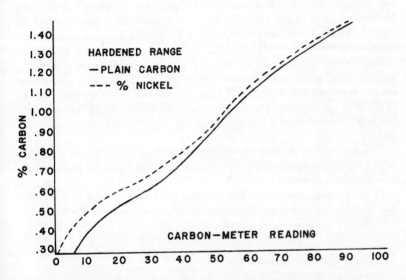

Figure 152. Typical calibration curve for Blosjo carbon meter.
Courtesy E. Leitz, Inc.

sample is then wiped to remove any water, and is inserted in the meter and tested.

For use with an alloy steel, the carbon meter must be calibrated over the full range of expected carbon content by making simultaneous chemical analyses and carbon meter tests on identical samples. A curve is then drawn.

Hardness Testing

Hardness testing of metals is based upon the principle of measuring the force required to penetrate the metal under standardized conditions.

Brinell Hardness Tests

The Brinell test is made by pressing into the metal under a specified load, a steel ball 10 millimeters in diameter, or a similar ball of tungsten carbide, known as *penetrators*, and measuring the diameter of the indentation produced. For measuring the hardness of steel a load of 3000 kilograms is applied, using the steel penetrator for hardnesses of 500 Brinell or less, and the tungsten carbide penetrator for hardnesses above 500 Brinell. The specifications of the American Society for Testing Materials provide that the load must be applied for a minimum of 10 seconds, but it customarily is applied for 30 seconds. For testing materials softer than steel, the steel penetrator is used with a load of 500 kilograms.

Smaller penetrators and correspondingly lighter loads sometimes are used for small articles. The ratio of load to size of penetrator is the same as for the standard penetrators: 30 times the square of the diameter of the penetrator for use with steel, and 5 times for use with soft metals, with the load expressed in kilograms and the diameter expressed in millimeters.

The diameter of the impression left in the work under test is measured with a microscope provided with a calibrated reticle, and the Brinell hardness number corresponding to that diameter is found in a table supplied with the testing instrument. This number is a figure obtained by dividing the load by the spherical area of the impression, expressed in kilograms and square millimeters, respectively.

Rockwell Hardness Tests

The Rockwell hardness test, like the Brinell test, is based upon penetration by pressure. The measure used, however, is the depth of penetration as indicated in Rockwell units on the dial of the machine. The two most commonly used ranges are the *Rockwell C hardness* and the *Rockwell B hardness* scales, used for steel and for softer metals, respectively.

For the C tests, the penetrator used is a diamond cone with a rounded point of 0.2 millimeter diameter; this is known by the trade-mark *brale*. In use, a *minor load* of 10 kilograms is applied first to seat the brale, after which a *major load* of 150 kilograms is applied. Then the *Rockwell C hardness* is read on the black scale of the dial; it represents the difference in penetration produced by the major load over that produced by the minor load. Brales must be replaced if chipped or cracked.

Rockwell B hardness is obtained in the same manner, except that the penetrator used is a hardened steel ball $\frac{1}{16}$ inch in diameter, and used

Figure 153. Rockwell hardness tester.
Courtesy American Chain & Cable Co., Inc.

with a major load of 100 kilograms. The steel ball is held in a chuck and is readily replaceable when it becomes flattened by use.

Other special scales are used, designated *A*, *D*, *E*, *F*, *G*, *H*, and *K*, each differing only in the loads applied and the types of penetrators used.

For hardness tests on thin material of high surface hardness, another model called the *superficial hardness tester* is used; it operates on the same principle but uses a minor load of 3 kilograms and major loads of 15, 30, or 45 kilograms.

The Rockwell tester is somewhat easier to use than the Brinell because readings are made on a dial, requiring no use of accessory apparatus such as the measuring microscope.

Scleroscope Hardness Tests

Scleroscope hardness is measured by the height of rebound of a diamond-faced hammer dropped on the work from a specified height. In the better model of the instrument the hardness number is read on a dial. Operation of the scleroscope requires more skill and experience on the part of the operator than does operation of either the Brinell or Rockwell instruments, but the scleroscope has the advantage of easy portability and may be taken to the work when large objects are to be tested for hardness.

Microhardness Tester

The microhardness tester has made it possible to determine the hardness of single crystals of steel. It can take hardness readings at three points within the width of a human hair. Under a carefully controlled load, a four-sided diamond pyramid marks a polished sample. The hardness of the steel is indicated by the depth of penetration measured in microns (one micron = 0.000039 inch). This test is very effective in studying the segregation of elements, the effects of heat treatment, and the effect of size and shape on mechanical properties.

Identification of the Constitution and Structure of Metals

Various test methods have been developed for identification of both the chemical constitution and the physical structures of metals and alloys.

In industrial plants it is necessary, for example, to determine readily the alloy content of metals for the purpose of classification and sorting. Metals frequently must be tested also to determine the microstructure produced by work or heat treatment in order to ascertain their suitability for stated applications, or to learn whether a given treatment has been effective.

Another important application of testing methods is to research work. The microstructures of metals are examined and studied, and the course of experimental work being performed to improve or develop a given metallurgical process is guided by the information so obtained.

Tests which may be classified as identification tests include a range from the simple spark tests used in the shop, to the precise x-ray diffraction tests performed in the laboratory.

The study of structures of metals by use of such tools as the metallographic microscope, x-ray diffraction equipment, the spectroscope and spectrograph, and equipment designed for magnetic analysis, constitutes the science of metallography.

Identification of Metals by the Spark Test

Steels and a few nonferrous metals and alloys can be identified and classified by examining the characters of the sparks emitted when they are ground on an abrasive wheel. The value of this method in any instance depends primarily upon the experience and judgment of the operator, and in no event is it capable of giving an exact result as is chemical analysis. It is, however, an economical and rapid method of approximate classification.

The operator uses a clean grinding wheel of any convenient type, and works under subdued light in order to see the sparks more readily. Pieces of black cardboard are so arranged that the spark stream is viewed against a black background for better contrast. In order that the spark stream may be sufficiently long for identification, the grinding wheel should have a surface speed of not less than about 4000 feet per minute.

Sparks from the metal to be identified are compared directly with sparks produced from standard test specimens made available for the purpose. A set of samples should include wrought iron, several carbon steels with different contents of carbon, and several samples of different important varieties of alloy steel.

Microscopic Examination

The crystalline structures of metals are studied by viewing them at magnifications as high as 3000 times. The shape and size of crystals so disclosed are indicative of the composition of an alloy, and of the work or heat treatment it has received.

Because metals are opaque, metallographic specimens must be examined by reflected light. In order that the light may fall upon the specimen in a direction perpendicular to its surface, it must pass first through

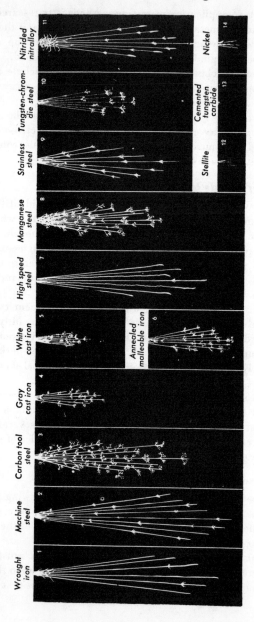

Figure 154. Spark test for metals.
Courtesy Norton Company.

the objective lens of the microscope; after reflection from the specimen, the light passes again through the objective and so to the eyepiece and through it to the eye of an observer or to a camera.

The microscope customarily is used in a vertical position, and the light, which enters horizontally, must be reflected through the objective by means of a reflector inside the microscope tube. This may be either a

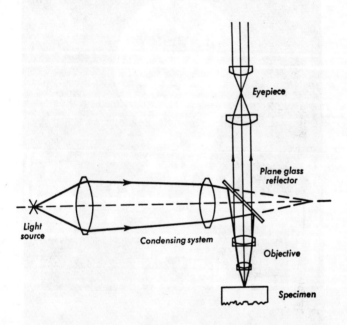

Figure 155. Diagram of bench type metallurgical microscope.
Courtesy Bausch and Lomb Optical Company.

totally reflecting prism that extends across one half the diameter of the tube, or a sheet of plane glass set at an angle of 45 degrees with the axis of the tube and extending entirely across the tube.

The prism allows more light to reach the eye and therefore produces a more brilliant image than does a plane glass reflector, but because the prism obscures one half the area of the objective, resolving power is diminished to such an extent that satisfactory images are unobtainable at high magnifications.

When a plane glass reflector is used, only a small portion of the light received from the illuminator is reflected through the objective; the remainder passes through the glass and is absorbed by the dead black internal surfaces of the microscope. The small portion which does pass

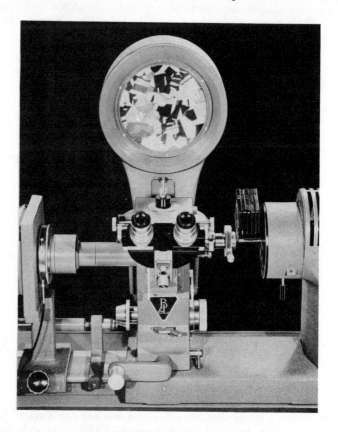

Figure 156. Bench type microscope.
Courtesy Bausch and Lomb Optical Company.

through the objective and is reflected from the specimen back through the objective strikes the plane glass plate again; here a further portion is lost, in this case by reflection, and the remainder passes to the eyepiece. Only a small proportion of the light originally received from the illuminator ever reaches the eyepiece, and a source of intense illumination therefore is essential. Despite this inefficiency of the plane glass reflector, however, it is used for high magnifications in order to secure adequate resolving power.

The microscope is used for either visual observation or for making photomicrographs.

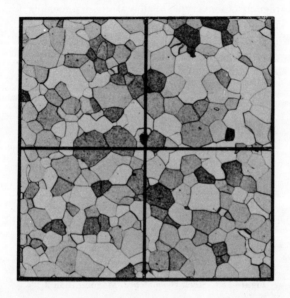

Figure 157. Estimating grain size with measuring eyepiece.
Courtesy Bausch & Lomb Optical Company.

PREPARATION OF SPECIMENS FOR MICROSCOPIC EXAMINATION   Proper choice and preparation of the specimen are essential to satisfactory results in microscopy. The optimum surface area of a specimen depends to some extent upon the nature of the investigation, but is ordinarily one square inch or less. If examination is to be made to discover the reason for a failure in a section of metal, the specimen should be taken from a spot close to where the failure originated. Both longitudinal and transverse sections of forged metal should be examined. The surface of a specimen cut from cold-worked metal should be parallel to the direction of working.

A portion of the outer skin should be retained in a specimen of heat-treated steel in order that decarburization or other surface conditions shall be disclosed.

The specimen is cut from the piece of metal by a saw or abrasive wheel, depending upon the nature of the metal. Cutting and all subsequent grinding and polishing operations should be done in such a way as to avoid heating to an extent which could result in tempering of a hardened steel, air hardening of some types of alloy steels, or annealing, for example.

Figure 158. Steel. Deep etched. Magnification 2 ×.
Courtesy Bausch & Lomb Optical Company.

The specimen preferably is mounted in a plastic material or in an alloy of low melting point, such as Woods metal; if means for such mounting are not available it may be clamped firmly between steel strips.

Rough grinding is done on a grinding wheel or belt grinder, using light pressure. Smooth grinding is accomplished by rubbing the specimen on emery paper, taking care that the surface is held flat on the paper at all times in order that the edges of the specimen shall not be rounded over. The specimen is moved across the abrasive paper in a straight line in one direction only, and always in a direction at right angles to the scratches left by the preceding grinding operation. This procedure is continued on successively finer grades of abrasive paper, grinding on each grade only long enough to remove the scratches left by the preceding grinding. After each grinding, the specimen is wiped clean of grit before starting to grind on the next finer grade of abrasive. Grades 1, 0, 00, and 000, of abrasive paper customarily are used.

When all scratches have been removed and the surface appears smooth upon examination at a magnification of about 7 times, the specimen is ready for polishing. Polishing is done on cloth mounted on a rotating disk. A suspension of an abrasive such as alundum or carborundum is applied to the cloth disk, and the specimen is held against it with firm pressure for about four revolutions, and thereafter with light pressure. The specimen may be moved slowly across the revolving disk in a radial direction, or may be turned slowly in a direction opposite the direction of rotation of the disk. Usually three cloth disks of successively increasing degrees of softness are used, with successively finer abrasives, the last of which is levigated alumina or jewelers' rouge.

When polishing is completed, the surface of the specimen is washed under a stream of hot water and is swabbed gently with absorbent cotton or facial tissue paper until all abrasive is removed. Finally it is dipped in ethyl alcohol and is dried in a stream of hot air.

Some specimens are ready for microscopic examination without further treatment, but etching usually is required to show the structure more clearly. The etchant is a solution which attacks some constituents or structures more readily than others, producing a contrast which makes grain boundaries more evident. Many different etchants are used, the choice depending upon the composition and structure of the metal being examined. A list of recommended etchants is contained in the recent edition of the *Metals Handbook* of the American Society for Metals. Two of the most commonly used are *picral* and *nital*. Picral is a solution of from 1 to 4 per cent of picric acid in alcohol; nital is a 1 to 5 per cent solution of strong nitric acid (specific gravity 1.42) in alcohol. Either ethyl or methyl alcohol may be used, of either the 95 per cent or the absolute grade. In electrolytic etching the specimen is etched by being used as an electrode in an electrolytic cell; both anodic and cathodic etching are used.

Examination by X-ray Diffraction

X-ray diffraction techniques, although still in an early stage of development, already have become some of the most powerful and important means of metallographic examination and testing. Examination of structures by the optical microscope is limited to those of dimensions little smaller than the wave length of light. Because the wave length of x-rays is so short, the parallel planes of atoms in the space lattice of a crystal act as diffraction gratings for this radiation somewhat as closely ruled parallel lines act as diffraction gratings for visible light. X-rays reflected from these atomic planes are directed to the surface of a photographic emulsion, on which their characteristic patterns are recorded.

Because every pure element and every compound forms a distinctive and unique x-ray diffraction pattern, an unknown mixture can be analyzed qualitatively. This method of qualitative analysis has the merit that the compounds present in an unknown substance can be identified,

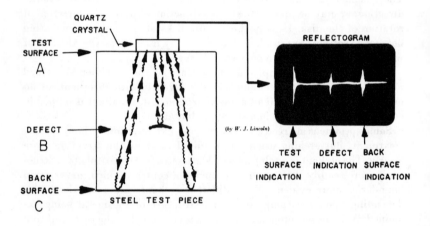

(by W. J. Lincoln)

Figure 159. Schematic representation of ultrasonic detection by the reflectoscope.

whereas chemical or spectrographic methods of qualitative analysis can identify only the individual elements of which the compounds are composed. The constituents present in an unknown mixture are identified by comparing the x-ray diffraction pattern produced by the mixture with the card index of x-ray diffraction patterns of compounds published by the American Society for Testing Materials. Because the pattern produced by a mixture is a composite of patterns produced by its constituent compounds, experience and skill are required for accurate interpretation in routine laboratory analytical work. Compounds of the same chemical composition but of different crystalline structures are differentiated by x-ray diffraction analysis. The relative proportion of each compound present can be estimated by comparing the relative densities of the lines of the diffraction patterns, but methods of exact quantitative analysis by x-ray diffraction have not been developed as yet.

Preferred orientation of crystals or distortion resulting from cold work can be identified by this method, and stresses can be measured with fair accuracy.

Physical structures of metals, and therefore the processing to which they have been subjected, can be identified. For routine industrial testing, x-ray diffraction patterns of standard samples are supplied to the operator, who compares with them the patterns obtained from the work under test. In this way the adequacy of annealing, tempering, hardening, case hardening, and like processes can be determined for purposes of quality control.

Examination by X-ray Spectrometry

X-ray spectrometry is used for rapid x-ray diffraction analysis; a Geiger counter is substituted for the photographic emulsion, and charts are produced automatically by associated electronic apparatus. This simplification, and elimination of film processing, make the procedure suitable for routine checking of material by relatively unskilled personnel.

Spectrographic Examination

When a chemical element is volatilized in an arc or in an electric spark, the individual atoms are caused to emit visible radiation that is characteristic of the element. If the radiation is directed through a spectrograph and is recorded on a photographic emulsion, one or more lines appear, each of which represents a single wave length of the spectrum. By comparing the spectrum so produced by an unknown substance with standard spectra of various elements, the coincidence of lines provides identification. Quantitative analysis for elements is accomplished by measuring with a densitometer the density of the spectral lines recorded on the film.

In industrial production testing by this method, a spark is struck on the surface of the metal under test, and the photographic record of the resulting spectrum is used for analysis. The National Bureau of Standards supplies a set of spectrographic standards for a number of steels; for steels not included in this set, or for other metals, standards are easily made.

Detection of Defects

The method to be chosen for detection of defects in metallic sections depends upon whether they are surface, subsurface, or internal. Nondestructive testing methods have superseded to a great extent the older, destructive methods which involved cutting up and so destroying the piece under investigation. For many varieties of metal products, methods have been developed which permit nondestructive testing of every unit produced instead of only a limited number of representative samples from each batch. Radiographic inspection of all welded pressure vessels such as boilers, for example, is not only customary but mandatory. Radiographic inspection is the most generally used method for detection of internal defects, and magnetic particle inspection is the most generally used method for detection of surface and subsurface defects.

Radiographic Inspection

Interest in radiography of metals started in 1912–13 when Dr. William D. Coolidge introduced an x-ray tube which could be operated continuously for long periods. This tube produced x-rays of greater penetrating power and greater intensity than those produced by medical x-ray apparatus heretofore used, because it operated at higher voltages and higher currents. During World War I, x-ray inspections were made of such relatively small objects as the fuse assemblies of high-explosive shells. In 1922 an x-ray tube of the new type was installed at the arsenal at Watertown, Massachusetts, for inspection of ordnance parts made there; this tube operated at 200,000 volts and 5 milliamperes. The first commercial laboratory for industrial x-ray testing was started in 1925 by Dr. Ancel St. John.

Industrial radiographic inspection by means of the gamma rays emitted by radium compounds started with the exhaustive experiments made by the United States Naval Research Laboratory in 1929.

RADIOGRAPHY BY X-RAYS In making an x-ray, a photographic film, enclosed in a cassette of material that is impervious to light rays but readily penetrated by x-rays, is placed below the object to be radiographed. X-rays emanating from the x-ray tube penetrate the metal and produce upon the film a shadowgraph of the interior of the object. The amount of radiation transmitted through a specimen depends upon the nature of the metal and upon its thickness. If, therefore, there are any cavities such as blowholes, gas cavities, shrinkage, pipe, porosity, and the like, the area of the film directly below a section in which such defects occur receives more exposure than do areas below sound metal; consequently those areas of the negative are darker. Inclusions such as oxide, slag, or segregations also appear as dark areas on the negative. Points of incomplete fusion in welds are more dense than the steel and therefore show as lighter areas on the negative.

The penetrating power of x-rays increases as their wave length decreases; this decrease is produced by applying a higher voltage to the anode of the tube. Intensity of radiation, and therefore the effect produced upon the photographic film, is increased by an increase in the flow of electrons in the tube, called *tube current;* tube current is increased by

Figure 160. Simultaneous gamma radiography of several castings.

Courtesy Metallurgy Branch, Material Laboratory, New York Naval Shipyard, Brooklyn, N.Y.

increasing the current supplied to the filament of the tube. Maximum penetration with minimum required exposure is therefore obtained by use of the highest voltage and highest filament current that the construction of the tube permits.

Equipment is available for as low as 3000 to 50,000 volts, for use with thin sections of such metals as aluminum and magnesium, and with non-metallic materials. Units of intermediate power ratings are available up to

2,000,000-volt units which are capable of making x-rays of steel as thick as 12 inches. An x-ray installation of moderate size used in many foundries for routine production inspection of castings is a 250,000-volt unit.

Figure 161. Reproduction of gamma graph of a cast monel sheave.

Courtesy Metallurgy Branch, Material Laboratory, New York Naval Shipyard, Brooklyn, N.Y.

Defects with thickness as little as 0.5 per cent of the thickness of the work can be detected, but a sensitivity of 2 per cent, as required by the boiler code of the American Society of Mechanical Engineers, usually is considered satisfactory.

Stereoscopic radiographs can be obtained by making two successive exposures with the x-ray tube placed at two different positions separated by about $2\frac{1}{2}$ inches. The stereographs are viewed in a suitable instrument to give a better impression of actual conditions, and to indicate the approximate depth of a defect. The method is used to a relatively small extent at present.

For rapid inspection of small objects, particularly those made of light metals, *fluoroscopy* is used. Instead of allowing the x-rays to produce a shadowgraph picture on a photographic film, they are caused to produce the same sort of picture for visual examination on a screen coated with a material that fluoresces with a brightness more or less proportional to the radiation that falls upon it. The operator sits in a dark room behind the fluorescent screen and examines the pictures as the pieces of work are passed through the x-ray beam; in many installations the pieces pass through the apparatus on an endless belt. X-ray tubes used for fluoroscopy usually are designed for operation at 150,000 volts or less because x-rays of greater penetrating power give an image of lower contrast, the details of which are less readily recognizable by the operator. The differences in density between adjacent areas of the image as seen on the fluorescent screen are less readily detectable than are the similar differences of portions of a photographic negative; as a result the smallest defect recognizable by fluoroscopy is about 5 per cent of the thickness of the material. Because of important individual differences in visual acuity, a fluoroscope observer must be selected by actual test of his ability to see the indications under the required conditions.

Fluoroscopic and photographic methods are combined in *photofluoroscopy*, which sometimes is used when a relatively permanent record is required. The image is formed on a fluoroscopic screen which then is photographed on a 70-millimeter film by means of a specially designed camera. The method is, of course, no more rapid than direct photographic recording, but the film size is so much smaller that economy in storage space and expense is effected.

GAMMA RADIOGRAPHY The gamma rays emitted by radium and its salts are of shorter wave length than x-rays, and therefore have greater power to penetrate metal. Because their intensity is less than that of x-rays emitted by tubes of high power ratings, however, longer exposures are required in gamma radiography. The source of gamma radiation for industrial uses is radium sulfate sealed in a silver capsule which in turn is enclosed in a container made of an aluminum alloy. Standard sizes range from 25 to 500 milligrams of radium sulfate, the choice for any stated application depending upon the intensity of radiation required.

The small size of the radium container makes it possible to make radiographs of boilers in position on ships, or portions of hulls of ships, and similar work that could not be done with the bulky equipment required for x-rays of high penetration. Gamma radiography is used economically also for smaller parts, because radiation is emitted from the capsule in all directions.

For protection of personnel, radium containers are kept in protective enclosures lined with thick walls of lead, and the containers when in use are placed in position by manipulation of cords from a distance.

Ultrasonic Inspection

Any sound wave of a frequency higher than about 16,000 cycles per second, which is the upper limit of audibility for the average human ear, is called *ultrasonic*. For ultrasonic testing, however, frequencies between 0.5 and 12 megacycles (500,000 to 12,000,000 cycles) per second are used in practice. The highest frequency permits detection of the smallest defects, but lower frequencies must be used when maximum penetration of considerable thickness of material is required. The waves are transmitted through the material under test just as are sound waves in a solid medium.

One model of ultrasonic test equipment operates on the principle of measuring the time required for an ultrasonic impulse applied to the surface of a metal to reach an existing defect and to return by reflection to the point from which it was transmitted. A quartz crystal is used both to apply the impulse obtained from the generator of ultrasonic pulses and to accept the echo and return it to a receiver that is also part of the equipment. The elapsed time is indicated on the screen of a cathode ray tube.

With this type of equipment it has been possible to detect defects at a distance of 24 feet through steel, and 28 feet through aluminum. Defects with diameters as small as 0.1 per cent of the distance from the quartz crystal have been detected. As compared with radiography, ultrasonic inspection with this equipment offers the advantages of greater penetrating power, ability to detect smaller defects, and absence of hazard to the operator. On the other hand, preparation of the work is required, because the surface to which the crystal is to be applied must be ground to approximate flatness.

Another model uses separate crystals for transmission and reception of the impulses. The two crystals are immersed in oil or water, with the object under test placed between them. Equipment of this type is used for production testing for defects because it provides a rapid and economical method that can be operated by semiskilled personnel after short training.

Magnetic Analysis Inspection

Equipment used for magnetic analysis testing includes an oscillator adjustable to any frequency, with a range of about 2000 to 200,000 cycles per second, and a test coil that forms part of the oscillating circuit. If a metal object is introduced into the field of the test coil, the frequency of the oscillator is changed as a result of losses in the metal. This change in frequency is indicated on the screen of a cathode ray tube incorporated in the apparatus. The apparatus can be calibrated to indicate differences in chemical composition or physical structure, stresses produced by cold work or heat treatment, and the like.

Magnetic Particle Inspection

Magnetic particle inspection, by the method known as *Magnaflux*, is a nondestructive method for the detection of discontinuities at or near

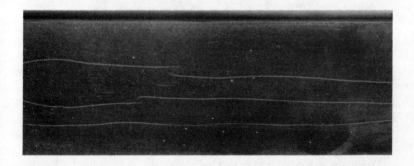

Figure 162. Cracks in a bar of hardened tool steel caused by improper heat treatment.
Courtesy Norton Company.

the surface of magnetic materials. Such a test reveals all types of surface defects such as quenching and fatigue cracks, grinding checks, and surface seams, and distinguishes between many of these types of defects. It also detects such subsurface defects as slag pockets in welds, and shrinkage cavities and pores in castings; such defects are usually detectable at a maximum depth of about one quarter inch in highly finished machine parts, and from three quarters inch to one inch in large castings and welds. The indication is a powder pattern magnification of the defect.

The method depends upon the fact that a magnetic field is distorted by a crack, a gas pocket, or an inclusion of slag or other nonmagnetic material. This distortion results in leakage fields and formation of small local magnetic poles. When powder of finely divided magnetic material is applied to the magnetized surface of a piece to be inspected, the particles are attracted by the leakage field, and form patterns that indicate the locations and shapes of any such discontinuities.

The work may be magnetized by passing an electric current through it or by placing it within the magnetic field of a coil of wire in which current flows, the choice of method depending upon the nature of the part to be inspected. For detection of subsurface defects, magnetization usually is produced by use of direct current, which penetrates the mass

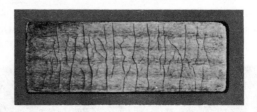

Figure 163. Cracks in a bar of hardened tool steel caused by too severe a grinding operation.

Courtesy Norton Company.

of the work. For surface defects, alternating current is used because of its *skin effect*, which concentrates its effect in a relatively thin layer at the surface of the part.

Dry magnetic powder may be blown over the work, a light oil suspension of the particles may be flowed over the work, or the work may be dipped in it. The dry method is suited particularly for surface and subsurface inspection of unmachined castings, welds, forgings, heavy machinery, and the like; the wet method is used more often for inspection of ground or polished parts with bright surfaces.

A variation of the wet method consists in coating the wet particles with material that is highly fluorescent under near ultraviolet light. This modification, called *Magnaglo*, is especially useful for detection of defects on interior surfaces where illumination is poor; it is useful also for surface work because of the added visibility that it provides.

Fluorescent Penetrant Inspection

Fluorescent penetrant inspection was developed by the Magnaflux Corporation for inspection of nonmagnetic materials such as brass, magnesium, aluminum, plastics, ceramics, and the like; the method is called *Zyglo*. A fluorescent penetrant applied to the work penetrates all surface

Figure 164. Magnaflux indication of surface cracks on a plug gauge.
Courtesy Norton Company.

defects. After subsequent processing, the parts are examined under near ultraviolet light of wave length between 3200 and 4000 angstrom units, and the defects are easily recognizable.

In the operation of the process, the parts to be inspected are dipped in or bathed with penetrant, excess penetrant is drained off, and penetration takes place for a period of time depending upon the material and the type of defect suspected. The work then is rinsed with water to remove the fluorescent material from the surface but leave it in the pores or cracks. From this point the operation is continued by either the dry method or the wet method. In the dry method the parts are dried after rinsing, by wiping, by heating in an oven, or by air blast, or by a combination of heat and air blast. They then are dusted over with a developing powder that draws the penetrant from the flaws, and by bringing it to the surface, provides a larger and more brilliant indication. It also provides a background that gives increased contrast. In the wet method, the parts after rinsing are dipped in a colloidal water suspension of a developer and are dried in a hot air recirculating drier; this method also produces increased brilliance and contrast in the indications. Finally the parts are inspected under near ultraviolet light in a darkened room, where all defects appear as brilliant white lines or spots against a dark background.

Questions

1 Describe a thermocouple pyrometer and explain the principle of operation.
2 Describe a radiation pyrometer and explain the principle of operation.
3 Describe the magnetic test for carbon content of steel, and explain the principle upon which it is based.
4 What is the principle upon which measurement of hardness is based? Describe the Rockwell hardness test.
5 Describe a metallurgical microscope. How does light reach the specimen?
6 Describe the preparation of a specimen for microscopic examination.
7 What information can be obtained from x-ray diffraction inspection?
8 What are the principal procedures used in radiographic examination? Describe each briefly and state under what conditions it is used.
9 Explain ultrasonic inspection.
10 Describe magnetic particle inspection procedure. For what purpose is it used?

24 / Foundry Practice

A casting is a metal object formed by solidification of metal introduced in the molten condition into a prepared mold of required shape and size.

Sand Casting

A casting is made by pouring molten metal into a mold formed by pressing molding sand around a pattern of the shape of the article to be cast. The pattern used for sand casting may be of wood, metal, or any other material that can be shaped readily.

Casting procedure is essentially the same with all metals used for sand casting, but with some modifications required by the nature of the metal. Steel is poured at such high temperatures that a mixture of silica and fire clay must be used instead of the molding sand used in making castings of aluminum, brass, cast iron, and magnesium. Steel castings require prompt annealing after cooling. Because at high temperatures it has a tendency to absorb gases, particularly hydrogen, the pouring temperature of aluminum is kept as low as is consistent with proper fluidity. Because aluminum has relatively high shrinkage, cores must be softer than those suitable for cast iron, for example, to avoid cracking of the casting as a result of resistance offered by hard cores. Multiple gating ordinarily is used for aluminum casting in order that the mold shall fill with metal as promptly as possible. The strong tendency of magnesium to oxidize makes it necessary to melt it under a flux containing such materials as fluorides, sulfur, and boric acid, and to mix inhibiting agents in the molding sand.

Pattern Making

The pattern must be slightly larger than the finished object, because all metals shrink upon cooling. Allowance is made for this required oversize by constructing the pattern from measurements made with a *shrink rule* that is so graduated as to include the proper allowance. The allowances vary from $\frac{1}{8}$ inch per foot for gray cast iron to as much as $\frac{1}{4}$ inch for some steels.

Wood patterns made of white pine, cherry, or mahogany are used if the number of castings to be made is not large. For larger production of castings a more durable pattern is required, and metal patterns of aluminum, brass, cast iron, or white metal are used. A metal pattern is made from a *master pattern* or *double shrink pattern* of wood, which must have double allowance to compensate for the shrinkage of the metal of which the pattern is to be made and for the metal to be used for the castings.

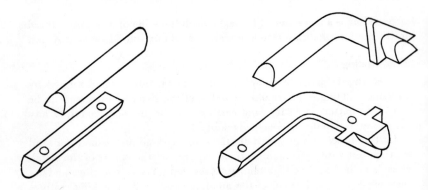

Figure 165. Split patterns.

In order that the pattern may be drawn from the sand after the mold is completed, without tearing the sand and without requiring excessive rapping, all draw surfaces of a pattern must have *draft* or *taper allowance*. An allowance of $\frac{1}{8}$ inch per foot is considered *standard draft*; some pattern shops use a taper of from 2 to 4 degrees.

Because castings usually require machining, an excess of metal is added at spots where the casting is to be machined. The allowance required is called *machine finish*.

A *solid pattern* is made in one piece, and is so constructed that it can be molded with a single joint. Use of a solid pattern requires more hand work in molding, with the result that cost of production of castings is increased.

A *split pattern* is made in two pieces joined at the *parting line* along which the two sections of a mold separate.

A *gated pattern* is actually a group of patterns, usually but not necessarily identical, so connected that all patterns of the group are molded at once. If such a gated pattern is mounted on a metal plate or on a wooden board, it is known as a *matchplate pattern*.

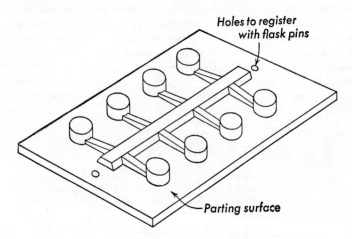

Holes to register with flask pins

Parting surface

Figure 166. Matchplate pattern.

Molds and Cores

The sand form containing the cavity into which molten metal is poured is the *mold*. A mold is made of molding sand pressed around the pattern in a metal or wooden container known as a *flask*. A flask is made in at least two parts, and often in three or more. The lower portion is known as the *drag*, and the upper portion as the *cope*. If a middle section is used, it is called a *cheek*. A flask is provided with lugs and pins on at least two opposite sides to assure accurate alignment when the mold is closed.

Molding sand must be sufficiently refractory to withstand the temperature of the molten metal without fusing, sufficiently cohesive to retain its form until the molten metal has solidified, sufficiently permeable to permit escape of gases and steam generated during pouring of the metal, and of uniform grain size.

GREEN SAND MOLDS Because they are the least expensive to make, and because they can be used immediately, *green sand* molds are used for a majority of castings. Green sand is molding sand in its natural state containing clay, and *tempered* by mixing with a suitable amount of water.

Because of their softness, green sand molds allow freedom of contraction of castings during solidification and cooling.

DRY SAND MOLDS   Large castings of intricate design usually are made in *dry sand* molds. The sand is mixed with a bonding agent such as dextrin, clay, pitch, or rosin, and the mold is baked before using, to dry out the water and to harden the bonding agent. Dry sand molds are stronger than those of green sand, and have less tendency to collapse during normal contraction of the casting. Baking the mold reduces the possibility of the face being washed away as the metal is poured over it, and precludes the possibility of distortion resulting from steam.

SKIN-DRIED MOLDS   For extremely large castings or for castings that must be made with maximum accuracy, *skin-dried* molds are used. Such

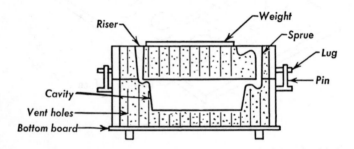

Figure 167. Mold reading for pouring.

a mold is made of green sand with a facing of sand containing a bonding material, and the skin is dried by rapid application of concentrated heat. Heating is done with a kerosene torch, by brief heating in a hot oven, or by spraying with flammable material and igniting. A skin-dried mold must be used immediately after drying before softening of the dry skin by absorption of moisture from the green sand can occur. A skin-dried mold combines the advantages of a firm face and a relatively collapsible body.

LOAM MOLDS   Large symmetrical castings such as large gears, for example, often are molded in *loam* molds made of bricks, with surfaces of loamy sand applied wet, and dried before use. Some forms can be molded by use of rotary sweeps instead of patterns.

CHILL MOLDS   In pouring a casting in which a thick section joins a thin section, distortion may occur as a result of the slower rate of cooling of the thick section; to compensate for the difference in normal cooling rates the thicker section may be chilled. Rapid chilling also helps to prevent formation of shrinkage cavities, and chilling sometimes is required to produce a hardened area on the surface of a casting. A metallic *chill* inserted in the mold conducts heat more rapidly from the molten metal at that spot, with the result that solidification and cooling take place more rapidly there than at other parts of the casting.

CORES   If a casting is of such design that an opening extends partly or completely through it, a *core* must be placed in the mold before casting. Cores also are used to assemble molds of such complicated structure that they would be difficult or impossible to make by use of patterns of the usual type. A core is made of sand combined with a binder such as linseed oil, pitch, dextrin, molasses, or a cereal; sometimes wires are used to strengthen a core. Before use, the core is baked in an oven to give the strength required to enable it to retain its shape until the casting has solidified. A core must be sufficiently permeable to permit steam and gases generated during pouring to escape readily, and it should crush when the casting shrinks during cooling.

In small foundries, cores are made by ramming sand by hand into *core boxes* containing cavities of required sizes and shapes. Additional provision for escape of gases sometimes is provided by piercing vents in cores. Cores in large numbers are made by extrusion or by a core-blowing machine that makes them automatically by ramming sand into core boxes.

Molding

In bench molding, the drag section of the flask is inverted on a flat *molding board*, and the lower section of the pattern is placed within it, large side down. A small amount of sand is *riddled* (sifted) over the pattern and is followed by enough sand to fill the drag. The sand is *peen-rammed* around the outer edges of flask and pattern, taking care to avoid ramming heavily over the pattern. Additional sand is added to fill the drag again, and is *butt-rammed* to uniform hardness sufficient to form a firm mold, but not so hard as to destroy permeability. Excess sand is scraped off level with the edges of the drag, a *bottom board* is clamped to the drag, the drag is rolled over, and the molding board is removed, exposing the top surface of the lower half of the pattern. The surface is smoothed off and parting dust is sprinkled over it to prevent the two halves of the mold from sticking together; parting dust then is blown from the surface of the pattern with a bellows.

The cope is fitted to the drag and the upper half of a pattern is fitted accurately to the lower half. A *sprue pin* is set on the sand in the drag, extending above the top edges of the cope, and the cope is filled with sand and rammed. The sprue pin then is removed and a funnel-shaped *pouring basin* is cut at the top of the sand in the cope around the hole left by removal of the sprue pin, to receive the molten metal. A similar opening is made elsewhere in the cope section of the mold to serve as a riser through which air and gases escape and into which the molten metal rises after the mold is full. Openings for sprue and riser may be cut with a sprue cutter instead of making them by use of a sprue pin. The cope section then is removed from the drag and inverted on a board, and the sand around the edges of both sections of the pattern is *swabbed* (dampened) in order that the sand may hold firmly when they are withdrawn. Both sections of the pattern are drawn from the mold after rapping to loosen them, and all loose sand is blown from the cavities. Channels called *gates* then are cut in the sand of the cope section, to connect the sprue and riser to the mold cavity. If a core is used, it is set in *core prints* provided in the mold for its support, and the entire surface of the mold is dusted with powdered graphite. Finally the cope section is placed again in position above the drag section, the flask is clamped together, and the metal is poured into the mold.

After the casting has cooled, the mold is broken up and the casting is removed. Gates, sprue, and the *fins* which appear at the parting lines or at junctions of cores with the mold are cut off, and the casting is cleaned by wirebrushing, sandblasting, pickling, or tumbling.

In large modern foundries nearly all of these operations are performed by machines. Their use accelerates production and reduces the amount of skilled labor required.

Permanent Mold Casting

Metal may be cast in permanent metal molds fitted with permanent metal cores, by pouring the molten metal by hand as in sand casting. Permanent mold casting is used principally for metals of low melting point, such as aluminum, lead, magnesium, and zinc. Advantages of this method of casting are: (1) finer grain structure produced by more rapid chilling; (2) relative freedom from shrink and blowholes; (3) greater accuracy of reproduction of the pattern, requiring less subsequent machining; (4) improved surface appearance.

Centrifugal Casting

Hollow objects can be cast by rotating a permanent mold at high speed while the metal is solidifying. The heavy metal is forced to the outside of the mold, leaving the lighter impurities, such as oxide and slag, at the inner surfaces; the process is less suitable for alloys of light metals that are little or no heavier than the inclusions likely to be present. Because all gas is squeezed out, the casting is free from blowholes.

Continuous Casting of Nonferrous Metals

Continuous casting is the continuous transformation of a flowing stream of molten metal into billets or bars ready for the extrusion press or the rolling mill, or into cast shapes ready for immediate mill processing. Advantages gained by continuous casting are uniformity of chemical composition and inherent soundness of metal structure. There are three basic types of the continuous casting process.

1. The Hunter-Douglas process, used by American Smelting and Refining Co., operates with the mold connected to a reservoir of molten metal. The process is entirely automatic and gravity-fed. Its short freezing zone eliminates porosity, insures feeding from the molten head, and permits solidification under controlled conditions. Except for the necessary periodic additions of molten metal, the process is completely mechanized. This process is the best suited to eliminate dirt and dross from the casting.

2. The Junghan-Rossi process, used by Scovill Manufacturing Co., employs a mold disconnected from the molten metal supply, with feed either by conduit or a falling stream. The water-cooled copper mold moves up and down, the down stroke being synchronized with the rate of discharge of the solidified bar; there is no relative movement between the mold and the solidified section during the major portion of the casting period.

During the pouring operation, molten metal from the electric melting furnace flows into a ladle, which transports it to the holding furnace that feeds the casting machine. The metal flows from the bottom of the holding furnace through a distributor into the water-cooled copper mold. It is discharged at an even and controlled rate below the surface of the pool of molten metal, which is maintained at an even level. A constant gas flame at the top of the mold prevents oxidation of the surface of the molten metal pool. Entrapment of dirt and dross in the molten metal is prevented by gravity-feeding the metal into the mold; also, any foreign matter that may enter the system floats on top of the mold, without turbulence to carry it into the product.

The metal descends through the water-cooled mold at constant speed and emerges as a solid slab of unusual density because (1) it freezes from

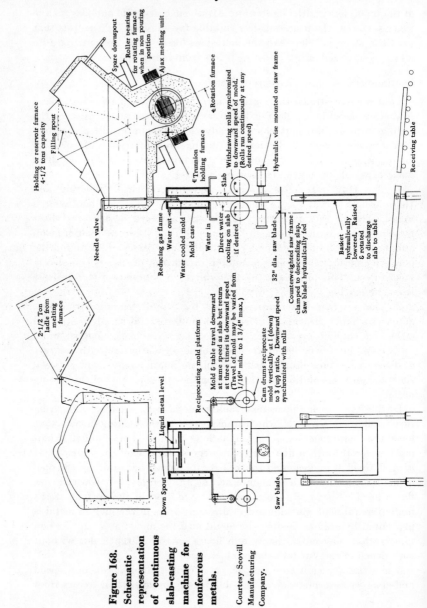

Figure 168. Schematic representation of continuous slab-casting machine for nonferrous metals.

Courtesy Scovill Manufacturing Company.

Spare downspout

Roller bearing for rotating furnace when in non pouring position

Ajax melting unit.

Holding or reservoir furnace 4-1/2 tons capacity

Filling spout

Rotation furnace

Needle valve

Trunnion holding furnace

Withdrawing rolls synchronized to downward speed of mold. (Rolls run continuously at any desired speed)

Reducing gas flame

Water out

Water cooled mold

Mold case

Water in

Slab

Direct water cooling on slab if desired

Hydraulic vise mounted on saw frame

32" dia. saw blade

Counterweighted saw frame clamped to descending slab, Saw blade hydraulically fed

Basket hydraulically lowered. Raised & rotated to discharge slab to table

Receiving table

2-1/2 Ton ladle from melting furnace

Liquid metal level

Reciprocating mold platform

Mold & table travel downward at same speed as slab but return at three times its downward speed (Travel of mold may be varied from 1/16" min. to 1 3/4" max.)

Cam drums reciprocate mold vertically at 1 (down) to 3 (up) ratio. Downward speed synchronized with rolls

Down Spout

Saw blade

the bottom and is therefore free of dissolved gases, and (2) the molten bath serves as a riser and head, thus preventing formation of shrinkage cavities. The dense solid slab passes through the withdrawing rolls, after which it is inspected and tested before being sent to the mill for cold rolling.

Because of its high initial cost, the process is profitable only when used to produce large tonnage of a few standardized shapes. It is especially well adapted for casting intermediate products such as extrusion billets and rolling slabs, which are of fairly large cross-sectional area.

3. The third type of continuous casting process employs a moving mold that preserves motionless contact with the casting during the freezing process.

Precision Casting

Precision casting involves use of a pattern made of wax or other plastic material that is destroyed by melting and burning it from the mold before casting the metal. The process sometimes is known as the *lost wax process.*

The wax pattern is formed in a die that usually is made of a soft metal or rubber, but which may be made of steel if a large number of patterns is to be made. When small objects are to be cast, several wax patterns are joined together by wax to form a multiple pattern corresponding to a gated pattern used in sand casting practice.

The pattern is placed in a cylindrical metal flask, a gate and sprue of the same wax are provided, and the flask is filled with the *investment*, which is a finely divided refractory material and a binder, mixed in water. To remove all air bubbles and to force the investment into all fine details of the pattern, the flask is vibrated rapidly in high vacuum. The investment hardens in much the same way as does a cement. After the investment has hardened, the flask is heated to dry out the remaining moisture; then it is heated to a temperature high enough to melt and burn out all wax, leaving a cavity of the exact size and shape of the pattern. Finally, the mold is heated to such a temperature that the cast metal will cool at the proper rate, and is filled with molten metal in a centrifugal air-pressure machine or a vacuum machine. (*See* Chapter 25, for vacuum casting techniques.) It is essential that temperatures of both mold and metal be correct.

The process permits casting to very close tolerances many intricate pieces that are too complex for successful production by ordinary casting and machining methods. The casting is free of dross and is of dense structure because of the centrifugal pressure under which it is cast, and

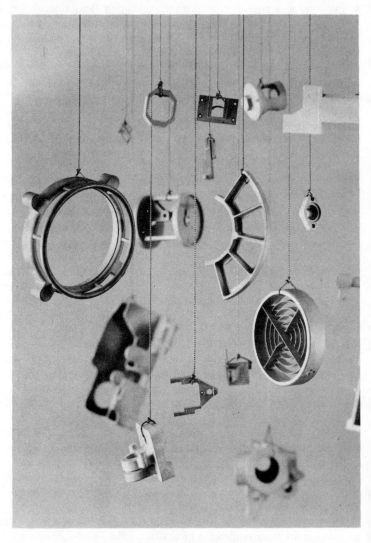

Figure 169. Precision castings.
Courtesy Arwood Precision Casting Corporation.

requires only removal of gates and sprue, and possibly light sandblasting, to make it ready for use.

Precision castings have been made in lengths to 18 inches, and weighing as much as 35 pounds.

Die Casting

Die castings are parts made by forcing nonferrous alloys, particularly those of aluminum, magnesium, and zinc, into metal molds by application

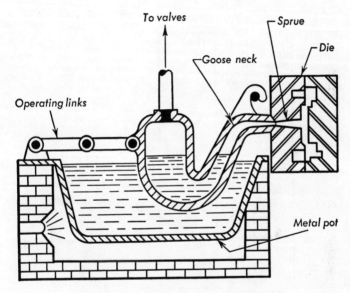

Figure 170. Air-injection gooseneck machine.
Courtesy New Jersey Zinc Company.

of high pressure. Die casting provides a method by which metal parts may be produced rapidly and to such close dimensional limits that only a small amount of subsequent machining is required. Because of the high pressure under which they are cast, die castings have greater density than sand castings, and thin sections of die castings have mechanical properties superior to those of sand castings of the same metals.

One of the early machines designed for die casting was a *gooseneck* machine operated by air pressure applied to the molten metal. The nozzle of the gooseneck is dipped below the surface of the metal in the pot for filling, and is raised to connect with the opening in the die before pressure is applied. This machine has the important disadvantage of a pressure limit of about 700 pounds per square inch, with resulting low density in the castings made with it. Aluminum alloys dissolve iron from the iron pot, adversely affecting the strength of the castings. Machines of this type

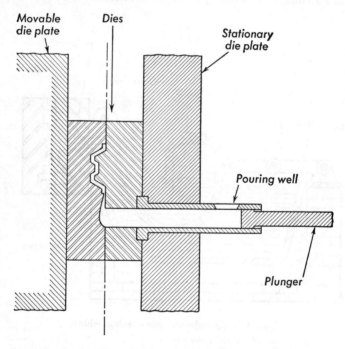

Figure 171. Cold-chamber die-casting machine.
Courtesy American Foundrymen's Association.

therefore are being superseded by cold-chamber machines for use with aluminum, although they still are being used for zinc alloys.

In another type of machine, pressure is applied by a plunger operated by compressed air or hydraulic pressure; such machines operate at a pressure of about 1500 pounds per square inch. The cylinder in which the plunger works is immersed in the pot of molten metal, and a port in the side of the cylinder admits a fresh charge of metal each time the piston is raised. This machine is used principally for zinc alloys.

A third type of machine, producing pressures as high as 20,000 pounds per square inch, is the cold-chamber machine used principally for aluminum and magnesium alloys, and also for copper alloys. Castings produced on a machine of this type have high density because of the extremely high pressure used. The charge is fed through the pouring well, and is forced into the die by the plunger. After the casting has solidified, the die

Figure 172. Zinc die-cast lawn mower parts.
Courtesy New Jersey Zinc Company.

is opened and the casting is ejected by further motion of the plunger. Cold-chamber die-casting machines have plungers and cylinders of a steel which is not attacked by aluminum alloys, as is the cast iron of the gooseneck type of machine.

Powder Metallurgy

Powder metallurgy is the art of producing metal parts or shapes by pressing together finely divided particles of one or more metals. The applied pressure produces such close contact between the particles that atoms diffuse from particle to particle, and atomic forces accordingly act to produce a continuous solid phase. When two or more metals are used, an actual alloy may result, but because it does not solidify from the liquid state, cored structures and segregations such as those found in cast alloys do not have an opportunity to form. The pressing operation constitutes cold work analogous to cold work on a metal that has solidified from the molten state, and plastic deformation therefore occurs. Subsequent heating in the sintering operation increases the mobility of the atoms and accelerates their diffusion between particles; simultaneously it produces recrystallization, just as does annealing a piece of cast metal, and leaves the part free from strain. Hot pressing in powder metallurgy is analogous to hot working of cast metal, and has the same effect of causing continuous recrystallization during working.

Parts that can be produced especially advantageously by powder metallurgy include:

1. Parts produced at low cost in large volume, that must be held to close tolerances with a minimum of finishing operations.

2. Parts having special properties not otherwise obtainable, such as porosity, controlled density, self-lubrication, resistance to wear, extreme hardness for cutting operations, resistance to burning by electric arc.

3. Parts impossible or impracticable to produce by other methods. Articles made by powder metallurgy can be produced economically only on a mass production basis, although stock shapes are available in some materials.

Products are classified as dense materials comparable to cast metal, porous materials including filters, oil-impregnated bearings and the like, magnetic materials, electric contacts, products made of refractory metals that cannot be melted in metallurgical furnaces, cemented carbides, and materials with high coefficient of friction for use in brake linings and similar applications.

Preparation of Powders

Important characteristics of powders used for powder metallurgy are density, shape, size, and size distribution of the grains; all are affected by the method by which a powder is produced. Many methods are in use, the choice of which depends upon the chemical and physical natures of the metal and upon its intended use.

Metal powders can be produced by grinding in mills, but the process has the important disadvantages that the surfaces of the particles become oxidized, and that material from the grinding surfaces of the mill may contaminate the metal powder. The method seldom is used for purposes of powder metallurgy, except for some iron and steel powders.

Fairly fine particles may be pulverized further by grinding in an *eddy mill*, in which the disadvantages of the ordinary grinding process have been overcome. The mill may be operated under an atmosphere of inert or reducing gas to prevent surface oxidation of the particles. Two high-velocity jets of gas blow the particles about with such violence that they are pulverized by repeated collisions with each other. Powders of aluminum, copper, and iron sometimes are prepared by this method.

A large proportion of all powder used in powder metallurgy is produced by direct reduction of an oxide or other compound of a metal. Reduction may be accomplished by exposing the solid compound of the metal to action of a reducing gas such as hydrogen or carbon monoxide, or by means of carbon. Direct reduction is used for preparation of powder of cobalt, copper, iron, molybdenum, and tungsten.

Powders of some metals can be produced by electrolysis of aqueous solutions of their salts if such conditions are chosen that the cathodic deposit is coarse grained and only loosely adherent to the cathode. These conditions are attained in general by low concentration of the metal ions, high concentration of acid in the bath, and high current density. The resulting deposit of granular metal is scraped from the cathode, washed thoroughly to remove all traces of electrolyte, and dried. In many cases further pulverization is required. The electrolytic method is used to a considerable extent for production of powders of copper, iron, nickel, and other less frequently used metals.

Powders of tantalum, thorium, and uranium sometimes are produced by fused salt electrolysis.

In a process called *atomization*, molten metal is forced through a small orifice, and the resulting stream is forced into contact with a jet of steam, air, or other gas. The metal is broken up into particles the size and shape of which depend upon the temperatures and pressures of both the stream of molten metal and the jet of steam or gas. The method appears particularly suited for the production of prealloyed powders such as brass and bronze powders.

Powders of iron and of nickel sometimes are prepared by the carbonyl process. The metal first is treated with carbon monoxide at a temperature between 210°F and 390°F and at a pressure of 3000 pounds per square inch to form iron carbonyl [$Fe(CO)_5$] or nickel carbonyl [$Ni(CO)_4$]. The carbonyls subsequently are decomposed by heating to a temperature between 300°F and 750°F at atmospheric pressure; the metal powder that results is of exceptional purity. Carbonyls of some other metals, notably chromium, niobium, molybdenum, and tungsten, can be formed, and powdered alloys of iron or of nickel with any of these metals can be made by mixing the carbonyls and decomposing them simultaneously. Powders so made have exceptionally favorable characteristics, but the high cost precludes the use of the process for many applications.

Powders of the austenitic stainless steels can be prepared by a process of intergranular corrosion. This steel can be *sensitized* by heating for several hours at temperatures between 930°F and 1650°F and subsequently cooling slowly to room temperature. This treatment changes the concentration of carbon in the grain boundaries, and when the metal is boiled in a solution of copper sulfate acidified with sulfuric acid, it is disintegrated. The size of the particles can be varied by variations in the sensitization methods.

The size of powder used in any stated application is controlled by screening it through a sieve of the proper mesh; standard sizes of sieves cover a range of from 100 to 325 mesh per linear inch. Powders are prepared for use by mixing and blending in tumbling or blending machines to produce a uniform mixture.

Pressing

Powders are formed into desired shapes by compression in dies. The pressure applied must be such as to force the particles into sufficiently intimate contact to produce enough molecular cohesion to hold the part together until it is strengthened by sintering. The pressure also must be adjusted to produce the required density in the finished part; pressures in industrial work may be as high as 50 tons per square inch. In the laboratory, much higher pressures are sometimes used for experimental work. Pressing operations and die design must be such as to pack the powder with maximum possible uniformity.

The powder is fed into the die cavity of the press, usually automatically from a hopper that is part of the press, but manually in the case of some larger presses. The punch or punches then are pressed into the die, and pressure is applied for a suitable time. The compressed article, known as a *briquette*, is ejected by the punch.

Some parts, such as rods and tubes of cemented carbide material, and some types of electric contact material, are compressed by extrusion.

Hot pressing has been found useful in many applications, and its use probably will increase as improved techniques are developed. A hot-pressed briquette is partially sintered when it is ejected from the press, and therefore requires a shorter period in the sintering furnace; it is also stronger than a cold-pressed briquette and therefore less liable to damage

Figure 173. Briquetted gears.
Courtesy Moraine Products Division General Motors Corp.

during the period before it is sintered. In some cases hot pressing produces a fully sintered briquette. Hot pressing is used successfully for cemented carbide parts.

Sintering

The green briquettes produced in the press are sintered for the purpose of causing recrystallization and grain growth, with consequent relief of strains produced by cold pressing. The new crystals so formed replace the crystals of the original particles, which no longer can be identified.

The sintering furnace is usually of the continuous type, provided with a belt conveyor that moves the briquettes through at a uniform rate. A

Figure 174. Miscellaneous iron-base oilite finished machine parts.
Courtesy Amplex Division, Chrysler Corporation.

controlled atmosphere sufficiently reducing to prevent oxidation of the type of material being sintered, or even to reduce existing oxides, is maintained throughout the furnace. Some furnaces operate under vacuum, which serves both to prevent oxidation and to remove from the briquettes any occluded gas. The charge passes successively through a preheating zone, a soaking zone, and a cooling zone. The soaking temperature varies with the material of the metal powders used, and in the case of an alloy is well below the melting point of the major constituent.

Sizing

Sintered parts that must be kept to close dimensional tolerances are brought to exact size after sintering by a re-pressing operation sometimes called *coining*. It ordinarily is done at room temperature and therefore causes work hardening which must be removed by subsequent annealing. Re-pressing increases the density of the article.

Products of Powder Metallurgy

One class of products includes those of highly refractory metals the melting points of which are so high that they cannot be melted in refractory-lined furnaces. These include molybdenum, tantalum, tungsten, and their alloys. Tungsten anodes for x-ray tubes, molybdenum electric contacts and parts for electronic tubes, tantalum sheets for repairing human skulls, tantalum chemical equipment, and applications of some alloys as thermoemissive elements in electronic equipment are examples. Alloys of these refractory metals with metals of lower melting points have been used for special purposes.

Another class of products includes mixtures of elements that do not form alloys with each other. These include electric contacts for heavy-duty service in which considerable arcing occurs, which would burn out contacts of silver or copper rapidly. Contacts made of mixtures of molybdenum or tungsten with silver, or of tungsten with copper, combine the satisfactory refractory properties of molybdenum or tungsten with the high electric conductivity of silver or copper. Bearings made from mixtures of copper powder with lead powder in proportions higher in lead than can be obtained in a lead-copper alloy are another example of this class of products.

Cemented carbide tips for cutting tools form another important class of products made from materials that do not form alloys. A cemented carbide is made from a mixture of the powdered carbide of a refractory metal, and a metal powder which acts as a cement. The most important products in this class are titanium carbide or tungsten carbide cemented with cobalt, and tantalum carbide cemented with nickel. The mixtures are pressed to shape and are sintered at about the melting point of the cementing metal. Particle size and distribution are important in quality control of cemented carbides.

A product consisting of a metal incorporated with a nonmetal can be made only by powder metallurgy. An important application is the manufacture of copper-graphite brushes for electric generators and motors, and of moving parts of other electric equipment. Silver-graphite products also are used to some extent.

One of the most important applications of powder metallurgy is production of articles having a high degree of porosity, for use as filters or to be impregnated with oil to serve as bearings for machinery. The required degree of porosity is obtained by suitable choice of particle size and by use of relatively light pressure in forming. After sintering, the product has a system of interconnecting pores that serves as a filter. In making porous bearings, graphite and a lubricant such as stearic acid or a stearate are mixed with the metal powder.

Questions

1 What is meant by *draft* of a pattern used in sand molding?
2 State the meaning of the following:

    a. Green sand mold   c. Skin-dried mold
    b. Dry sand mold   d. Loam mold

3 What is the purpose of a *chill*?
4 Describe in detail the process of bench molding with a split pattern.
5 What are the advantages of permanent mold casting?
6 Define continuous casting.
7 What advantages are gained by continuous casting?
8 What are the advantages of precision casting?
9 Describe the process of precision casting.
10 What metals are used principally for die casting?
11 Why is the cold-chamber machine used for die casting aluminum in preference to the air-injection gooseneck machine?
12 State some of the advantages of production by powder metallurgy over production by casting, rolling, forging, and machining.

25 / Forming and Finishing

Brazing

Joints that are to be subjected to moderate stress can be joined by brazing. Brazing is done by applying to the base metal a molten alloy of lower melting point than that of the workpiece. The bond is believed to result from atomic forces that come into play as a result of the intimate contact obtained between the base metal and the molten alloy. Because the surfaces to be joined are not fused together, the strength of the joint depends principally upon the strength of the brazing alloy used.

Brazing is more economical than welding and makes a neater joint that requires less finishing. Sections that are too thin for welding can be joined successfully by brazing. Many metals can be joined with a minimum of filler material, and the required temperatures are relatively low; therefore less heat is required for brazing than for welding. Brazing should not be used on finished work at temperatures above 500°F, nor where stresses above 15,000 psi will be encountered.

Brazing is frequently used for plain carbon steels, cast iron, copper, nickel, monel, inconel, brass, and bronze. Dissimilar metals with widely different melting points are readily brazed.

Since the melting points of the filler metals are lower than the temperatures encountered in welding, lower preheating temperatures are required in brazing, therefore less distortion results. In brazing such metals as cast iron and medium- to high-carbon steel, the critical temperatures are not exceeded so that the heated zones are not hardened seriously, as they are when welded.

Brazing Alloys

Silver-base alloys are most commonly used for brazing because they can be used for brazing all metals except aluminum and magnesium. These alloys have melting points of 1250–1600°F, and are free-flowing. They have high strength and are naturally resistant to corrosion.

Copper and copper alloys can be brazed either with silver-base alloys or with lower-melting and less costly bronze brazing alloys.

Flux and Tinning

Surfaces that are to be brazed must be free of oxide. The surfaces are cleaned of oxide and dirt by the chemical action of a *flux*, and are protected from oxidation during the heating, brazing, and cooling periods. In addition to cleaning the workpiece and protecting it from oxidation, the flux reduces the surface tension of the molten alloy, permitting it to flow freely over the work surfaces. When the molten brazing metal is brought into contact with the hot base metal, it should spread over the surface in a thin layer that is equivalent to plating the base metal with the brazing metal. Care must be taken to avoid entrapment of flux in the joint, as otherwise corrosion might develop in the joint while it is in service.

Soldering

Soft Soldering

Soldering is possible because the metallic bond is not selective. Soft soldering is used to join, without melting, metals that melt at temperatures above 600°F. No strain should be applied to a soft-soldered joint until the solder has solidified completely, because the solder has no strength at temperatures above its melting point. Soldered joints have relatively poor mechanical properties and poor corrosion resistance. Depending on the type of service, the joint should be designed to provide interlocking strength, such as lock joints, twisting of wires, and the like. Corrosion can be minimized by thorough cleaning with flux, and painting or otherwise protecting the surface.

Soft solders are used primarily because they are easy to handle, and require only comparatively low temperatures for successful application. Each solder melts at a definite temperature, which depends upon its composition. After the solder has melted at its *melting point*, the temperature rises as additional heat is applied and the solder reaches its *flowing point*. In order that the solder shall flow, it is necessary that the base metal be heated to a temperature higher than the flowing temperature.

Soldering processes are classified according to the manner in which heat is applied: (1) machine soldering; (2) dip soldering; (3) iron, or bit, soldering, which is the most widely used because of its maneuverability and heat localization; (4) blowpipe or torch soldering; and (5) sweat soldering.

TABLE 12   TEMPERATURE CHARACTERISTICS OF COMMONLY USED SOLDERS

Composition (per cent)		Melting Point (°F)	Flowing Point (°F)
Tin	60	361	372
Lead	40		
Tin	50	361	421
Lead	50		
Tin	40	361	453
Lead	60		
Tin	30	361	486
Lead	70		
Lead	95	579	689
Silver	5		
Lead	97.5	579	579
Silver	2.5		

Courtesy: Linde Company

A flux is required in soft soldering to retard formation of an oxide film on the work by protecting the cleaned metal from the air, and to remove any oxide that may be present on the surface of the metal when the solder is applied.

TABLE 13   CHARACTERISTICS AND USES OF VARIOUS FLUXES

Flux	Characteristics	Use
Rosin or rosin in alcohol	Noncorrosive	Electrical
Zinc chloride mixtures	Corrosive	Nonferrous
Zinc chloride and hydrochloric acid mixture	Corrosive	Brasses and bronzes containing silicon, aluminum, and manganese

Courtesy: Linde Company

Silver Soldering

Silver soldering, known also as *silver brazing* or *hard soldering*, is used for making strong joints. Silver-soldering alloys contain 10–80 per cent silver, with copper and zinc as the other principal components; melting points are 1200–1500°F, depending on the composition. A silver-soldered joint has the following characteristics: (1) it is strong, with tensile strength of 40,000–50,000 psi; (2) ductility is 5–35 per cent in 2 inches; (3) electric conductivity is relatively high, 14–77 per cent that of copper; (4) it offers high resistance to vibration, shock, and corrosion; (5) its color ranges from yellow to white. Silver soldering is most economically carried out with the oxyacetylene blowpipe.

During the heating operation, the joint must not be heated beyond the temperature required for free flowing and bonding of the soldering alloy used, since otherwise the benefit gained by use of this low-temperature process will be lost. The heat must be applied to the workpiece, never to the alloy in the joint itself.

A suitable flux must be used to dissolve any oxide that forms on the metal during heating. The flux also protects the heated surface from atmospheric oxidation, and facilitates tinning.

TABLE 14   STANDARD SPECIFICATIONS FOR SILVER-SOLDERING ALLOYS

Grade[1]	Ag	Cu	Zn	Cd	Max. Per Cent Impurities	Melt. Pt. (°F)	Flow Pt. (°F)	Color
	(percentage composition)							
1	10	52	38	Not over 0.15	0.15	1510	1600	Yellow
2	20	45	35	Not over 0.15	0.15	1430	1500	Yellow
3	30	45	30	5.	0.15	1430	1500	Yellow
4	40	30	25	0	0.15	1250	1370	Nearly white
5	50	34	16	0	0.15	1280	1425	Nearly white
6	60	20	15	0	0.15	1280	1325	White
7	70	20	10	0	0.15	1335	1390	White
8	80	16	4	0	0.15	1310	1460	White

Courtesy: Linde Company

[1] Grade 1 is used where a temperature of 1600°F will not damage the base metal. Grades 2 and 3 are used on copper, brass, bronze, and other copper-base metals, as well as on steel and dissimilar metals. Grades 4 and 5 are used for general all-round applications. Grade 6 is used by silversmiths because of its white color and low melting point. It is excellent also for monel. Grades 7 and 8 have extreme ductility, and are used for joining copper rods that will later be drawn into wire.

Soldering Aluminum

An oxide film that instantaneously forms on aluminum when the metal is exposed to the air must be removed to make possible the soldering of aluminum.

In one method, very chemically active fluxes must be used, because the aluminum oxide is not readily attacked by mild fluxes. Unfortunately, these active fluxes emit copious fumes, and leave corrosive residues that are difficult to remove from the aluminum surface.

In a second method, the aluminum oxide is removed by application of ultrasonic vibration to the molten solder while it is in contact with the aluminum. This causes cavitation and rupturing of the oxide film, permitting the solder to wet the aluminum. The equipment required for the process is expensive, and the rapid attenuation of the ultrasonic impulses limits the use of the process.

In a third method that has found limited use, the surface of the aluminum is wetted by vigorously scratching it with a wire brush or sharp instrument while the molten solder is applied to the area. Once the surface is wet, soldering is readily accomplished.

G. M. Bouton and P. R. White of the metallurgical research staff of Bell Laboratories have developed a soldering technique for aluminum, employing zinc-base solders. The method requires no flux, abrasion, or ultrasonic vibration. A solder stick $\frac{1}{16}$ by $\frac{3}{16}$ inch in cross section and containing zinc-aluminum solder plus 0.05 per cent of magnesium is used to puncture the aluminum oxide film. The magnesium gives the solder greater stability, and the narrow edge of the stick permits the bar to touch the surfaces close to their intersection and to break up the oxide films within the fillet. The wide edge of the solder stick is excellent to finish off the fillets and produce smooth and clean surfaces. The workpiece is heated only to a temperature that will melt the end of the solder stick, and a very slight motion of the stick against the surface of the aluminum is sufficient to wet it. The molten aluminum solder does not dissolve the oxide, but floats it so that it can be readily swept away with the end of the solder stick while the solder is molten. Since the solder will not flow by itself into a capillary joint during the actual soldering, it is essential that the heated aluminum surfaces be in direct contact with the solder stick, and stroked slightly with it, in order to wet the aluminum.

This process has two important advantages: The joints made on aluminum and galvanized parts are stronger than joints made with other solders. The joints are stable in the presence of humidity.

Spinning

Spinning is a shaping method in which metal is held in a revolving chuck and is shaped over a form by pressure of a tool held against it. The tool may be of wood or of metal, and it may or may not revolve.

Cavitron

The cavitron, an ultrasonic machine tool, is a marvel of the machine age. It has upset the theory that cutting tools must be harder than the material being cut. The cavitron can bore through the hardest metals and even through diamond. It can readily machine a material to extreme accuracy at incredible speed and at low cost, however brittle or hard the material may be. It can bore a square hole and can cut, carve, dice, drill, tap, engrave, or machine materials into simple or intricate shapes— internal or external. It does not affect surface appearance nor the physical, chemical, electrical, magnetic, or metallurgical properties of the material being machined. The cavitron is capable of slicing from the solid work-piece the thin wafers (0.015 inch or less) required for making transistors.

The apparently motionless drill tip of the cavitron actually vibrates at the speed of 20,000 strokes per second, traveling up to 0.004 inch with each stroke. The cutting action of the drill tip is obtained by spraying a jet of water containing tiny abrasive particles against the tip of the drill. The rapid vibrations cause the abrasive particles to be driven against the work, cutting the material to exact cross-section dimension. The shape of the hole depends upon the design of the cutting tip.

The cavitron has a drill or die mounted at the end of a metal motion amplifier, which is driven by a nickel rod surrounded by a coil carrying alternating current at a frequency corresponding to the frequency of vibration of the mechanical system. The alternating current causes the rod to expand and contract minutely, activated by the magnetostrictive property of nickel, in conjunction with the changing magnetic flux in the nickel. This motion is amplified, providing the repeated small hammer blows that do the work.

Machinability

Machinability may be defined as "that characteristic possessed by the material which permits the ready removal of the material by a cutting tool at economic speeds, producing a satisfactory finish and providing normally expected tool life."

Some factors that determine machinability are: (1) the microstructure of the material being machined, that is, the size and distribution of the ferrite, pearlite, and spheroidized carbides; (2) the strength of the material; (3) the rubbing friction of the material against the tool; and (4) the plastic properties of the material. Factors other than the machinability of the metal that contribute to easy machining are: (1) the shape and material of the cutting tool; (2) the coolant used; (3) the speeds and feeds used; and (4) the skill of the machine operator.

Good machinability is very important for material that is to be used for making parts by automatic screw machines, which can turn out vast quantities of finished parts in a given time. If suitable steels were not available for use on these machines, our present-day mass production era would be somewhere in the future.

Cold drawn B1112 steel is the standard by which the machinability of all other metals is rated. When a material has a machinability rating of 70, this means that it machines 70 per cent as well and as fast as cold-drawn B1112 steel for equal tool life.

Vacuum Melting

Vacuum melting meets the demand of the aircraft industry for metals of very high strength-weight ratios. The conventional methods of heat treatment increase the strength of a metal, but decrease ductility in the center because of segregation. Since vacuum melting reduces inclusions and gases, it overcomes this advantage appreciably.

The major objectives of the vacuum melting processes are:

1. to lower gas content
2. to control composition more closely
3. to improve cleanliness
4. to obtain ingot structures free from center porosity and segregation
5. to produce metals and alloys which cannot be melted or, if air melted, cannot be hot worked economically by conventional techniques
6. to improve properties such as
   a. high-temperature rupture and creep strength
   b. room- and elevated-temperature ductility
   c. impact strength
   d. fatigue strength
   e. magnetic, electrical, and other properties
   f. improvements in stress-rupture performance due to the close control over increased amounts of titanium and aluminum, improved cleanliness, and low gas content

One of the most important advantages of vacuum melting is the substantial improvement in the hot workability of super alloys. When air melted in conventional furnaces, these alloys are difficult to hot work due to the presence of titanium nitrides and carbonitride stringers. Both vacuum induction melting and vacuum arc remelting greatly improve the hot workability of these alloys due to the relative freedom from stringers and to the increased elevated temperature ductility.

Three methods of vacuum melting are in commercial use: (1) vacuum induction melting, (2) consumable electrode melting, and (3) vacuum degassing. Each method produces metals of greater purity than does conventional air melting.

Vacuum Induction Melting

During vacuum induction melting the charge is always under a vacuum and the only source of contamination is the crucible. Since the melt contains no slag or air, there is no necessity for adding deoxidizers such as manganese or silicon. One disadvantage is that the amount of metal produced is limited. High-purity raw materials are used, and there is sufficient time to refine the molten metal. In addition, inductive stirring greatly aids the refining by continuously bringing fresh metal to the surface.

The vacuum induction melting process is the most flexible process as far as composition control is concerned, being due to:

(1) flexibility in selection of raw materials; high-purity virgin metals can be used

(2) there is sufficient time for removal of gaseous impurities

(3) no metallic deoxidizers, such as silicon and manganese, are needed

(4) molten metal can be sampled and analyzed and corrective additions made if necessary

(5) at relatively low temperatures, losses of volatile elements are small and uniform from heat to heat

Deoxidation is achieved by carbon and in some cases by hydrogen; in both cases, the deoxidation products are gaseous oxides which are pumped out continuously.

Since the ingots are poured in the same manner as in air melting, there is no improvement in center porosity and segregation. This, however, is not a problem for many alloys in the sizes that vacuum induction melting can produce at present.

Consumable Electrode Melting

In this method the alloy is remelted and reacts under a vacuum. Thus the advantages realized by original vacuum melting are obtained. In addition, larger ingots are produced, segregation is reduced, and there is no contamination from slag or oxygen.

The high temperatures and low pressures attained in vacuum melting are very favorable for dissociation, degassification, and deoxidation. Due to the short time the material is molten and due to the use of air melted electrodes the removal of gases may not be complete.

During vacuum arc melting it is impossible to take samples and analyze them or make corrective additions. Thus, the electrode must initially possess the correct composition; it may also have to contain higher than normal amounts of such volatile metals as manganese to compensate for vaporization during vacuum arc remelting.

In open air melting it is a normal practice to deoxidize melts by the addition of silicon, manganese, aluminum, and titanium. Although insoluble in molten metal, the deoxidation products can remain in the metal or alloy as nonmetallic inclusions. In vacuum arc remelting, on the other hand, deoxidation is carried out efficiently by the high temperatures and low pressures encountered during the remelting. Furthermore, large inclusions and stable compounds tend to float out during the remelting and concentrate at the top. As a result, small inclusions are left dispersed throughout the ingot.

Center porosity and segregation are common defects in conventionally solidified ingots. These defects become serious and lower the mechanical properties of the ingot as the ingot size and alloying content increase. Since the melt in vacuum arc remelting is built up incrementally, the tendency toward center porosity and segregation is eliminated or minimized. This is the feature that makes vacuum arc remelting ideally suited for the production of large ingots with no decreased segregation.

Vacuum Degassing

Degassing differs from the two methods previously described in that pouring only is performed under vacuum, whereas the remainder of the melting is carried out by conventional practices in an electric or open-hearth furnace. The primary objective of this process is the removal of hydrogen (the presence of which may develop cracks or weakening flaws without warning) and the subsequent elimination of "flake" in large sections caused by its presence. For this purpose vacuum degassing is extremely effective, and entire heats up to 250 tons can be

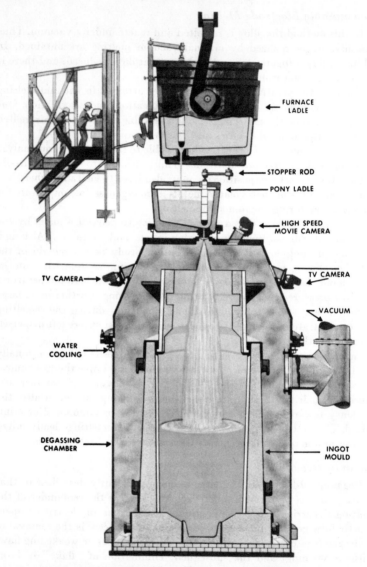

Figure 175. Cutaway diagram of vacuum-casting machine.

Courtesy *Steelways*, published by American Iron and Steel Institute.

processed at one time. To a lesser degree the contents of other gases, such as nitrogen and oxygen, are also reduced. Since normal deoxidation practice is still required, it will produce inclusions of silica and alumina. Entrapped slag is another source of contamination not eliminated.

The molten metal is tapped first into a conventional ladle in the usual manner. This ladle is positioned over a small ladle, called the "pony ladle," attached to the vacuum chamber that contains the ingot mold. When the small ladle is full, its stopper is opened and the molten metal melts an aluminum seal and flows into the vacuum chamber where it separates into many droplets. The droplets expel the harmful gases because atmospheric pressure has been practically removed. Powerful pumps quickly draw the expelled gases out of the chamber. Hence, before the molten steel flows into the ingot mold at a controlled rate, it is free of gases. As a result of the removal of the gases, a sounder steel with a strong internal structure and greater ductility is produced.

After the last ingot has been poured, the tank is repressurized with an inert gas, the seal of the chamber broken, and the ingot removed. Before further processing, large ingots must be cooled for at least two days. No alloys can be added during the degassing method.

Vacuum Casting

Vacuum casting techniques are similar to those employed in the vacuum degassing process. *See also* Chapter 11, section on vacuum induction process.

Questions

1 Define brazing.
2 How does brazing differ from welding and soldering?
3 Why is a brazed joint weak?
4 Define soldering.
5 How does soldering differ from welding?
6 Why is a hard soldered joint stronger than a brazed joint?
7 Why is it difficult to solder aluminum?
8 What is the importance of the cavitron?
9 Explain how the cavitron works.
10 Define machinability.
11 State three factors that effect machinability.
12 How has the use of free-cutting metals made mass production possible?
13 What industry demands the use of vacuum melted metals? Why?
14 Describe the degassing method of vacuum melting.

26 / Ultrahigh-Purity Metals

Zone Refining

Zone refining refers to a class of solidification techniques, all of which involve the movement of one or more liquid zones through an elongated charge of meltable material. It may be compared to distillation, the essen-

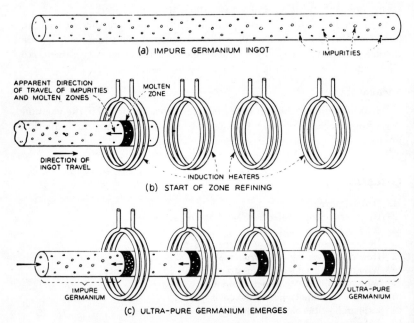

Figure 176. Schematic representation of zone refining.

Courtesy Bell Telephone Laboratories, Inc.

tial difference being that the change in phase is from solid to liquid to solid, instead of from liquid to vapor to liquid.

Zone refining produces a single crystal of ultrahigh purity with a minimum of effort. It has been used to obtain ultrapure germanium, silicon, molybdenum, tungsten, and other semiconductors, metals, organic compounds, and inorganic compounds. It can be used to produce any crystalline substance which can be safely melted and in which there exists an appreciable difference in impurity concentration between the freezing solid and the liquid from which it freezes.

Zone refining is based on the principle that most impurities are more soluble in molten than in solid material. During zone refining a series of rather narrow molten zones pass through a relatively long charge of alloy or relatively impure metal. The zones can be produced by external, ring-shaped heaters. As each molten zone advances, impure solid melts at its leading interface and purified solid freezes at its trailing interface. As each molten zone passes through the charge, the liquid picks up additional impurities and leaves a purer solid behind. Hence, each successive zone causes further refining. The molten zone always travels through the ingot in the same direction until the desired purity is obtained. Since the impurities collect in the liquid zones and are carried with them to the end of the ingot, most of the impurities concentrate there because that is the last volume to freeze. This impure section can be cut off.

The degree of purification depends on:

1. The number of passes
2. The initial impurity level
3. The difference in solubility of the impurities in the solid and those in the liquid
4. The length of the molten zone. Short zones are most suitable for many passes; long zones are most suitable for few passes

Automatic Floating Zone Refining

Bell Telephone Laboratories has developed an automatic refining device that reduces electrically active impurities in silicon to less than one part per billion. The device makes use of the floating zone refining technique, in which a molten zone is passed through the silicon, carrying impurities with it. The method can also be used to purify germanium, molybdenum, tungsten, and many other materials. The floating zone process requires no crucible, and is therefore free from contamination by impure molten silicon. The molten zone, supported only by surface tension, is passed along a vertical silicon rod held rigidly at both ends.

Refining takes place in a controlled atmosphere obtained by surrounding the rod with a water-cooled gas-tight envelope. The molten zone is produced by inductive heating, and the motion of either the heating coil or the rod sweeps the zone along. The process is repeated, with the molten zone always traversing the rod in the same direction until the desired purity is obtained. The rod is moved up and down by mechanical means as it rotates inside the gas-tight silica enclosure that automatically recycles the mechanism as long as required.

Figure 177. Refining of a crystal of silicon by the automatic floating zone method.
Courtesy Bell Telephone Laboratories, Inc.

Questions

1   What does zone refining refer to?
2   Compare zone refining and distillation.
3   Explain the principle involved in zone refining.
4   What is the floating zone refining technique?

Bibliography

Aluminum Company of America, *Alcoa Aluminum and Its Alloys*, 1947.

——, *The Story of Aluminum*, 2d ed., 1946.

American Foundrymen's Association, *Cast Metals Handbook*, 3d ed., Chicago, 1944.

American Iron and Steel Institute, *The Picture Story of Steel*, New York.

American Machinist Editorial Staff, "The Working of Aluminum Alloys," *American Machinist*, **84**, 617–632 (Aug. 21, 1940).

American Society for Metals, *Metal—Inside Out*, Cleveland, 1941.

——, *Metals Handbook*, Cleveland, 1948.

American Zinc Institute, *The Zinc Industry*, 1943.

Aston, J., and Story, E. B., *Wrought Iron—Its Manufacture, Characteristics, and Applications*, 2d ed., Pittsburgh, A. M. Byers Co., 1944.

Bancroft, W. E., "Salt Baths for Hardening High Speed Steel," *Metal Progress*, November 1946, 941–947.

Basch, D., "Metal Mold Castings," *Metal Progress*, Oct. 1945, 761–768.

Bausch & Lomb Optical Company, *The Educational Focus*, Rochester, N.Y. Vol. XII, No. 4, 1941; Vol. XV, No. 1, 1944.

Bell Laboratories Record, September 1957, "Automatic Floating—Zone Refining."

Bell Telephone System, Monograph 2000, "Principles of Zone Melting."

——, Monograph 2147, "Ultrapure Metals Produced by Zone Melting Technique."

——, Monograph 2388, "Continuous Multistage Separation by Zone Melting."

——, Monograph 2626, "Single Crystals of Exceptional Quality by Zone Leveling."

Bethlehem Steel Company, *Properties of Frequently Used Carbon and Alloy Steels*, Bethlehem, Pa., 1944.

——, *Steel in the Making*, Bethlehem, Pa., 1942.

——, *Tool Steel Treaters' Guide*, Bethlehem, Pa., 1942.

Blake, E. A., "Fundamentals of Molding Machines," *The Foundry*, September 1947.

Bouton, G. M., and White, P. R., "A Method for Soldering Aluminum," *Bell Laboratories Record*, May 1958.

Bray, J. L., *Non-Ferrous Production Metallurgy*, 2d ed., New York, John Wiley and Sons, Inc., 1947.

Bristol Company, *Pyrometers*, Bulletins P1200, P1202, P1213.

Bullens, D. K., *Steel and Its Treatment*, New York, John Wiley and Sons, Inc., Vol. 1, 1938; Vol. 2, 1939.

Camp, J. M., and Francis, C. B., *The Making, Shaping, and Treatment of Steel*, 5th ed., Pittsburgh, Carnegie-Illinois Steel Corporation, 1940.

Campbell, H. L., *Metal Castings*, New York, John Wiley and Sons, Inc., 1936.

Carnegie-Illinois Steel Corporation, *Suiting the Heat Treatment to the Job*, 1946.

Chambersburg Engineering Company, *Impact Die Forging, Publication No. 4401* Chambersburg, Pa., 1944.

Cohen, Morris, "Tempering of Toolsteels," *Metal Progress*, May 1947, 781–788; June 1947, 962–968.

Coonan, F. L., *Principles of Physical Metallurgy*, New York, Harper and Bros., 1943.

Denver Equipment Company, *Denver Flotation, Bulletin No. F 11–B*.

Dietert, H. W., *Modern Core Practices and Theories*, Chicago, American Foundrymen's Association, 1942.

Doan, G. E., and Mahla, E. M., *Principles of Physical Metallurgy*, 2d ed., New York, McGraw-Hill Book Co., Inc., 1941.

Dow Chemical Company, *Dow and Magnesium*.

———, *The Metal of Motion*.

———, *Dow Metal Magnesium Alloys*.

———, *Dow Metal Data Book*.

———, *Vital Materials from Sea Water*.

Dowdell, R. L., Jerabek, H. S., Forsyth, A. C., and Green, C. H., *General Metallography*, New York, John Wiley and Sons, Inc., 1943.

Drop Forging Association, *Metal Quality*, Cleveland, 1944.

E. I. du Pont de Nemours Company (Inc.), *Molten Salt Baths*, 1946.

Eastman Kodak Company, *Radiography in Modern Industry*, Rochester, N.Y., 1947.

Ess, T. J., "The Modern Arc Furnace," *Iron and Steel Engineer*, February 1944.

Federated Metals Division, American Smelting and Refining Company, *Aluminum Casting Alloys*, New York, 1946.

Frey, M. L., Shepherd, B. F., and Elmendorf, H. J., "Incomplete Quenches," *Metal Progress*, August 1944, 308–314.

Frier, W. T., *Elementary Metallurgy*, New York, McGraw-Hill Book Co., Inc., 1942.

Gathmann, E., *The Ingot Phase of Steel Production*, 2d (revised) ed., Catonsville, Baltimore, Md., The Gathmann Engineering Co., 1942.

Gladding, S. D., and Bigge, H. C., "The Basic Electric Furnace for Steelmaking," *Metal Progress*, October 1945, 642–651.

Henning, C. C., *Manufacture and Properties of Bessemer Steel*, Pittsburgh, Pa., Jones and Laughlin Steel Corp., 1938.

Henry, O. H., and Claussen, G. E. (rev. by G. E. Linnert), *Welding Metallurgy—Iron and Steel*, 2d ed., American Welding Society, 1940.

Heppenstall Company, *Effects of Alloying Elements and the Physical Properties of Steel in Forged Sections*, Pittsburgh, Pa., 1936.

Heyer, R. H., *Engineering Physical Metallurgy*, New York, D. Van Nostrand Co., Inc., 1939.

Houghton, E. F. and Co., *Houghton on Quenching*, 1943.

———, *Interrupted Quenching in Salt*, 1943.

International Welding Association, *Braze Welding of Iron and Steel by the Oxy-Acetylene Process*.

Jeffries, Z., and Archer, R. S., *The Science of Metals*, New York, McGraw-Hill Book Co., Inc., 1924.

Juppenlatz, J., "Acid Electric Process for Steel Castings," *Metal Progress*, October 1945, 638–641.

Kaiser Aluminum and Chemical Corporation, *Heavy Press Extrusions*.

Kerr Dental Manufacturing Company, *Fundamentals of Industrial Precision Casting*.

Lake Superior Iron Ore Association, *Lake Superior Iron Ores*, Cleveland, 1938.

Langhammer, A. S., *Oilite: Powdered Metal Parts*, Detroit, Chrysler Corporation, Amplex Division, 1945.

Leeds and Northrup Company, *Pyrometers*, Catalogs N–33A, N–33D.

Leitz, E., Inc., *Leitz-Blosjo Carbon Meter*.

Leland, J. F., "Tooling for Cold Steel Extrusion," American Society of Tool Engineers Meeting, March 1955.

Leland, J. F., and Helms, J. W., "The Influence of Proper Lubrication on the Design of Cold Extruded Components," Reprint, Society of Automotive Engineers Meeting, March, 1954.

Liddell, D. M., and Doan, G. E., *Principles of Metallurgy*, New York, McGraw-Hill Book Co., Inc., 1933.

Lincoln, R. F., "Fundamentals of Core Blowing," *The Foundry*, February and March 1940.

Linde Air Products Company, *How to Soft-Solder and Silver-Braze*.

Lippert, T. W., "Continuous Casting of Semifinished Steel," *The Iron Age*, Aug. 14, 1948.

Loughrey, D. R., "Design, Operation and Construction of a Bessemer Converting Mill," *Iron and Steel Engineer*, March 1942.

Magnaflux Corporation, *Magnaflux*.

———, *Zyglo*.

Magnesium Association, *Magnesium Base Alloys, Alloy Nomenclature and Temper Designation*, November 1956.

Malleable Founders' Society, *American Malleable Iron, A Handbook*, Cleveland, 1944.

Materials in Design Engineering, "Materials for High Temperature Service," April 1954.

Materials in Design Engineering, "The New Stainless Steels," April 1956.

Metal Progress, "Vacuum Melting Today," August 1958.

Moldenke, R. G. G., *Principles of Iron Founding,* New York, McGraw-Hill Book Co., Inc., 1938.

Moraine Products Division, General Motors Corporation, *Durex: Self-lubricating Bearings,* Dayton, Ohio, 1945.

Najarian, H. K., "Weaton-Najarian Vacuum Condenser," "Reduction and Refining of Non-Metals," *Trans. A.I.M.E.,* **159,** 161–175 (1944).

New Jersey Zinc Company, *Die Casting for Engineers,* New York, 1946.

Norton Company, *Characteristics of Sparks Generated by the Grinding of Metals,* Worcester, Mass.

Parker, C. M., *The Metallurgy of Quality Steels,* New York, Reinhold Publishing Corp., 1946.

———, "Selecting Steels by Hardenability Bands," *Materials and Methods,* April 1957, 68–72.

Pfann, W. G., and Olsen, K. M., "Zone Melting," *Bell Laboratories Record,* June 1955.

Republic Steel Corporation, *National Emergency Steels; Properties, Treatment, Application,* 2d ed., Cleveland, 1944.

Revere Copper and Brass, Inc., *Revere Copper and Copper Alloys,* 1943.

Reynolds Metals Company, Inc., *Reynolds Aluminum Alloys and Mill Products Data Book,* Louisville, Ky., 1946.

Rustless Iron and Steel Corporation, *Heat Treatment of Stainless Steels,* Baltimore, Md., 1944.

Sachs, G., and Van Horn, K. R., *Practical Metallurgy,* Cleveland, American Society for Metals, 1940.

Schwarzkopf, P., *Powder Metallurgy, Its Physics and Production,* New York, The Macmillan Company, 1947.

Seasholtz, A. P., "Interrupted Quenching in Salt Baths," *Metal Progress,* October 1944, 730–738.

Spalding, S. C., "Why Tool Steels Do Not Act Alike," *American Machinist,* March 14, 1946.

Stimpson, W. C., Gray, B. L., and Grennan, J., *Foundry Work,* Chicago, American Technical Society, 1944.

St. John, A., and Isenburger, H. R., *Industrial Radiography,* 2d ed., New York, John Wiley and Sons, Inc., 1943.

Strauss, Jerome, "The Ugine-Sejournet Extrusion Process," American Iron and Steel Institute Meeting, May 21, 1952.

Thum, E. E., *Modern Steels,* Cleveland, American Society for Metals, 1939.

United States Steel Corporation, *U.S. Steel News,* June 1937, September 1937, October 1937, October 1938, January 1942, July 1943, January 1944.

———, *Atlas of Isothermal Transformation Diagrams,* 1943.

Vilella, J. R., *Metallographic Technique for Steel*, Cleveland, American Society for Metals, 1938.

———, "The Grain Size of Steel," *Mechanical Engineering*, April 1940.

Weaton, G. F., Najarian, H. I., and Long, C. C., "Production of Electrothermic Zinc at Josephtown Smelter," *Trans. A.I.M.E.* **159**, 141–160 (1944).

Wendt, R. E., *Foundry Work*, 4th ed., New York, McGraw-Hill Book Co., Inc., 1942.

White, A. H., *Engineering Materials*, New York, McGraw-Hill Book Co., Inc., 1939.

Williams, R. S., and Homerberg, V. O., *Principles of Metallography*, 4th ed., New York, McGraw-Hill Book Co., Inc., 1939.

Wilson Mechanical Instrument Company, Inc., *Rockwell Hardness Tester, Catalog No. 24*.

Wissman, C. C., "Working a Heat of Acid Electric Steel," *Metal Progress*, December 1944, 1277–1284.

———, "Deoxidizing a Heat of Acid Electric Steel," *Metal Progress*, September 1945, 499–504.

Tables

MELTING POINTS AND SPECIFIC GRAVITIES

	Chemical Symbol	Melting Point Centigrade	Melting Point Fahrenheit	Specific Gravity
Aluminum	Al	660	1220	2.7
Antimony	Sb	630	1166	6.7
Arsenic	As	sublimes		5.7
Beryllium	Be	1350	2462	1.8
Bismuth	Bi	271	520	9.8
Boron	B	2300	4172	2.5
Cadmium	Cd	321	610	8.6
Carbon (graphite)	C	3500	6330	2.3
Chromium	Cr	1615	2939	7.1
Cobalt	Co	1480	2696	8.8
Copper	Cu	1083	1981	8.9
Gold	Au	1063	1943	19.3
Iridium	Ir	2350	4262	22.4
Iron	Fe	1535	2795	7.9
Lead	Pb	327	621	11.3
Magnesium	Mg	651	1204	1.7
Manganese	Mn	1260	2300	7.2
Molybdenum	Mo	2625	4757	10.2
Nickel	Ni	1452	2646	8.9
Niobium	Nb	1950	3542	8.4
Phosphorus (yellow)	P	44	111	1.8
Potassium	K	62	144	0.86
Selenium	Se	220	428	4.8
Silicon	Si	1420	2588	2.4
Silver	Ag	961	1762	10.5
Sodium	Na	98	208	0.97
Sulfur	S	113	235	2.1
Tantalum	Ta	2850	5162	16.6

MELTING POINTS AND SPECIFIC GRAVITIES—*Continued*

	Chemical Symbol	Melting Point Centigrade	Melting Point Fahrenheit	Specific Gravity
Tellurium	Te	452	846	6.2
Thallium	Tl	304	579	11.9
Thorium	Th	1845	3353	11.2
Tin	Sn	232	450	7.3
Titanium	Ti	1800	3272	4.5
Tungsten	W	3370	6098	19.3
Uranium	U	1850	3362	18.7
Zinc	Zn	419	786	7.1
Zirconium	Zr	1700	3092	6.4

HARDNESS CONVERSION TABLE

The following values are to be considered only as rough approximations due to variables of size, shape, and mass:

    (1) Brinell values over 578.   (3) Rockwell C scale figures under 20.
    (2) All Shore Scleroscope values.   (4) Rockwell B scale figures over 104.

Approximate Relations between Brinell, Rockwell, and Shore Hardnesses and the Tensile Strengths of AISI Carbon and Alloy Constructional Steels

Brinell		Rockwell			
Dia. in mm, 3000 kg load 10 mm ball	Hardness No.	C 150 kg load 120° Diamond Cone	B 100 kg load $\frac{1}{16}$ in. dia. ball	Shore Scleroscope No.	Tensile Strength psi
2.05	898	..	..	..	440,000
2.10	857	..	..	..	420,000
2.15	817	..	..	..	401,000
2.20	780	70	..	106	384,000
2.25	745	68	..	100	368,000
2.30	712	66	..	95	352,000
2.35	682	64	..	91	337,000
2.40	653	62	..	87	324,000
2.45	627	60	..	84	311,000
2.50	601	58	..	81	298,000
2.55	578	57	..	78	287,000
2.60	555	55	120	75	276,000
2.65	534	53	119	72	266,000

HARDNESS CONVERSION TABLE—*Continued*

Brinell Dia. in mm, 3000 kg load 10 mm ball	Hardness No.	Rockwell C 150 kg load 120° Diamond Cone	Rockwell B 100 kg load $\frac{1}{16}$ in. dia. ball	Shore Scleroscope No.	Tensile Strength psi
2.70	514	52	119	70	256,000
2.75	495	50	117	67	247,000
2.80	477	49	117	65	238,000
2.85	461	47	116	63	229,000
2.90	444	46	115	61	220,000
2.95	429	45	115	59	212,000
3.00	415	44	114	57	204,000
3.05	401	42	113	55	196,000
3.10	388	41	112	54	189,000
3.15	375	40	112	52	182,000
3.20	363	38	110	51	176,000
3.25	352	37	110	49	170,000
3.30	341	36	109	48	165,000
3.35	331	35	109	46	160,000
3.40	321	34	108	45	155,000
3.45	311	33	108	44	150,000
3.50	302	32	107	43	146,000
3.55	293	31	106	42	142,000
3.60	285	30	105	40	138,000
3.65	277	29	104	39	134,000
3.70	269	28	104	38	131,000
3.75	262	26	103	37	128,000
3.80	255	25	102	37	125,000
3.85	248	24	102	36	122,000
3.90	241	23	100	35	119,000
3.95	235	22	99	34	116,000
4.00	229	21	98	33	113,000
4.05	223	20	97	32	110,000
4.10	217	18	96	31	107,000
4.15	212	17	96	31	104,000
4.20	207	16	95	30	101,000
4.25	202	15	94	30	99,000
4.30	197	13	93	29	97,000
4.35	192	12	92	28	95,000
4.40	187	10	91	28	93,000
4.45	183	9	90	27	91,000
4.50	179	8	89	27	89,000

HARDNESS CONVERSION TABLE—*Continued*

| Brinell | | Rockwell | | | |
Dia. in mm, 3000 kg load 10 mm ball	Hardness No.	C 150 kg load 120° Diamond Cone	B 100 kg load $\frac{1}{16}$ in. dia. ball	Shore Scleroscope No.	Tensile Strength psi
4.55	174	7	88	26	87,000
4.60	170	6	87	26	85,000
4.65	166	4	86	25	83,000
4.70	163	3	85	25	82,000
4.75	159	2	84	24	80,000
4.80	156	1	83	24	78,000
4.85	153	..	82	23	76,000
4.90	149	..	81	23	75,000
4.95	146	..	80	22	74,000
5.00	143	..	79	22	72,000
5.05	140	..	78	21	71,000
5.10	137	..	77	21	70,000
5.15	134	..	76	21	68,000
5.20	131	..	74	20	66,000
5.25	128	..	73	20	65,000
5.30	126	..	72	..	64,000
5.35	124	..	71	..	63,000
5.40	121	..	70	..	62,000
5.45	118	..	69	..	61,000
5.50	116	..	68	..	60,000
5.55	114	..	67	..	59,000
5.60	112	..	66	..	58,000
5.65	109	..	65	..	56,000
5.70	107	..	64	..	55,000
5.75	105	..	62	..	54,000
5.80	103	..	61	..	53,000
5.85	101	..	60	..	52,000
5.90	99	..	59	..	51,000
5.95	97	..	57	..	50,000
6.00	95	..	56	..	49,000

Reprinted by permission of American Society for Metals

TEMPERATURE CONVERSION TABLE

Albert Sauveur type of table. Values revised.

−459.4 to 0			0 to 100						100 to 1000					
C	C/F	F	C	C/F	F	C	C/F	F	C	C/F	F	C	C/F	F
−273	−459.4		−17.8	0	32	10.0	50	122.0	38	100	212	260	500	932
−268	−450		−17.2	1	33.8	10.6	51	123.8	43	110	230	266	510	950
−262	−440		−16.7	2	35.6	11.1	52	125.6	49	120	248	271	520	968
−257	−430		−16.1	3	37.4	11.7	53	127.4	54	130	266	277	530	986
−251	−420		−15.6	4	39.2	12.2	54	129.2	60	140	284	282	540	1004
−246	−410		−15.0	5	41.0	12.8	55	131.0	66	150	302	288	550	1022
−240	−400		−14.4	6	42.8	13.3	56	132.8	71	160	320	293	560	1040
−234	−390		−13.9	7	44.6	13.9	57	134.6	77	170	338	299	570	1058
−229	−380		−13.3	8	46.4	14.4	58	136.4	82	180	356	304	580	1076
−223	−370		−12.8	9	48.2	15.0	59	138.2	88	190	374	310	590	1094
−218	−360		−12.2	10	50.0	15.6	60	140.0	93	200	392	316	600	1112
−212	−350		−11.7	11	51.8	16.1	61	141.8	99	210	410	321	610	1130
−207	−340		−11.1	12	53.6	16.7	62	143.6	100	212	413.6	327	620	1148
−201	−330		−10.6	13	55.4	17.2	63	145.4	104	220	428	332	630	1166
−196	−320		−10.0	14	57.2	17.8	64	147.2	110	230	446	338	640	1184
−190	−310		−9.4	15	59.0	18.3	65	149.0	116	240	464	343	650	1202
−184	−300		−8.9	16	60.8	18.9	66	150.8	121	250	482	349	660	1220
−179	−290		−8.3	17	62.6	19.4	67	152.6	127	260	500	354	670	1238
−173	−280		−7.8	18	64.4	20.0	68	154.4	132	270	518	360	680	1256
−169	−273	−459.4	−7.2	19	66.2	20.6	69	156.2	138	280	536	366	690	1274
−168	−270	−454	−6.7	20	68.0	21.1	70	158.0	143	290	554	371	700	1292
−162	−260	−436	−6.1	21	69.8	21.7	71	159.8	149	300	572	377	710	1310
−157	−250	−418	−5.6	22	71.6	22.2	72	161.6	154	310	590	382	720	1328
−151	−240	−440	−5.0	23	73.4	22.8	73	163.4	160	320	608	388	730	1346
−146	−230	−382	−4.4	24	75.2	23.3	74	165.2	166	330	626	393	740	1364
−140	−220	−364	−3.9	25	77.0	23.9	75	167.0	171	340	644	399	750	1382
−134	−210	−346	−3.3	26	78.8	24.4	76	168.8	177	350	662	404	760	1400
−129	−200	−328	−2.8	27	80.6	25.0	77	170.6	182	360	680	410	770	1418
−123	−190	−310	−2.2	28	82.4	25.6	78	172.4	188	370	698	416	780	1436
−118	−180	−292	−1.7	29	84.2	26.1	79	174.2	193	380	716	421	790	1454
−112	−170	−274	−1.1	30	86.0	26.7	80	176.0	199	390	734	427	800	1472
−107	−160	−256	−.6	31	87.8	27.2	81	177.8	204	400	752	432	810	1490
−101	−150	−238	0	32	89.6	27.8	82	179.6	210	410	770	438	820	1508
−96	−140	−220	.6	33	91.4	28.3	83	181.4	216	420	788	443	830	1526
−90	−130	−202	1.1	34	93.2	28.9	84	183.2	221	430	806	449	840	1544
−84	−120	−184	1.7	35	95.0	29.4	85	185.0	227	440	824	454	850	1562
−79	−110	−166	2.2	36	96.8	30.0	86	186.8	232	450	842	460	860	1580
−73	−100	−148	2.8	37	98.6	30.6	87	188.6	238	460	860	466	870	1598

TEMPERATURE CONVERSION TABLE—*Continued*

Albert Sauveur type of table. Values revised.

-459. to 0			0 to 100						100 to 1000					
C	$\frac{C}{F}$	F	C	$\frac{C}{F}$	F	C	$\frac{C}{F}$	F	C	$\frac{C}{F}$	F	C	$\frac{C}{F}$	F
− 68	− 90	−130	3.3	38	100.4	31.1	88	190.4	243	470	878	471	880	1616
− 62	− 80	−112	3.9	39	102.2	31.7	89	192.2	249	480	896	477	890	1634
− 57	− 70	− 94	4.4	40	104.0	32.2	90	194.0	254	490	914	482	900	1652
− 51	− 60	− 76	5.0	41	105.8	32.8	91	195.8				488	910	1670
− 46	− 50	− 58	5.6	42	107.6	33.3	92	197.6				493	920	1688
− 40	− 40	− 40	6.1	43	109.4	33.9	93	199.4				499	930	1706
− 34	− 30	− 22	6.7	44	111.2	34.4	94	201.2				504	940	1724
− 29	− 20	− 4	7.2	45	113.0	35.0	95	203.0				510	950	1742
− 23	− 10	14	7.8	46	114.8	35.6	96	204.8				516	960	1760
− 17.8	0	32	8.3	47	116.6	36.1	97	206.6				521	970	1778
			8.9	48	118.4	36.7	98	208.4				527	980	1796
			9.4	49	120.2	37.2	99	210.2				523	990	1814
						37.8	100	212.0				538	1000	1832

Look up reading in middle column. If in degrees Centigrade, read Fahrenheit equivalent in right-hand column; if in degrees Fahrenheit, read Centigrade equivalent in left-hand column.

Courtesy Bethlehem Steel Company

1000 to 2000						2000 to 3000					
C	$\frac{C}{F}$	F	C	$\frac{C}{F}$	F	C	$\frac{C}{F}$	F	C	$\frac{C}{F}$	F
538	1000	1832	816	1500	2732	1093	2000	3632	1371	2500	4532
543	1010	1850	821	1510	2750	1099	2010	3650	1377	2510	4550
549	1020	1868	827	1520	2768	1104	2020	3668	1382	2520	4568
554	1030	1886	832	1530	2786	1110	2030	3686	1388	2530	4586
560	1040	1904	838	1540	2804	1116	2040	3704	1393	2540	4604
566	1050	1922	843	1550	2822	1121	2050	3722	1399	2550	4622
571	1060	1940	849	1560	2840	1127	2060	3740	1404	2560	4640
577	1070	1958	854	1570	2858	1132	2070	3758	1410	2570	4658
582	1080	1976	860	1580	2876	1138	2080	3776	1416	2580	4676
588	1090	1994	866	1590	2894	1143	2090	3794	1421	2590	4694
593	1100	2012	871	1600	2912	1149	2100	3812	1427	2600	4712
599	1110	2030	877	1610	2930	1154	2110	3830	1432	2610	4730
604	1120	2048	882	1620	2948	1160	2120	3848	1438	2620	4748
610	1130	2066	888	1630	2966	1166	2130	3866	1443	2630	4766
616	1140	2084	893	1640	2984	1171	2140	3884	1449	2640	4784

TEMPERATURE CONVERSION TABLE—*Continued*

Albert Sauveur type of table. Values revised.

1000 to 2000						2000 to 3000					
C	$\dfrac{C}{F}$	F	C	$\dfrac{C}{F}$	F	C	$\dfrac{C}{F}$	F	C	$\dfrac{C}{F}$	F
621	1150	2102	899	1650	3002	1177	2150	3902	1454	2650	4802
627	1160	2120	904	1660	3020	1182	2160	3920	1460	2660	4820
632	1170	2138	910	1670	3038	1188	2170	3938	1466	2670	4838
638	1180	2156	916	1680	3056	1193	2180	3956	1471	2680	4856
643	1190	2174	921	1690	3074	1199	2190	3974	1477	2690	4874
649	1200	2192	927	1700	3092	1204	2200	3992	1482	2700	4892
654	1210	2210	932	1710	3110	1210	2210	4010	1488	2710	4910
660	1220	2228	938	1720	3128	1216	2220	4028	1493	2720	4928
666	1230	2246	943	1730	3146	1221	2230	4046	1499	2730	4946
671	1240	2264	949	1740	3164	1227	2240	4064	1504	2740	4964
677	1250	2282	954	1750	3182	1232	2250	4082	1510	2750	4982
682	1260	2300	960	1760	3200	1238	2260	4100	1516	2760	5000
688	1270	2318	966	1770	3218	1243	2270	4118	1521	2770	5018
693	1280	2336	971	1780	3236	1249	2280	4136	1527	2790	5036
699	1290	2354	977	1790	3254	1254	2290	4154	1532	2790	5054
704	1300	2372	982	1800	3272	1260	2300	4172	1538	2800	5072
710	1310	2390	988	1810	3290	1266	2310	4190	1543	2810	5090
716	1320	2408	993	1820	3308	1271	2320	4208	1549	2820	5108
721	1330	2426	999	1830	3326	1277	2330	4226	1554	2830	5126
727	1340	2444	1004	1840	3344	1282	2340	4244	1560	2840	5144
732	1350	2462	1010	1850	3362	1288	2350	4262	1566	2850	5162
738	1360	2480	1016	1860	3380	1293	2360	4280	1571	2860	5180
743	1370	2498	1021	1870	3398	1299	2370	4298	1577	2870	5198
749	1380	2516	1027	1880	3416	1304	2380	4316	1582	2880	5216
754	1390	2534	1032	1890	3434	1310	2390	4334	1588	2890	5234
760	1400	2552	1038	1900	3452	1316	2400	4352	1593	2900	5252
766	1410	2570	1043	1910	3470	1321	2410	4370	1599	2910	5270
771	1420	2588	1049	1920	3488	1327	2420	4388	1604	2920	5288
777	1430	2606	1054	1930	3506	1332	2430	4406	1610	2930	5306
782	1440	2624	1060	1940	3524	1338	2440	4424	1616	2940	5324
788	1450	2642	1066	1950	3542	1343	2450	4442	1621	2950	5342
793	1460	2660	1071	1960	3560	1349	2460	4460	1627	2960	5360
799	1470	2678	1077	1970	3578	1354	2470	4478	1632	2970	5378
804	1480	2696	1082	1980	3596	1360	2480	4496	1638	2980	5396
810	1490	2714	1088	1990	3614	1366	2490	4514	1643	2990	5414
			1093	2000	3632				1649	3000	5432

Look up reading in middle column. If in degrees Centigrade, read Fahrenheit equivalent in right-hand column; if in degrees Fahrenheit, read Centigrade equivalent in left-hand column.

Courtesy Bethlehem Steel Company

STANDARD CLASSIFICATION FOR COPPER AND COPPER ALLOYS
Wrought Products Only

COPPER NUMBER	PREVIOUS COMMONLY ACCEPTED TRADE NAME	Copper plus Silver(%min.)	Silver (oz./ton)	COMPOSITION, PER CENT MAXIMUM (Unless shown as a range or minimum)							
				Arsenic	Antimony	Phosphorus	Tellurium	Nickel	Bismuth	Lead	Other named elements
101 ⓐ	Oxygen Free Certified.....	99.96	–	–	–	.0003	–	–	–	–	.004 Sulfur / .0003 Zinc / .0001 Mercury
102 ⓐ	**Oxygen Free.............**	99.95	–	–	–	–	–	–	–	–	–
104 ⓐ	Oxygen Free with Silver....	99.95	8(min)	–	–	–	–	–	–	–	–
105 ⓐ	Oxygen Free with Silver....	99.95	10(min)	–	–	–	–	–	–	–	–
110 ⓐ	**Electrolytic Tough Pitch ...**	99.90	–	–	–	–	–	–	–	–	–
111 ⓐ	Electrolytic Tough Pitch Anneal Resistant........	99.90	–	–	–	–	–	–	–	–	©
113 ⓐⓑ	Tough Pitch with Silver.....	99.90	8(min)	–	–	–	–	–	–	–	–
114 ⓐⓑ	Tough Pitch with Silver.....	99.90	10(min)	–	–	–	–	–	–	–	–
116 ⓐⓑ	Tough Pitch with Silver.....	99.90	25(min)	–	–	–	–	–	–	–	–
120	Phosphorus Deoxidized Low Residual Phosphorus.	99.90	–	–	–	.004-.012	–	–	–	–	–
121	....	99.90	4(min)	–	–	.005-.012	–	–	–	–	–
122 ©	**Phosphorus Deoxidized High Residual Phosphorus**	99.90	–	–	–	.015-.040	–	–	–	–	–
123	....	99.90	4(min)	–	–	.015-.025	–	–	–	–	–
125 ⓖ	Fire Refined Tough Pitch...........	99.88	–	.012	.003	–	.025 ©	.05	.003	.004	–
127 ⓖ	Fire Refined Tough Pitch with Silver	99.88	8(min)	.012	.003	–	.025 ©	.05	.003	.004	–
128 ⓖ	Fire Refined Tough Pitch with Silver ..	99.88	10(min)	.012	.003	–	.025 ©	.05	.003	.004	–
130 ⓖ	Fire Refined Tough Pitch with Silver ..	99.88	25(min)	.012	.003	–	.025 ©	.05	.003	.004	–
141	Arsenical Tough Pitch. ...	99.40	–	.15-.50	–	–	–	–	–	–	–
142	Phosphorus Deoxidized Arsenical......	99.40	–	.15-.50	–	.015-.040	–	–	–	–	–
145 ⓗ	Phosphorus Deoxidized Tellurium Bearing.......	99.90 ①	–	–	–	.004-.012	.40-.60	–	–	–	–
147	Sulfur Bearing............	99.90 ②	–	–	–	–	–	–	–	–	.2-.5 Sulfur
150	Zirconium Copper........	99.80	–	–	–	–	–	–	–	–	.10-.15 Zirconium

COPPER ALLOY NUMBER	PREVIOUS COMMONLY ACCEPTED TRADE NAME	Copper+Silver +elements with specific limits(%min.)	COMPOSITION, PER CENT MAXIMUM (Unless shown as a range or minimum)									
			Iron	Tin	Nickel	Cobalt	Chromium	Silicon	Beryllium	Lead	Cadmium	Other named elements
162	Cadmium Copper.	99.75	.02	–	–	–	–	–	–	–	.7-1.2	–
164	....	99.75	.02	.2-.4	–	–	–	–	–	–	.6-.9	–
165	....	99.75	.02	.5-.7	–	–	–	–	–	–	.6-1.0	–
172	Beryllium Copper.	99.5	ⓡ	–	–	ⓡ	ⓡ	–	1.80-2.05	–	–	–
182	Chromium Copper.	99.5	.10	–	–	–	.6-1.2	.10	–	.05	–	–
184	Chromium Copper.	99.75	.15	–	–	–	.40-1.20	.10	–	–	–	.005 Arsenic / .005 Calcium / .05 Lithium / .05 Phos. / .70 Zinc
185	Chromium Copper.	99.75	–	–	–	–	.40-1.00	–	–	.015	–	.04 Phos. / .08-.12 Silver
190	....	99.5	.10	–	.9-1.3	–	–	–	–	.05	–	.75 Zinc / .15-.35 Phos.
191	....	99.5	.2	–	.9-1.3	–	–	–	–	.1	–	.5 Zinc / .15-.35 Phos. / .35-.65 Tellurium

COPPER ALLOY NUMBER	PREVIOUS COMMONLY ACCEPTED TRADE NAME	COMPOSITION, PER CENT MAXIMUM (Unless shown as a range or minimum)										
		Copper	Iron	Tin	Nickel	Cobalt	Chromium	Zinc	Aluminum	Lead	Cadmium	Total other elements ⓡ
193	....	92.0-94.0	2.05-2.60	.03	–	–	–	Rem.	.02	.03	–	.05

See page 358 for notes

STANDARD CLASSIFICATION FOR COPPER AND COPPER ALLOYS—
Continued

COPPER ALLOY NUMBER	PREVIOUS COMMONLY ACCEPTED TRADE NAME	COMPOSITION, PERCENT MAXIMUM (Unless shown as a range or minimum)												
		Copper	Lead	Iron	Tin	Zinc	Nickel	Aluminum	Phosphorus	Arsenic	Antimony	Manganese	Silicon	Total other elements ⑭
210	Gilding, 95%	94.0-96.0	.05	.05	–	Rem.	–	–	–	–	–	–	–	.10
220	Commercial Bronze, 90%	89.0-91.0	.05	.05	–	Rem.	–	–	–	–	–	–	–	.10
226	Jewelry Bronze 87½%	86.0-89.0	.05	.05	–	Rem.	–	–	–	–	–	–	–	.15
230	Red Brass, 85%	84.0-86.0	.05	.05	–	Rem.	–	–	–	–	–	–	–	.15
234		81.0-84.0	.05	.05	–	Rem.	–	–	–	–	–	–	–	.15
240	Low Brass, 80%	78.5-81.5	.05	.05	–	Rem.	–	–	–	–	–	–	–	.15
260	Cartridge Brass, 70%	68.5-71.5	.07	.05	–	Rem.	–	–	–	–	–	–	–	.15
261		68.5-71.5	.05	.05	–	Rem.	–	–	.02-.05	–	–	–	–	.15
262		67.0-70.0	.07	.05	–	Rem.	–	–	–	–	–	–	–	.15
268	Yellow Brass 66% (Sheet)	64.0-68.5	.15	.05	–	Rem.	–	–	–	–	–	–	–	.15
270	Yellow Brass 65% (Rod and Wire)	63.0-68.5	.10	.05	–	Rem.	–	–	–	–	–	–	–	.15
274	Yellow Brass 63%	61.0-64.0	.10	.05	–	Rem.	–	–	–	–	–	–	–	.20
280	Muntz Metal 60%	59.0-63.0	.30	.07	–	Rem.	–	–	–	–	–	–	–	.20
298	Brazing Alloy	49.0-52.0	.50	.10	–	Rem.	–	.10	–	–	–	–	–	–
310	Leaded Commercial Bronze (Low Lead)	89.0-91.0	.3-.7	.10	–	Rem.	–	–	–	–	–	–	–	.50
314	Leaded Commercial Bronze	87.5-90.5	1.3-2.5	.10	–	Rem.	–	–	–	–	–	–	–	.50
316	Leaded Commercial Bronze (Nickel Bearing)	87.5-90.5	1.3-2.5	.10	–	Rem.	.7-1.2	–	–	–	–	–	–	.50
320	Leaded Red Brass	83.5-86.5	1.5-2.2	.10	–	Rem.	–	–	–	–	–	–	–	.50
325		72.0-74.5	2.5-3.0	.10	–	Rem.	–	–	–	–	–	–	–	.50
330	Low Leaded Brass (Tube)	65.0-68.0	.2-.8	.07	–	Rem.	–	–	–	–	–	–	–	.50
331		65.0-68.0	.7-1.2	.06	–	Rem.	–	–	–	–	–	–	–	.50
332	High Leaded Brass (Tube)	65.0-68.0	1.3-2.0	.07	–	Rem.	–	–	–	–	–	–	–	.50
335	Low Leaded Brass	62.5-66.5	.3-.7	.10	–	Rem.	–	–	–	–	–	–	–	.50
340	Medium Leaded Brass 64½%	62.5-66.5	.8-1.4	.10	–	Rem.	–	–	–	–	–	–	–	.50
342	High Leaded Brass 64½%	62.5-66.5	1.5-2.5	.10	–	Rem.	–	–	–	–	–	–	–	.50
344		62.0-66.0	.5-1.0	.10	–	Rem.	–	–	–	–	–	–	–	.50
347		62.5-64.5	1.0-1.8	.10	–	Rem.	–	–	–	–	–	–	–	.50
348		61.5-63.5	.4-.8	.10	–	Rem.	–	–	–	–	–	–	–	.50
350	Medium Leaded Brass 62%	61.0-64.0	.8-1.4	.10	–	Rem.	–	–	–	–	–	–	–	.50
353	High Leaded Brass 62%	59.0-64.5	1.3-2.3	.10	–	Rem.	–	–	–	–	–	–	–	.50
356	Extra High Leaded Brass	60.0-64.5	2.0-3.0	.10	–	Rem.	–	–	–	–	–	–	–	.50
360	Free Cutting Brass	60.0-63.0	2.5-3.7	.35	–	Rem.	–	–	–	–	–	–	–	.50
362		60.0-63.0	3.5-4.5	.15	–	Rem.	–	–	–	–	–	–	–	.50
365	Leaded Muntz Metal, Uninhibited	58.0-61.0	.4-.9	.15	.25	Rem.	–	–	–	–	–	–	–	.10
366	Leaded Muntz Metal, Arsenical	58.0-61.0	.4-.9	.15	.25	Rem.	–	–	–	.02-.10	–	–	–	.10
367	Leaded Muntz Metal, Antimonial	58.0-61.0	.4-.9	.15	.25	Rem.	–	–	–	–	.02-.10	–	–	.10
368	Leaded Muntz Metal, Phosphorized	58.0-61.0	.4-.9	.15	.25	Rem.	–	–	.02-.10	–	–	–	–	.10
370	Free Cutting Muntz Metal	59.0-62.0	.9-1.4	.15	–	Rem.	–	–	–	–	–	–	–	.50
371		58.0-62.0	.6-1.2	.15	–	Rem.	–	–	–	–	–	–	–	.50
377	Forging Brass	58.0-62.0	1.5-2.5	.30	–	Rem.	–	–	–	–	–	–	–	.50
385	Architectural Bronze	55.0-60.0	2.0-3.8	.35	–	Rem.	–	–	–	–	–	–	–	.50
405		94.0-96.0	.05	.05	.7-1.3	Rem.	–	–	–	–	–	–	–	.15
408		94.0-96.0	.05	.05	1.8-2.2	Rem.	–	–	–	–	–	–	–	.15
410		91.0-93.0	.05	.05	2.0-2.8	Rem.	–	–	–	–	–	–	–	.15
411		89.0-92.0	.10	.05	.3-.7	Rem.	–	–	–	–	–	–	–	.15
413		89.0-93.0	.10	.05	.7-1.3	Rem.	–	–	–	–	–	–	–	.15
415		89.0-93.0	.10	.05	1.5-2.2	Rem.	–	–	–	–	–	–	–	.15
419		89.0-92.0	.10	.05	4.8-5.5	Rem.	–	–	–	–	–	–	–	.15
420		88.0-91.0	–	.05	1.5-2.0	Rem.	–	–	.25	–	–	–	–	.15
422		86.0-89.0	.05	.05	.8-1.4	Rem.	–	–	–	–	–	–	–	.15
425		87.0-90.0	.05	.05	1.5-2.2	Rem.	–	–	–	–	–	–	–	.15
430		85.0-89.0	.10	.05	1.7-2.7	Rem.	–	–	–	–	–	–	–	.15
432		85.0-87.0	.05	.05	.4-.6	Rem.	–	–	–	–	–	–	–	.15
434		84.0-86.0	.05	.05	.5-1.0	Rem.	–	–	–	–	–	–	–	.15
435		79.0-83.0	.10	.05	.6-1.2	Rem.	–	–	–	–	–	–	–	.15
438		79.0-82.0	.05	.05	1.0-1.5	Rem.	–	–	–	–	–	–	–	.15
442	Admiralty Uninhibited	70.0-73.0	.07	.06	.8-1.2	Rem.	–	–	–	–	–	–	–	.10
443	Admiralty Arsenical	70.0-73.0	.07	.06	.8-1.2	Rem.	–	–	–	.02-.10	–	–	–	.10
444	Admiralty Antimonial	70.0-73.0	.07	.06	.8-1.2	Rem.	–	–	–	–	.02-.10	–	–	.10
445	Admiralty Phosphorized	70.0-73.0	.07	.06	.8-1.2	Rem.	–	–	.02-.10	–	–	–	–	.10
462	Naval Brass 63½%	62.0-65.0	.20	.10	.5-1.0	Rem.	–	–	–	–	–	–	–	.10
464	Naval Brass	59.0-62.0	.20	.10	.5-1.0	Rem.	–	–	–	–	–	–	–	.10

STANDARD CLASSIFICATION FOR COPPER AND COPPER ALLOYS— *Continued*

COPPER ALLOY NUMBER	PREVIOUS COMMONLY ACCEPTED TRADE NAME	Copper	Lead	Iron	Tin	Zinc	Nickel	Aluminum	Phosphorus	Arsenic	Antimony	Manganese	Silicon	Total other elements Ⓐ
465	Naval Brass Arsenical	59.0-62.0	.20	.10	.5-1.0	Rem.	–	–	–	.02-.10	–	–	–	.10
466	Naval Brass Antimonial ...	59.0-62.0	.20	.10	.5-1.0	Rem.	–	–	–	–	.02-.10	–	–	.10
467	Naval Brass Phosphorized .	59.0-62.0	.20	.10	.5-1.0	Rem.	–	–	.02-.10	–	–	–	–	.10
470	Naval Brass Welding & Brazing Rod	57.0-61.0	.05	–	.25-1.0	Rem.	.01	–	–	–	–	–	–	.50 Ⓒ
472	Brazing Alloy	49.0-52.0	.50	.10	3.0-4.0	Rem.	–	–	–	–	–	–	–	–
482	Naval Brass Medium Leaded	59.0-62.0	.4-1.0	.10	.5-1.0	Rem.	–	–	–	–	–	–	–	.10
485	Naval Brass High Leaded .	59.0-62.0	1.3-2.2	.10	.5-1.0	Rem.	–	–	–	–	–	–	–	.10

COPPER ALLOY NUMBER	PREVIOUS COMMONLY ACCEPTED TRADE NAME	Copper+Tin+ Phosphorus (% min.)	Lead	Iron	Tin	Zinc	Nickel	Aluminum	Phosphorus	Arsenic	Antimony	Manganese	Silicon
502	Phosphor Bronze E	99.5	.05	.10	1.0-1.5	–	–	–	.04	–	–	–	–
505	....	99.5	.05	.10	1.0-1.7	.30	–	–	.03-.35	–	–	–	–
507	....	99.5	.05	.10	1.5-2.0	–	–	–	.04	–	–	–	–
508	....	99.5	.05	.10	2.6-3.4	–	–	–	.01-.07	–	–	–	–
509	....	99.5	.05	.10	2.5-3.8	–	–	–	.15-.30	–	–	–	–
510	Phosphor Bronze A	99.5	.05	.10	3.5-5.8	.30	–	–	.03-.35	–	–	–	–
518	Phosphor Bronze	99.5	.02	–	4.0-6.0	–	–	.01	.10-.35	–	–	–	–
521	Phosphor Bronze C	99.5	.05	.10	7.0-9.0	.20	–	–	.03-.35	–	–	–	–
524	Phosphor Bronze D	99.5	.05	.10	9.0-11.0	.20	–	–	.03-.35	–	–	–	–

COPPER ALLOY NUMBER	PREVIOUS COMMONLY ACCEPTED TRADE NAME	Copper+Tin+ Phosphorus+ Lead (% min.)	Lead	Iron	Tin	Zinc	Nickel	Aluminum	Phosphorus	Arsenic	Antimony	Manganese	Silicon
532	Phosphor Bronze B	99.5	2.5-4.0	.10	4.0-5.5	.20	–	–	.03-.35	–	–	–	–
534	Phosphor Bronze B-1	99.5	.8-1.2	.10	3.5-5.8	.30	–	–	.03-.35	–	–	–	–
544	Phosphor Bronze B-2	99.5 Ⓡ	3.5-4.5	.10	3.5-4.5	1.5-4.5	–	–	.01-.50	–	–	–	–
546	Phosphor Bronze B-2 (P 0.50 max.)	99.5 Ⓡ	3.5-4.5	.10	3.5-4.5	1.5-4.5	–	–	.50	–	–	–	–

COPPER ALLOY NUMBER	PREVIOUS COMMONLY ACCEPTED TRADE NAME	Copper+ Elements with specific limits (% min.)	Lead	Iron	Tin	Zinc	Nickel	Aluminum	Phosphorus	Arsenic	Antimony	Manganese	Silicon
606	....	99.5	–	.50	–	–	–	4.0-7.0	–	–	–	–	–
608	....	99.5	.10	.10	–	–	–	5.0-6.5	–	.35	–	–	–
610	....	99.5	.02	–	–	.20	–	6.0-9.0	–	–	–	–	.10
612	....	99.5	–	.50	–	–	–	7.0-9.0	–	–	–	–	–
614	Aluminum Bronze D	99.5	–	1.5-3.5	–	–	–	6.0-8.0	–	–	–	1.0	–
616	....	99.5	–	4.0	.6	1.0	1.0	6.5-11.0	–	–	–	1.5	.25
618	....	99.5	.02	1.5	–	.02	–	9.0-11.0	–	–	–	–	.10
620	....	99.5	–	3.2-3.7	–	–	–	9.8-10.5	–	–	–	–	–
622	....	99.5	.02	3.0-4.25	–	.02	–	11.0-12.0	–	–	–	–	.10
624	....	99.5	–	2.0-4.0	.20	–	–	9.0-11.0	–	–	–	.30	–
626	....	99.7	–	2.0-4.5	–	–	3.0-4.5	9.7-10.7	–	–	–	1.5	–
628	....	99.5	–	1.5-3.5	–	–	4.0-7.0	8.0-11.0	–	–	–	.5-2.0	–
630	....	99.5	–	2.0-4.0	.20	–	4.0-5.5	9.0-11.0	–	–	–	1.5	.25
637	....	99.5	.05	.3	.6	1.0	.25	6.5-8.5	–	–	–	–	1.2-2.2
639	....	99.5	.05	.10	–	–	.25	6.5-8.0	–	–	–	–	1.5-3.0
642	....	99.5	–	4.0	.6	1.0	.25	6.5-11.0	–	–	–	1.5	2.2
647	....	99.5	.1	.1	–	.5	1.6-2.2	–	–	–	–	–	.4-.8
651	Low Silicon Bronze B	99.5	.05	.8	–	1.5	–	–	–	–	–	.7	.8-2.0
653	....	99.7	.05	.8	–	–	–	–	–	–	–	–	2.0-2.6
655	High Silicon Bronze A	99.5	.05	1.6	–	1.5	.6	–	–	–	–	1.5	2.8-3.5
656	....	99.5	.02	.5	1.5	1.5	–	.01	–	–	–	1.5	2.8-4.0
658	....	99.5	.05	.5	–	–	–	.01	–	–	–	1.3	2.8-3.8
661	....	99.5	.2-.8	.25	–	1.5	–	–	–	–	–	1.5	2.8-3.5

STANDARD CLASSIFICATION FOR COPPER AND COPPER ALLOYS—

Continued

COPPER ALLOY NUMBER	PREVIOUS COMMONLY ACCEPTED TRADE NAME	COMPOSITION, PERCENT MAXIMUM (Unless shown as a range or minimum)												
		Copper	Lead	Iron	Tin	Zinc	Nickel	Aluminum	Phosphorus	Arsenic	Antimony	Manganese	Silicon	Total other elements ⓐ
665	....	80.0-82.0	.05	.10	–	Rem.	–	–	–	–	–	.7-1.5	–	.15
667	Manganese Brass	68.5-71.5	.07	.10	–	Rem.	–	–	–	–	–	.8-1.5	–	.50
670	Manganese Bronze B	63.0-68.0	.20	2.0-4.0	.50	Rem.	–	3.0-6.0	–	–	–	2.5-5.0	–	.10
675	**Manganese Bronze A**	57.0-60.0	.20	.80-2.0	.5-1.5	Rem.	–	.25	–	–	–	.05-.5	–	.10
680	Bronze, Low Fuming (Nickel)	56.0-60.0	.05	.25-1.25	.75-1.10	Rem.	.2-.8	.01	–	–	–	.01-.50	.04-.15	.50 ⓒ
681	Bronze, Low Fuming	56.0-60.0	.05	.25-1.25	.75-1.10	Rem.	–	.01	–	–	–	.01-.50	.04-.15	.50 ⓒ
685	....	85.0-89.0	.05	1.5-2.5	.10	Rem.	–	3.5-4.5	–	–	–	–	–	.10
687	**Aluminum Brass Arsenical**	76.0-79.0	.07	.06	–	Rem.	–	1.8-2.5	–	.02-.10	–	–	–	.10
692	Silicon Brass	89.0-91.0	.05	.05	–	Rem.	–	–	–	–	–	–	.8-1.5	.50
694	Silicon Red Brass	80.0-83.0	.20	.10	–	Rem.	–	–	–	–	–	–	3.5-4.5	.50
697	....	75.0-80.0	.5-1.5	.10	–	Rem.	–	–	–	–	–	.40	2.5-3.5	.50

COPPER ALLOY NUMBER	PREVIOUS COMMONLY ACCEPTED TRADE NAME	Copper+Elements with specific limits (% min.)	COMPOSITION, PERCENT MAXIMUM (Unless shown as a range or minimum)										
			Lead	Iron	Tin	Zinc	Nickel	Aluminum	Phosphorus	Arsenic	Antimony	Manganese	Silicon
702	....	99.7	.05	.10	–	–	2.0-3.0	–	–	–	–	.40	–
703	....	99.5	–	.05	–	–	4.7-5.7	–	–	–	–	.5	–
704	Copper Nickel 5%	99.6	.05	1.3-1.7	–	–	4.8-6.2	–	–	–	–	.30-.80	–
705	Copper Nickel 7%	99.5	.05	.10	–	.20	6.0-8.0	–	–	–	–	.15	–
706	**Copper Nickel 10%**	99.5	.05	.5-2.0	–	1.0	9.0-11.0	–	–	–	–	1.0	–
707	....	99.5	–	.05	–	–	9.5-10.5	–	–	–	–	.5	–
708	Copper Nickel 11%	99.5	.05	.10	–	.20	10.5-12.5	–	–	–	–	.15	–
710	Copper Nickel 20%	99.5	.05	1.0	–	1.0	19.0-23.0	–	–	–	–	1.0	–
715	**Copper Nickel 30%**	99.5	.05	.40-.70	–	1.0	29.0-33.0	–	–	–	–	1.0	–
720	Copper Nickel 40%	99.5	.05	1.5-2.5	–	.3	40.0-43.0	–	–	–	–	.8-1.7	–

COPPER ALLOY NUMBER	PREVIOUS COMMONLY ACCEPTED TRADE NAME	COMPOSITION, PERCENT MAXIMUM (Unless shown as a range or minimum)												
		Copper	Lead	Iron	Tin	Zinc	Nickel	Aluminum	Phosphorus	Arsenic	Antimony	Manganese	Silicon	Total other elements ⓐ
732	....	70.0(min)	.05	.6	–	3.0-6.0	19.0-23.0	–	–	–	–	1.0	–	.50
735	....	70.5-73.5	.10	.25	–	Rem.	16.5-19.5	–	–	–	–	.50	–	.50
740	....	69.0-73.5	.10	.25	–	Rem.	9.0-11.0	–	–	–	–	.50	–	.50
745	**Nickel Silver 65-10**	63.5-68.5	.10	.25	–	Rem.	9.0-11.0	–	–	–	–	.50	–	.50
752	**Nickel Silver 65-18**	63.0-66.5	.10	.25	–	Rem.	16.5-19.5	–	–	–	–	.50	–	.50
754	**Nickel Silver 65-15**	63.5-66.5	.10	.25	–	Rem.	14.0-16.0	–	–	–	–	.50	–	.50
757	**Nickel Silver 65-12.**	63.5-66.5	.05	.25	–	Rem.	11.0-13.0	–	–	–	–	.50	–	.50
762	....	57.0-61.0	.10	.25	–	Rem.	11.0-13.5	–	–	–	–	.50	–	.10
764	....	58.5-61.5	.05	.25	–	Rem.	16.5-19.5	–	–	–	–	.50	–	.50
766	....	55.0-58.0	.10	.25	–	Rem.	11.0-13.5	–	–	–	–	.50	–	.50
770	**Nickel Silver 55-18**	53.5-56.5	.10	.25	–	Rem.	16.5-19.5	–	–	–	–	.50	–	.50
773	....	46.0-50.0	.05	–	–	Rem.	9.0-11.0	–	.25	–	–	–	.04-.25	.50
774	....	43.0-47.0	.20	–	–	Rem.	9.0-11.0	–	–	–	–	–	–	.50
782	....	63.0-67.0	1.50-2.25	.35	–	Rem.	7.0-9.0	–	–	–	–	.50	–	.10
784	....	60.0-63.0	.8-1.4	.25	–	Rem.	9.0-11.0	–	–	–	–	.50	–	.50
786	....	60.0-63.0	1.25-1.75	.35	–	Rem.	8.5-11.0	–	–	–	–	.50	–	.10
788	....	63.0-67.0	1.5-2.0	.25	–	Rem.	9.0-11.0	–	–	–	–	.50	–	.10
790	....	63.0-67.0	1.50-2.25	.35	–	Rem.	11.0-13.0	–	–	–	–	.50	–	.50
792	....	60.0-63.0	.8-1.4	.25	–	Rem.	11.0-13.0	–	–	–	–	.50	–	.50
794	....	59.0-66.5	.8-1.2	.25	–	Rem.	16.5-19.5	–	–	–	–	.50	–	.50

NOTES

ⓐ Analysis is regularly made for the elements for which limits are listed except Zinc. If, however, the presence of other elements is suspected or indicated in the course of routine analysis further analysis shall be made to determine that the total of these "other" elements is not in excess of the limits specified.

ⓑ These are high conductivity coppers which have in the annealed condition a minimum conductivity of 100% IACS.

ⓒ Small amounts of Cadmium or other elements may be added by agreement to improve resistance to softening at elevated temperature.

ⓓ This includes Low Resistance Lake Copper and Electrolytic Copper.

ⓔ This includes Oxygen Free Copper to which Phosphorus has been added in an amount agreed upon.

ⓕ This includes High Resistance Lake Copper.

ⓖ This includes permissible Selenium.

ⓗ This includes Oxygen Free Tellurium Bearing Copper to which Phosphorus has been added in an amount agreed upon.

ⓘ This includes Copper plus Silver plus Tellurium.

ⓙ This includes Copper plus Silver plus Sulfur.

ⓚ Nickel and/or Cobalt .20 minimum. Nickel plus Cobalt plus Iron .60 maximum.

ⓛ Including Aluminum and Lead.

ⓜ Including Zinc

Appendix 1 / Maraging Steels

During the past fifteen years precipitation hardening has been used by the steel industry, and high strength steels with toughness previously impossible to attain were achieved. One precipitation hardening technique used is the *maraging process*, the name *maraging* being derived from a combination of the two major reactions involved in the hardening: *martensite + aging = maraging*.

A typical maraging alloy contains 18 per cent nickel, plus 8–9 per cent cobalt, 3–3.5 per cent molybdenum, smaller amounts of titanium and aluminum, and the balance low carbon (.03 per cent maximum) iron. When the alloy is heated above its transformation temperature the alloying elements dissolve in the austenite; upon cooling, the austenite transforms directly to practically 100 per cent martensite. Because the carbon content of maraging steels is exceedingly low this martensite is very tough and relatively soft and formable. To increase the strength of the steel further, the martensite is aged by reheating to 900°F, causing the precipitates to form a fine dispersion of the nickel compounds: Ni_3Mo, Ni_3Ti, and Ni_3Al. These precipitates harden the material but appreciable toughness is retained.

Advantages of 18 Per Cent Nickel Maraging Steels

1. Excellent mechanical properties:
   a. high tensile strength combined with good toughness
   b. high strength-to-weight ratio.

2. Good processing characteristics:
   a. excellent castability
   b. high resistance to hot tearing
   c. good hot and cold workability.

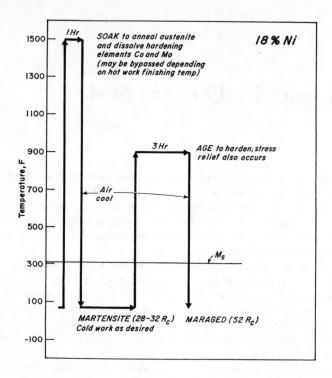

Figure I. Heat treating the maraging steels. Starting temperatures for martensite transformation (M_s) shown in charts are based on air cooled $\frac{1}{4}$-inch round specimens. The martensite transformation in these steels is not only temperature dependent but also highly time dependent—similar to the transformation in the precipitation hardening stainless steels.

From *Materials in Design Engineering*, courtesy The International Nickel Co., Inc.

3. Simple heat treatment:

    a. no decarburization
    b. no quenching
    c. deep hardening
    d. good dimensional stability.

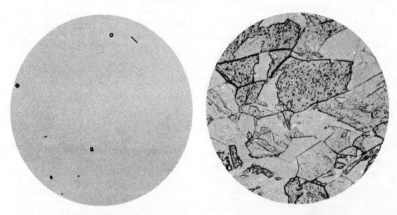

Mag. 100x. Typical field of inclusions; as-polished.　　Mag. 500x. Condition, as-rolled; etchant, 10 per cent ferric chloride.

Figure II. Photomicrographs of the 250 grade 18% Nickel maraging steel showing the typical cleanliness and the as-rolled microstructure.

Courtesy International Nickel Co., Inc.

Mag. 500x. Annealed at 1500°F; etchant, 10 per cent ferric chloride.　　Mag. 500x. Annealed at 1500°F, aged at 900°F; etchant, modified Fry's reagent.

Figure III. Typical microstructures of the 250 grade 18% Nickel maraging steel.

Courtesy International Nickel Co., Inc.

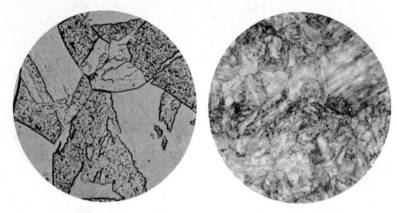

Mag. 500x. Annealed at 1800°F; etchant 10 per
cent ferric chloride.

Mag. 500x. Annealed at 1800°F, aged at 900°F;
etchant, modified Fry's reagent.

Figure IV. Typical microstructures of the 250 grade 18% Nickel maraging steel.
Courtesy International Nickel Co., Inc.

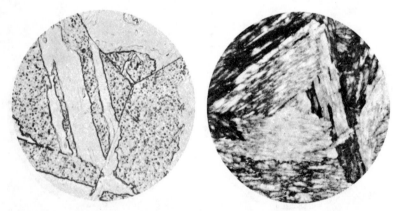

Mag. 500x. Annealed at 2100°F; etchant 10 per
cent ferric chloride.

Mag. 500x. Annealed at 2100°F, aged at 900°F;
etchant, modified Fry's reagent.

Figure V. Typical microstructures of the 250 grade 18% Nickel maraging steel.
Courtesy International Nickel Co., Inc.

TABLE 1—COMPOSITION OF MARAGING STEELS (%)[a]

Alloy Addition ↓	18% Ni 200,000 Psi	18% Ni 250,000 Psi	18% Ni 300,000 Psi	20% Ni	25% Ni
Ni	17-19	17-19	18-19	18-20	25-26
Ti	0.15-0.25	0.3-0.5	0.5-0.7	1.3-1.6	1.3-1.6
Al	0.05-0.15	0.05-0.15	0.05-0.15	0.15-0.35	0.15-0.35
Co	8-9	7-8.5	8.5-9.5	—	—
Mo	3.0-3.5	4.6-5.1	4.7-5.2	—	
Cb	—	—	—	0.3-0.5	0.3-0.5

[a]Other elements: C, 0.03 max; Mn, 0.10 max; Si, 0.10 max; S, 0.01 max; P, 0.01 max; B, 0.003 added; Zr, 0.02 added; Ca, 0.05 added.

TABLE 2—MECHANICAL PROPERTIES OF ANNEALED MARAGED STEELS

Alloy →	18% Ni[a]	20% Ni[b]	25% Ni[a]
Yld Str (0.2% offset), 1000 psi	110	115	40
Ten Str, 1000 psi	140	152	132
Elong (in 1 in.), %	17	8	30
Red. of Area, %	75	—	72
Hardness, R_c	28-32	26-35	10-15

[a]Bar specimen 0.252 in. dia, gage length 1 in.
[b]Flat specimen 0.145 in. thick, gage length 2 in.

TABLE 3—PHYSICAL AND ENVIRONMENTAL PROPERTIES OF MARAGING STEELS

Grade →	18% Ni[a]	20% Ni	25% Ni
Density, lb/cu in.	0.290	0.284	0.286
Mod of Elast, 10^6 psi	26.5	25.5	24.5
Mod of Rigidity, 10^6 psi	10.2	—	—
Poisson's Ratio	0.30	—	—
Coef of Ther Exp (70-900 F), 10^{-6}/°F	5.6	6.2	6.2
Nil Ductility Trans Temp, F	<−80	−80 to RT	300-400

Grade →	18% Ni[a]	20% Ni	25% Ni
Chg in Length During Maraging, %	Nil	−0.12	−0.10
ELECTRICAL, MAGNETIC			
Elec Res, μohm-cm			
As Annealed[b]	60-61	—	—
Ann. + Maraged[c]	38-39	—	—
Intrinsic Induction, gauss			
When H=250 oe	16,550	17,100	13,500
When H=5000 oe	18,500	18,375	14,750

Grade →	18% Ni[a]	20% Ni	25% Ni
Remanence, gauss	5,500	5,000	3,700
Coercive Force, oe	28.1	15.6	25.0
STRESS COR-ROSION			
Min Life (days) at Yield Strength			
U-Bends in Sea Water	35[d]	1	<1
3-Point Loaded in Bayonne Atm.	>360	—	>360

[a]Properties for 250,000 psi yield strength grade. Elastic modulus for 200,000 and 300,000 psi grades is 27.5 × 10^6 psi; other data not available.
[b]At 1500 F for 1 hr. [c]At 900 F for 3 hr. [d]200,000 psi yield strength grade showed no failures in six months.

Figure VI. Composition and properties of maraging steels.

From Materials in Design Engineering, courtesy The International Nickel Co., Inc.

4. Good fabrication characteristics:

   a. excellent weldability
   b. good machinability
   c. low rate of work hardening.

Some uses of maraging steels are: aircraft forgings, landing-gear components, jet engine starters, impellers, dies for aluminum extrusion, and aircraft arrester hooks.

Questions

1 Define maraging steel.
2 Describe the metallurgy and heat treatment involved in the hardening of 18 per cent nickel maraging steel.
3 Draw schematic sketches showing the heat treatment of 18 per cent nickel maraging steel.
4 How does a maraging steel differ from a conventional steel?
5 State ten advantages of a maraging steel over conventional steel.
6 How does 18 per cent nickel maraging steel attain its remarkable properties?
7 Why is 18 per cent nickel maraging steel readily welded and machined?
8 Explain precipitation hardening.
9 What is the structure of 18 per cent nickel maraging steel in the annealed condition? Why is it soft and tough?
10 Why does 18 per cent nickel maraging steel become stronger during aging?

Appendix 2 / The Plasma Arc

The conductor in a high pressure arc discharge is called *plasma*. Plasma is composed of electrons and positive or negative ions which are formed when a gas is heated to a temperature high enough to ionize it. Only at the very center of the high pressure arc discharge can one expect to have 100 per cent ionized gas (plasma). The other portions of the arc conductor consist of impure plasma, i.e., ions and electrons mixed with excited atoms and molecules. Plasma may be considered the fourth state of matter, the other three being solid, liquid, and gas.

In order to change water at 212°F into steam at 212°F latent heat must be supplied. Similarly, the plasma torch must supply energy to a gas in order to convert it into a plasma at the same temperature. It is this tremendous load of heat, carried by the plasma, which melts the metal in its path.

Practically any gas can be used to form the plasma. The rate at which the gas transfers heat is more important than the temperature the gas reaches. For example, argon and hydrogen produce arcs of similar temperature range, but hydrogen transfers more heat because of its higher thermal conductivity.

One feature of the plasma arc is that it combines the directional stability of a combustion flame of high gas velocity with the high temperature of an electric arc.

There are four ways to control the temperature and heat output of the plasma arc:

1. Varying the current.
2. Varying the flow or composition of the gas.
3. Modifying the orifice size.
4. Mode of operation: the plasma arc can be operated as a transferred or nontransferred arc.

365

Comparison of Transferred and Nontransferred Arcs

The transferred-arc and nontransferred-arc plasma arc torches differ in their electrical connections. In the transferred-arc torch the arc occurs between the cathode contained in the torch and the anode located on the workpiece. In the nontransferred-arc torch the nozzle itself is an electrode. The arc ends at the nozzle so that only the hot discharged stream reaches the workpiece.

The transferred-arc torch is normally used in cutting and produces a much higher plasma than that normally obtained from nontransferred operations. It transfers more heat and does it more efficiently than the nontransferred arc.

The nontransferred arc can concentrate its heat inside the torch. Its greatest use is for plating. During the plating operation the plating material is heated to a plastic or molten state inside the torch and blasted onto the work. The nontransferred-arc torch is important in applying

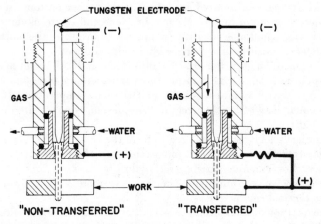

Figure VII. Schematic diagram showing the differences in electrical connections for nontransferred and transferred-arc torches.

Courtesy Linde Company, Division of Union Carbide Corporation.

platings of refractory materials, particularly to parts for missiles and rockets. Both metallic and nonmetallic materials can be coated. The coating material may be almost any inorganic solid that can be melted without decomposing. By depositing layer on layer, coatings can be made any thickness, without sacrifice of mechanical strength.

Advantages of Cutting with Plasma

1. Because of its extremely hot flame it can cut two-inch thick steel about twice as fast as the oxygen torch, and it leaves a narrower heat-affected zone. However, it is limited at present to about four-inch thick steel.

2. It does not require iron powder to cut stainless steel as does oxygas cutting (the oxygen-fuel-gas torch requires iron, which must be fed into the oxygen stream at a carefully controlled rate).

Weld Surfacing with the Plasma

Plasma (arc-welding surfacing) is the first method which makes it possible to deposit a very thin fused coating with controlled dilution at high speeds. The deposit is a true weld fused to the workpiece, and it resists impact and high stresses. In addition it is free of porosity.

Plasma Welding

Plasma welding produces no combustion products. Uniformly penetrated, high quality welds result because of the nature of the energy transfer.

As the torch moves along the path of the pieces to be joined, it melts a seam which cools and solidifies within the inert gas shield. The absolute arc stability and uniform penetration permit close tolerances and thereby make the process ideal for automated welding.

Questions

1 Technically, what is a plasma?
2 Explain why plasma may be considered the fourth state of matter.
3 What is latent heat?
4 Why does the hydrogen arc transfer more heat than the argon arc?
5 State four ways of controlling the temperature and heat output of the plasma arc.
6 Compare the transferred-arc and nontransferred-arc plasma torches.
7 By means of schematic illustrations show the differences in electrical connections for nontransferred- and transferred-arc torches.
8 State and explain two advantages of the plasma torch.
9 What advantages are gained by weld surfacing with the plasma?
10 State the advantages of plasma welding.

Appendix 3 / The Laser

The *laser*—abbreviation for *light amplification by stimulated emission of radiation*—is one of the most important developments of modern technology. It provides a tool for which new uses are daily being discovered.

Simply, the laser can be described as an extremely intense light source. It can produce light beams with energy densities ranging up to hundreds of millions of calories per square centimeter per second. These light outputs can be focused in small spots of high intensities of radiant energy and thereby produce temperatures high enough to melt any metal. This property makes the laser very promising as a source of welding energy; laser beams can also machine very small, close-tolerance holes in metals.

The Principle of the Laser

The basic idea behind the laser is not difficult to understand. Laser light is a kind of *fluorescence*. A *fluorescent material* is one which emits light upon absorption of some form of energy—light, electrons, etc. This phenomenon is explained by modern atomic physics as follows. The electrons surrounding the nucleus of any atom can occupy many different energy levels. Ordinarily they occupy the lowest possible energy level, but they can be excited to higher levels by the absorption of external energy. These electrons will then return spontaneously to their original level by dissipating their excess energy as radiation. This radiation is impractical as a source of energy because:

1. it is made up of light of many different wave lengths. This causes interference, with resulting loss of total energy.

2. the light is *incoherent*, i.e., not in phase. This results in interference; also, there is no way to amplify incoherent light.

By *stimulating* emission of light the limitations inherent in spontaneous

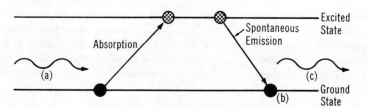

Figure VIII. In thermal equilibrium, the internal energy of atoms rests at ground level, but it can be raised to another level by absorption of a photon of optical energy (a). In time, the internal energy spontaneously drops to ground level (b), with the emission of a photon of light (c)—this is the phenomenon of fluorescence.

From *Metal Progress*, November 1962, courtesy American Society for Metals.

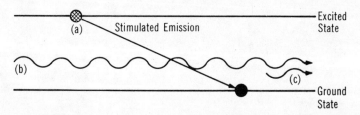

Figure IX. The laser phenomenon occurs when excited atoms (a) are struck by photons (b) which have precisely the right energy—that is, the energy of the photon which would otherwise have been emitted spontaneously (see Figure X). This "triggered" photon falls into phase with the incoming photon to produce an augmented wave (c). Meanwhile, the internal energy of the originally excited atom returns to the ground state.

From *Metal Progress*, November 1962, courtesy American Society for Metals.

emission can be overcome. In the laser, an atom is first "pumped" to an excited state, and then bombarded with a photon of just the right energy to stimulate the emission of another photon. This photon is *coherent* with the inducing radiation, i.e., it is in phase with it and has the same wave length; therefore it *amplifies* the original energy, instead of causing interference. This new, amplified, light energy can in turn be used as inducing energy. This is done in the laser by the use of reflecting surfaces: the amplified light energy is bounced back into the excited material, thus stimulating the emission of more and more coherent light energy. In this way tremendous energies surpassing that of the sun can be built up.

The idea is simple, but the technology making its implementation possible has been only recently developed. It is expected that in not too many years commercial laser welders and metal cutting devices will be available to solve many of the unusual problems of metal technology.

Questions

1 What does the word laser mean?
2 Why are lasers considered excellent potential sources of welding energy?
3 Define coherent light.
4 What is fluorescence?
5 How does the laser principle make use of fluorescence?

Glossary

ABRASION RESISTANCE resistance of a material to being worn away by friction.

ACID in metallurgical terminology, the oxide of a nonmetal.

ACID BOTTOM AND LINING the inner bottom and lining of a melting furnace composed of materials having an acid reaction in the melting process. The materials may be sand, siliceous rock, or silica bricks.

ACID PROCESS a steel-making process, either bessemer, open hearth, or electric, in which the furnace is lined with a siliceous refractory, and for which the raw materials to be used are low in phosphorus and sulfur, because those elements are not removed by this process.

ACID STEEL steel melted under a slag that has an acid reaction, and in a furnace with an acid bottom or lining.

AGE HARDENING hardening caused by a change in the physical properties of a low-carbon alloy steel at room or elevated temperatures. It is an attempt to overcome an unstable condition by restoring real equilibrium.

AGING change in properties (for example, increase in tensile strength and hardness) that occurs in certain metals at relatively low temperature after a final heat treatment (as in duralumin) or after a final cold working (as in mild steel). The method employed to bring about aging consists of exposure to a favorable temperature subsequent to (a) a relatively rapid cooling from some elevated temperature (quench aging) or (b) a limited degree of cold work (strain aging).

AIR FURNACE a form of reverberatory furnace for melting ferrous and nonferrous metals and alloys. Flame from fuel burning at one end of the hearth passes over the bath to the stack at the opposite end of the furnace.

AIR-HARDENING STEEL an alloy steel that does not require quenching from a high temperature to harden, but which is hardened by simply cooling in air from above its critical range.

ALLOTRIOMORPHIC CRYSTAL a crystal the regular growth of which has been obstructed and distorted, causing it to lack external symmetry.

ALLOTROPY occurrence of an element in two or more modifications. For example, carbon occurs in nature as the hard crystalline diamond, soft

SOURCES OF GLOSSARY: The Bethehem Steel Company booklets. *Metals Handbook*, American Society for Metals. "Glossary of Foundry Terms," *The Foundry*.

crystalline graphite, and amorphous coal.

ALLOY a metallic material formed by mixing (in the molten state) two or more chemical elements. It usually possesses properties different from those of the components.

ALLOY STEEL a carbon steel to which is added a definite amount of one or more elements other than carbon, in order to impart special properties to the steel so that it can be used for specific purposes.

ALPHA-BETA BRASS brass which has zinc content of between about 39 per cent and about 45.5 per cent and contains the alpha solid solution and another known as beta.

ALPHA BRASS brass that contains about 38 per cent or less of zinc and consists of a solid solution designated alpha.

ALPHA IRON a magnetic allotropic form of iron; it crystallizes in the body-centered cubic structure.

AMORPHOUS SUBSTANCES substances that exhibit no crystalline structure because the atoms of which they are composed are not arranged in the geometrical structure of a space lattice. They may be considered solid solutions which can be supercooled to any degree without occurrence of crystallization.

ANISOTROPIC SOLIDS solids that possess directional properties because their atoms are arranged in a definite geometric order.

ANNEALING a heating and cooling operation implying usually a relatively slow cooling. In annealing, the temperature of the operation and the rate of cooling depend upon the material being heat treated and the purpose of the treatment.

ARC FURNACE an electric furnace in which an electric arc between carbon or graphite electrodes and the furnace charge is used to produce the heat required to melt metals in a confined space.

ARREST POINTS the various temperatures at which pauses occur in the rise or fall of temperature when steel is heated from room temperature or cooled from the molten state.

ATOM the smallest particle of an element, and the fundamental unit from which the grain structure of a metal is formed.

ATOMIC PLANES The layers of atoms or planes along which the atoms are arranged within the crystal.

AUSTEMPERING an interrupted quenching process the purpose of which is to produce a bainitic structure.

AUSTENITE a solid solution of iron carbide in gamma iron. It is stable at temperatures above the Ac transformation and has a face-centered cubic lattice structure. In some special steels it exists at room temperature.

AUSTENITIC STEELS steels containing sufficient amounts of nickel, nickel and chromium, or manganese to retain austenite at atmospheric temperature; for example, austenitic stainless steel, and Hadfield's manganese steel.

BABBIT a term applied to white metals having a tin base.

BAINITE a structure intermediate between pearlite and martensite, which is formed when steel is cooled rapidly to about 800°F and is held at any temperature between 800°F and about 400°F for a sufficient length of time. The structure depends upon the temperature at which transformation occurs.

BANDED STRUCTURE a segregated structure of nearly parallel bands which run in the direction of working.

BAR MILL a mill consisting of one or more stands of grooved rolls for reducing blooms or billets to bars.

BASE in metallurgical terminology, the oxide of a metal.

BARS rounds, squares, flats, hexagons, octagons, half ovals, half rounds, special sections, and small shapes. Angles, channels, tees, or zees are *bar* size when their greatest diameter is under 3 inches. Flats are classified as bars when they are 6 inches or less in width and 0.250 inch or more in thickness.

BASIC BOTTOM AND LINING the inner lining and bottom of a melting furnace composed of materials having a basic reaction in the melting process. The materials may be crushed burnt dolomite, magnesite, magnesite bricks, or basic slag.

BASIC STEEL a steel melted in a furnace with a basic bottom and lining, and under a slag that is mainly basic. The raw materials used contain appreciable amounts of phosphorus and sulfur.

BAUXITE a residual clay, consisting essentially of aluminum hydroxide. It is the most important ore of aluminum.

BAYER PURIFICATION PROCESS a process for purification of bauxite, as the first stage in production of aluminum.

BEARING METALS alloys used for that part of a bearing which is in contact with the journal; for example, bronze or white metal, used because of their low coefficient of friction when used with a steel shaft.

BED CHARGE the charge of iron placed on the coke bed in a cupola.

BED COKE coke placed in the cupola well to support the following iron and coke charges.

BENEFICIATION the washing out of free sand and free clay from an ore.

BENTONITE a widely distributed and peculiar type of clay which is considered to be the result of devitrification and chemical alteration of the glassy particles of volcanic ash or tuff. Used in the foundry to bond sand.

BESSEMER ORE an iron ore containing not more than 0.045 per cent of phosphorus.

BESSEMER PROCESS a process for making steel by blowing air through molten pig iron contained in a converter. The process depends upon rapid oxidation of silicon and carbon to secure the necessary rise in temperature.

BETA BRASS copper-zinc alloys containing 46 to 49 per cent of zinc, which consists (at room temperature) of the intermediate constituent known as beta.

BILLET an ingot or bloom that has been reduced through rolling or hammering to an approximate square ranging from $1\frac{1}{2}$ to 6 inches square, or to an approximate rectangular cross section of an equivalent area. Billets are classified as semifinished products for rerolling or forging.

BINDER material to hold the grains of sand together in molds or cores. May be cereal oil, clay, resin, pitch, and the like.

BLACK-HEART American type of malleable iron. The normal fracture shows a velvety black appearance having a mouse-gray rim.

BLAST FURNACE a brick-lined cylindrical shell supplied with air blast, usually preheated in stoves, for producing pig iron by reduction of iron ore.

BLIND RISER an internal riser that does not reach to the exterior of the mold.

BLISTER a defect in a metal produced by gas bubbles either on the surface or beneath the surface.

BLOOM (SLAB, BILLET, SHEET BAR) semifinished products of rectangular cross section with rounded corners, hot rolled from ingots. The chief differences are in cross-sectional area and in their intended use.

BLOOMING MILL   a mill that rolls ingots usually to blooms, billets, and slabs. Sometimes called a *cogging mill*, and when so called in United States, refers to a mill producing shaped blooms used as blanks for subsequent rolling of I-beams, channels, and the like.

BLOW   the forcing of air through the molten pig iron contained in a converter.

BLOWHOLE   a casting defect caused by trapping of gas in molten or partially molten metal.

BODY-CENTERED CUBIC SPACE LATTICE (B.C.C.)   in crystals, an arrangement of atoms in which the atomic centers are disposed in space in such a way that they may be presumed to be situated at the corners and centers of a set of cubic cells.

BOND   cohesive material in sand.

BOSH   the combustion zone of a blast furnace. It is the tapered zone just above the hearth.

BOTT (BOD)   a piece of clay or other material to stop the flow of metal from the taphole.

BOTT-STICK   a stick or rod on which the bott is mounted so that it may be forced into the taphole.

BOTTOM BOARD   board supporting the mold.

BOX ANNEALING   softening steel by heating it, usually at a subcritical temperature, in a suitable closed metal box or pot to protect it from oxidation, employing a slow heating and cooling cycle; also called *closed annealing* or *pot annealing*.

BRASS   primarily an alloy of copper and zinc, but sometimes containing also small amounts of other elements such as aluminum, iron, lead, manganese, and tin.

BRAZING   a method of joining by use of an alloy of lower melting point than that of the metal to be joined. Brazing is similar to soldering except that a higher temperature is required.

BRIDGING (in a blast furnace)   sticking or arching of fine ore against the walls.

BRIDGING (in a cupola furnace)   material adhering to the wall that retards or prevents descent of the stock charges.

BRIDGING (in an electric furnace)   formation by the upper layers of scrap of a bridge over the pool of steel.

BRINELL HARDNESS   value of hardness of a metal or alloy, tested by measuring the diameter of an impression made by a ball of given diameter applied under a known load. Values are expressed in Brinell hardness numbers obtained from tables.

BRIQUETTES   compact blocked formed of finely divided materials by incorporation of a binder, by pressure, or both.

BRIQUETTING   the process of feeding a predetermined quantity of metal powders into hardened steel dies and pressing the powders into a *slug* or *green briquette* of the desired shape.

BRITTLENESS   the tendency to fracture without appreciable deformation and under low stress.

BROACHING   a machining operation in which a long cutting tool having a series of cutting teeth (broaches) of continuously changing dimensions, is forced through a roughly finished hole or over a surface to produce the desired shape or size of the article.

BRONZE   primarily an alloy of copper and tin, but the name is now applied to alloys with aluminum, silicon, and some other metals.

BRUCITE   a naturally occurring magnesium hydrate $Mg(OH)_2$.

BUNG   a section of the removable roof of an air furnace or reverberatory furnace. Also used to indicate choking or plugging up.

BURNING the heating of a metal to temperatures sufficiently close to the melting point to cause permanent injury. Such injury may be caused by the melting of the more fusible constituents, by the penetration of gases such as oxygen into the metal with consequent reactions, or by the segregation of elements already present in the metal.

BUSTLE PIPE in a blast furnace, the large pipe that encircles the bosh and receives the hot blast of air from the stoves in order to distribute it to the tuyeres.

BUTT the large flat round end of the rammer used in the foundry.

BUTT-RAMMING ramming the molding sand with the large round end of the molder's rammer.

BUTT WELD the welding of two abutting edges. Used in the manufacture of steel pipe, the pipe so made being called *butt-weld pipe*. Also applied to butt-welding of ends of two bars.

CALCINING the removal of chemically held water by the prolonged heating of a material at fairly high temperatures.

CAPPED STEEL a rimmed steel in which the rimming action is intentionally stopped shortly after the mold is filled.

CARBIDE a compound of carbon with a more positive element such as iron.

CARBIDE FORMER an alloying element that reacts chemically with the carbon present in a steel to form a carbide.

CARBOMETER TEST testing a steel for carbon (on a carbometer) by determining its magnetic properties. These properties bear a definite relationship to the carbon content.

CARBON STEEL a steel in which carbon is the only alloying element added to the iron to control its properties; also

known as *ordinary steel*, or *straight carbon steel*, or *plain carbon steel*.

CARBURIZING adding carbon to the surface of iron-base alloys by heating the metal below its melting point in contact with carbonaceous solids, liquids, or gases.

CASE the surface layer of a steel that has been made substantially harder than the interior by a process of case hardening.

CASE HARDENING a heat treatment or combination of heat treatments by which the surface layer of steel is made harder than the interior. The processes of carburizing, nitriding, and cyaniding accomplish this result by changing the composition of the case.

CASTING the metal shape that is obtained as a result of pouring metal into a mold.

CASTING the act of pouring molten metal into a cavity and allowing it to harden so that it will assume and retain the size and shape of the cavity when cold.

CAST IRON alloys of iron containing so much carbon that, as cast, they usually are not appreciably malleable at any temperature. Usually from 1.7 to 4.5 per cent carbon is present and in most cases an important percentage of silicon.

CEMENTED or SINTERED CARBIDES powdered carbides of tungsten, tantalum, or titanium cemented into solid masses by mixing with powdered cobalt or nickel, then compressing and sintering. Used instead of high-speed steel to form cutting tips of cutting tools, and in parts subjected to heavy wear.

CEMENTITE a chemical compound of iron and carbon, also known as iron carbide (Fe_3C), which contains about 6.8 per cent carbon. It occurs as grain envelopes or as needles within a grain of hypereutectoid steel. It occurs as lamellae in pearlite. It may also occur as spheroids in annealed steel. It is extremely hard and brittle.

CENTRIFUGAL CASTING   process of filling molds by pouring the metal into a sand or metal mold revolving about either its horizontal or vertical axis, or pouring the metal into a mold that subsequently is revolved before solidification of the metal is complete. Molten metal is removed from the center of the mold to the periphery by centrifugal action.

CHAPLETS supports of the same metal or alloy as the casting used to secure a core in its proper position in the mold, which fuse in with and become part of the casting.

CHARPY TEST   a notched-bar or impact test in which a notched specimen, fixed at both ends, is struck behind the notch by a striker carried on a pendulum.

CHEEK   the side walls of the mold cavity. Also the intermediate sections of flask between the cope and drag section numbered 1, 2, 3, etc., from the drag section. Necessitated by difficulty of molding unusual shapes or in cases where more than one parting line is required.

CHILL   a metal object placed on the outside or inside of a mold cavity to induce more rapid cooling at that point.

CHILLED CASTING   a casting that was cooled very rapidly and therefore possesses a very hard surface and soft, tough interior.

CINDER NOTCH   the opening through which the slag is removed from the blast furnace; also called *slag hole*.

CLASSIFIER   a machine for separating the product of an ore-crushing plant into two portions consisting of particles of different sizes. In general, the finer particles (the overflow or slime) are carried off by a stream of water, while the larger or coarse particles (underflow or sand) settle.

CLEAVAGE PLANE   a surface of fracture that is smooth and plane, and always parallel to some definite crystallographic plane. A substance may cleave on more than one crystallographic plane.

COARSE GRAIN   the grain formed during slow cooling because few nuclei are formed.

COARSE-GRAIN STEELS   steels which begin to coarsen near the transformation temperature and continue to coarsen with increasing temperature.

COGGING   rolling or forging ingots to reduce them to blooms or billets in a bloom or cog mill.

COKE BED   first layer of coke placed in the cupola. Also the coke used as the foundation in constructing a large mold in a flask or pit.

COLD DRAWING   permanent deformation of metal below its recrystallization temperature, by drawing the bar through one or more dies.

COLD ROLLING   permanent deformation of metal below its recrystallization temperature by rolling. This process is frequently applied in finishing rounds, sheets, strip, and tin plate. Cold rolling of metal sheets results in a smooth surface finish.

COLD SHORTNESS   brittleness when metal is at a low temperature.

COLD WORKING   permanent deformation of a metal below its recrystallization temperature.

COLD WORKING   plastic deformation of a metal at a temperature low enough to insure strain (work) hardening.

COLUMNAR CRYSTALS   elongated crystals formed by growth taking place at right angles to the surface of the mold.

COMBINED CARBON   the carbon in iron or steel combined with other elements and therefore not in the free state as graphite or temper carbon.

COMMINUTION   size reduction by breaking, crushing, or grinding, as in ore dressing, for example.

COMPRESSION TEST   a test for ductility and malleability of an iron or steel bar, in which a specimen, of length 1.5 diameters, is compressed to half its length, without cracking.

CONCENTRATION   elimination of the waste material or gangue present in an ore in order to increase the metal content of the product.

CONSTITUTIONAL DIAGRAM   see EQUILIBRIUM DIAGRAM.

CONTINUOUS MILL   a succession of roll stands, usually in single line formation. The reductions and speeds of rolls in the various stands are progressively related so the bar, strip, etc. can be in engagement with all stands of rolls at the same time.

CONTRACT or STOPE   the place where the ore is being mined.

CONTRACTION   the decrease in size of a casting as it cools.

COOLING CURVE   the curve obtained by plotting time against temperature for a metal cooling under constant conditions.

COPE   the upper or topmost section of a flask, mold, or pattern.

CORE   a separable part of the mold, usually made of sand and generally baked, to create openings and various shaped cavities in the casting.

CORE   the interior portion of an iron-base alloy casting; it is substantially softer than the surface layer as the result of case hardening.

CORE   the inner portion of a rolled section of rimmed steel as distinct from the rimmed portion or rim.

CORE (DRY SAND)   a dry sand core is a body of molded sand baked to a consistency sufficient to maintain its shape, allow for handling while being placed in the mold, and sustain the impact of flowing metal in the mold. It may be used in any position in the mold cavity to form holes through the casting, recesses in either the interior or exterior surfaces of castings, to fill voids made by coreprints used to facilitate the molding of a pattern, or to form the mold cavity itself.

CORE BINDER   any material used to hold the grains of core sand together.

CORE BOX   a wooden or metal box shaped internally for molding sand cores in the foundry.

CORE PRINT   a projection on a pattern for locating the core or an extension of the mold cavity for locating and supporting the core.

CORE VENT   a wax product round or oval in form used to form the vent passage in a core.

CORED STRUCTURE (CRYSTAL)   a grain structure having composition gradients caused by progressive freezing of the components in different proportions; the term *zonal structure* is in general preferable. In dendrites the cored or zonal structure is manifested by solvent-rich crystal axes and solute-rich interstices.

CORROSION   chemical combination of and consequent wasting away of metals by chemical attack, usually at ordinary temperatures. It may be classified as atmospheric, submarine, subterranean, or electrolytic.

COTTRELL ELECTRIC PRECIPITATOR   a device used to remove fine dust particles and moisture from gas or air.

CREEP   slow impermanent deformation in a metallic specimen produced by a relatively small steady force, below the elastic limit, acting for a long period of time.

CRITICAL COOLING RATE   the slowest rate of cooling which results in transformation of austenite into martensite without production of any pearlite.

CRITICAL POINTS   the temperatures at which changes of structure take place in a metal as it is heated from room temperature to its melting point or is cooled from the molten state to room temperature.

CRITICAL RANGE the range of temperature in which occurs the reversible change from austenite, which is stable at high temperature, to ferrite, pearlite, and cementite, which are stable at low temperature.

CRITICAL RANGE the range between the recalescence point and the decalescence point.

CRITICAL TEMPERATURE the temperature at which some change occurs in a metal or alloy during heating or cooling; for example, the temperature at which an arrest or critical point occurs on heating or cooling curves.

CROP the end or ends of a rolled or forged product containing the pipe or other defects that are cut off and discarded.

CRYOLITE (ICE STONE) sodium aluminum fluoride, a mineral found in Greenland or produced synthetically. It is used as the electrolyte and solvent in the electrolytic process of obtaining pure aluminum from alumina.

CRYSTAL a homogeneous solid of regular geometrical structure peculiar to the element, compound, or isomorphous mixture of which it is composed. Within each crystal the atoms are spaced in characteristic pattern.

CRYSTAL BOUNDARIES the surfaces of contact between adjacent crystals in a metal.

CRYSTAL NUCLEI the minute crystals the formation of which is the beginning of crystallization.

CUPOLA FURNACE a stack-type melting unit in which metal is melted in direct contact with the fuel.

CYANIDING surface hardening by carbon and nitrogen absorption of an iron-base alloy article or portion of it by heating at a suitable temperature

in contact with a molten cyanide salt, followed by quenching.

DAMPING CAPACITY the ability to absorb vibration. More accurately defined as the mount of work dissipated into heat by a unit volume of material during a completely reversed cycle of unit stress.

DECALESCENCE POINT the first critical point, 1333°F, at which a change of structure occurs in steel. The steel absorbs a considerable amount of heat as the structure changes in part to the face-centered cubic form.

DECARBURIZATION removal of carbon from the surface of solid steel by the (normally oxidizing) action of media that react with carbon. The media may be gaseous, solid, or liquid.

DEFORMATION change of form under mechanical stress (see ELASTIC DEFORMATION, PERMANENT DEFORMATION).

DEGASIFIER a material employed for removing gases from metals and alloys.

DENDRITE a crystal formed during solidification of a metal, or in any other way, having many branches and a treelike pattern; also termed *pine tree* and *fir tree* crystals.

DEOXIDATION the process of elimination of oxygen from molten metal before casting by adding elements with a high oxygen affinity, which form oxides that tend to rise to the surface.

DEOXIDIZER a material used to remove oxygen or oxides from metals and alloys.

DEPRESSANT a chemical that causes a finely powdered sulfide mineral to sink through a froth, in froth flotation. The mineral so sunk is said to be depressed.

DEZINCIFICATION that corrosive action which acts on brass, removing the surface zinc, leaving the copper exposed.

DIE   a solid or split block of metal used for cold drawing.

DIE   a set of metal blocks used for blanking, coining, or forging various shapes.

DIE CASTING   a method of casting which makes use of external pressure to shape the molten metal.

DIFFUSION migration of atoms through a solid metallic lattice. It takes place between cores and encasements of solid-solution alloys and is the basis for heat treating many alloys.

DISSOLVED CARBON   carbon in solution in either the liquid or solid state.

DOLOMITE   a mixed carbonate of magnesium and calcium. Calcined dolomite is used as a basic refractory for withstanding high temperatures and attack by basic slag in metallurgical furnaces.

DOUBLE BELL AND HOPPER   a contrivance located in the furnace top of a blast furnace to prevent the escape of gas during the charging of the furnace. It also permits even distribution of the charge into the stack.

DOWNCOMER   the pipe attached to the top of the stack of the blast furnace which conducts the hot gases to the auxiliary units.

DRAFT   taper allowed on the vertical faces of a pattern to permit removal from the sand mold without excessive rapping and tearing of the mold walls.

DRAG   the lower or bottom section of a mold or pattern.

DRAWING TEMPER or DRAWING BACK the operation of tempering hardened steel by heating it to some specific temperature and quenching in order to obtain some definite degree of hardness.

DRIFT   a level or tunnel pushed forward underground in a metal mine, for purposes of exploration or exploitation.

DROP FORGING   the process of shaping metal parts by forging between two dies, one fixed to the hammer and the other to the anvil of a steam or mechanical hammer. It is used for mass production of such parts as connecting rods, crankshafts, and similar articles.

DROSS   similar to slag but consisting of metallic oxides that rise to the surface in metallurgical oxidation processes.

DRY SAND MOLD   a mold made of prepared molding sand dried thoroughly before being filled with molten metal.

DUCTILITY   the property permitting permanent deformation by stress in tension without rupture.

DUPLEX STEEL   a steel which is refined in two stages. In one method the steel is first refined partially in a bessemer converter before final refinement in an open-hearth furnace. In another method the steel is refined partially in either the bessemer converter or open-hearth furnace before final refinement in an arc furnace.

DUST CATCHER   a cylindrical chamber, part of the auxiliary apparatus of the blast furnace, in which the direction of flow of the gases of combustion and of decomposition is reversed suddenly, causing some of the dust to settle on the bottom.

ELASTIC   DEFORMATION   deformation which is removed with removal of load, permitting the body to revert to its original form.

ELASTIC LIMIT   the limit of an applied stress, which, if exceeded will cause permanent deformation. For commercial purposes elastic limit is considered as yield strength.

ELECTRIC FURNACE STEEL   steel produced in electrically heated furnaces of arc, induction, or resistance type.

ELECTRON   a negatively charged particle that revolves about the heavy positive nucleus of an atom.

ELEKTRON   the trade name for magnesium alloys in Europe.

ELONGATION the amount of permanent extension in the vicinity of the fracture in the tension test; usually expressed as a percentage of the original gage length, such as 25 per cent in 2 inches. It may also refer to the amount of extension at any stage in any process which continuously elongates a body, as in rolling.

ENDOTHERMIC REACTION a chemical reaction accompanied by the absorption of heat.

ENDURANCE LIMIT a limiting stress, below which metal will withstand without fracture an indefinitely large number of cycles of stress. Above this limit failure occurs by the generation and growth of cracks until fracture results in the remaining section.

ENERGIZER a chemical mixed with charcoal to increase the speed with which steel will absorb carbon in the carburizing process.

EQUILIBRIUM DIAGRAM (CONSTITUTIONAL DIAGRAM) a curve that shows all relationships between composition and temperature of the various constituents in an alloy when that alloy is in equilibrium, its most stable and permanent condition.

ETCHING the process of revealing the structure of metals and alloys by attacking a highly polished surface with a reagent that has a differential effect on different crystals or different constituents.

EUTECTIC ALLOY the composition in an alloy system at which two descending liquidus curves in a binary system intersect at a point. Such an alloy has thus a lower melting point than neighboring compositions.

EUTECTOID a solid solution of any series which cools without change to its temperature of final composition.

EUTECTOID STEEL a steel of the eutectoid composition. Composition S on the iron-carbon diagram. This composition in pure iron-carbon alloys is 0.83 per cent carbon, but variations from this composition are found in commercial (impure) steels, and particularly in alloy steels in which the eutectoid composition is usually lower.

EXOTHERMIC REACTION a chemical reaction accompanied by evolution of heat.

EXTRUSION a method of shaping solid metal by forcing it through a die.

FACE-CENTERED CUBIC STRUCTURE an arrangement of atoms in crystals in which the atomic centers are disposed in space in such a way that they may be supposed to be situated at the corners and the middle of the faces of a set of cubic cells.

FACING refractory material applied to the face of a mold.

FATIGUE the phenomenon of progressive fracture of metal by a crack that spreads under repeated cycles of stress.

FERRITE nearly pure iron which contains less than 0.05 per cent of carbon and minor amounts of other elements. In its exact meaning, a solid solution of any element in alpha iron.

FERROMANGANESE an alloy of iron and manganese consisting of 80 per cent manganese, 12 per cent iron, $6\frac{1}{2}$ per cent carbon, $1\frac{1}{2}$ per cent silicon, which is added to the ladle to deoxidize steel and to increase the content of manganese and carbon in the steel.

FIBER a characteristic of wrought metal manifested by a fibrous or woody appearance of fractures and indicating directional properties. Fiber is caused principally by extension in the direction of working of the constituents of the metal, both metallic and non-metallic.

FILLET an arc connecting two surfaces converging at less than 180 degrees.

FILM TEST the time required for a spoonful of metal to skin over.

FIN a thin piece of metal projecting from a casting at the parting line or at the junction of cores, or of cores and mold, etc.

FINE GRAIN the grain obtained when simultaneous crystallization occurs on many nuclei during solidification.

FINE GRAIN STEELS steels which resist grain growth over a considerable temperature range, when held at temperature for a reasonable length of time as is customary in heat treatment of steel.

FINISH. See machine finish allowance.

FIRE REFINING subjecting a liquid melt to slow, limited oxidation so that the impurities are oxidized but only a minimum of the metal oxidizes.

FLAME HARDENING a surface hardening method in which a highly concentrated oxyacetylene flame is passed rapidly over the portions of the pieces of work which are to be hardened. The flame is followed immediately by a jet of water or an air blast to quench the heated surface layer.

FLASH a protrusion or overfill of excess metal in the form of a fin, usually occurring on forgings made in dies and sometimes on semifinished rolled products. It is the result of excess metal forcing out at the parting of dies and rolls.

FLASK a wooden or metal frame in which a mold is made.

FLOTATION an ore concentration process in which air is blown into a mixture of ore pulp, water, oil, and various chemicals. The oil forms a film on the mineral particles and air bubbles adhere to this. Thus the mineral particles are floated while other matter sinks.

FLUORESCENCE the unique property of emitting light during activation by ultraviolet light.

FLUORSPAR commercial grade of calcium fluoride.

FLUSH SLAG see runoff slag.

FLUX a material added to a charge to react with the gangue and to reduce its melting point to such a value that the entire charge melts and produces a slag sufficiently fluid to be handled readily.

FLUX a material which dissolves oxides that form on the surface of a metal being heated preparatory to soldering, brazing, galvanizing, etc., thus affording increased contact of the metals. The surfaces being joined must be wet by it.

FOLLOW BOARD board shaped to fit the pattern and form the joint.

FOREHEARTH a reservoir, connected to the cupola by a short spout, which holds a supply of the molten metal.

FORGING a mechanical method for hot working steel into specific shapes or parts. It may be accomplished by forcing plastic metal to flow in dies and conform to the shape of the impressions which have been sunk in the dies.

FORGING STEELS steels that contain between 0.3 per cent and 0.6 per cent carbon, and which may be heat treated to increase hardness and strength.

FRACTURE TEST a test for carbon in which a specimen of metal is drawn off, cooled rapidly in water, and broken with a sledge hammer. The appearance of the metal exposed in the fracture permits a fairly accurate estimate of carbon content or the presence of internal defects.

FRANKLINITE a zinc ore, practically free of sulfur, which is mined in Franklin, New Jersey.

FREE FERRITE ferrite in steel or cast iron other than that associated with cementite in pearlite.

FROTH FLOTATION separating of finely crushed minerals from one another by causing some to float and others to sink in a froth. Oils and various chemicals are used to activate, make floatable, or to depress the minerals.

FULL ANNEALING heating iron-base alloys above the critical temperature range, holding the alloy above that range for a proper period of time and then slowly cooling it to below the range either in the furnace or in some thermal insulating material.

GAGGER (JAGGER) an L-shaped rod used for reinforcing sand in the cope mold.

GALVANIZING a process used to coat iron or steel with zinc.

GAMMA IRON one of the allotropic forms of iron which crystallizes in the face-centered cubic lattice form. Its range of stability when pure is 2552 to 1670°F. It dissolves carbon up to 1.7 per cent. Its range of stability is lowered by carbon, nickel, and manganese, and it is the basis of the solid solution known as austenite.

GANGUE a claylike mixture of oxides of aluminum and silicon, found as an impurity in iron ore.

GAS CARBURIZING a method of carburizing carried out in an atmosphere of carburizing gases, including carbon monoxide and such hydrocarbons as butane, ethane, methane, and propane.

GATE specifically, the point where molten metal enters the casting cavity. Sometimes employed as a general term to indicate the entire assembly of connected columns and channels carrying the metal from the top of the mold to that forming the casting cavity proper. Term also applied to the pattern parts which form the passages, or to the metal that fills them.

GATE CUTTER a U-shaped piece of thin metal used to cut a shallow trough in a mold which will serve as a passage for the hot metal.

GATED PATTERNS one or more patterns with gating systems attached. They are so assembled that they can be molded at one time.

GRAINS crystals in metals.

GRAIN BOUNDARY a layer only a few atoms thick which constitutes the zone of contact between adjacent grains having differently oriented space lattices.

GRAIN GROWTH an increase in the average grain size resulting from some crystals absorbing adjacent ones when the metal is raised to a temperature above that necessary for recrystallization and kept at that temperature for a sufficient length of time.

GRAIN SIZE this is usually defined as the average diameter of the grains or by the number of grains per unit area in any cross section under examination. In steel, grain size affects the critical rate of cooling; in alloys, the ductility.

GRANULAR PEARLITE (GLOBULAR and DIVORCED PEARLITE) a structure formed from ordinary lamellar pearlite by long annealing of steel or at a temperature below but near to the lower critical point, causing the cementite to spheroidize in a ferrite matrix.

GRAPHITIZATION decomposition of carbide to give free carbon as graphite or as temper carbon.

GRAVITY CASTING a casting process in which the metal flows into the mold by gravity.

GRAY CAST IRON cast irons which as cast have combined or cemetitic carbon not in excess of a eutectoid percentage—the balance of the carbon occurring as graphite flakes. The term *gray iron* is derived from the characteristic gray fracture of this metal.

GREEN SAND prepared molding sand in the moist or *as-mixed* condition.

It is sand in its natural condition that contains just enough clay and water to hold it together.

GREEN SAND MOLD   a mold made from moist refractory· sand that is not dried before use.

GUIDE   the strip or other suitable device used to locate the cope in the proper place on the drag.

GYRATORY CRUSHER   a device which has a fixed crushing surface in the form of a frustrum of an inverted cone, within which is a moving crushing surface in the form of a frustrum of an erect cone. The ore is crushed in the downward converging annular space between the two crushing surfaces.

H-STEELS   steels made under specifications that include hardenability tolerances.

HALL-HÉROULT PROCESS   an electrolytic process in which alumina is dissolved in cryolite and then reduced to metallic aluminum by means of electricity.

HARDENABILITY   the depth to which steel can be hardened to martensite under stated conditions of cooling.

HARDNESS   resistance to plastic deformation by penetration, scratching or bending.

HEARTH   in the blast furnace, the zone or basin at the bottom of the furnace into which molten pig iron and slag trickle and are stored until tapped off.
In other furnaces, the basin which holds the charge during the refining period.

HEAT-RESISTANT STEEL   steel that resists scaling (oxidation) and creep when heated to a high temperature.

HEAT TREATMENT   an operation or combination of operations involving the heating and cooling of a metal or an alloy in the solid state for the purpose of obtaining desirable conditions or properties. Heating and cooling for the sole purpose of mechanical working are excluded from the meaning of this definition.

HEMATITE   the iron ore mined in greatest quantity in the United States; it is essentially ferric oxide.

HETEROGENEOUS FIELD or STRUCTURE   the field between the liquidus and solidus curves in an equilibrium diagram. In this field any alloy is partly liquid and partly solid and is said to be in the mushy state and not uniform in chemical composition.

HIGH-SPEED TOOL STEEL   a hard steel used to make tools that can cut while red hot and running at high speeds. It permits fast working and deep cuts while hot.

HOMOGENIZING   a heat treating process to make cast solid solution alloys uniform. It is carried out at a temperature below the melting point.

HOT COLLAR   see SHRINKHEAD.

HOT JUNCTION   the measuring junction or junction of the thermocouple placed at the point of measurement.

HOT ROLLING   see HOT WORKING.

HOT SHORTNESS   brittleness in metal when hot. Steels that are hot short have a tendency to crack and tear during rolling.

HOT TOP   see SHRINKHEAD.

HOT WORKING (HOT ROLLING or HOT FORGING)   work carried out at a temperature at or above the recrystallization temperature of the metal in question. During hot working the grains are constantly being deformed and broken up but new ones are constantly forming to take their place.

HYDROMETALLURGY   extraction of metals by leaching with an aqueous solvent that dissolves the valuable metal without attacking the gangue and permits subsequent recovery of the dissolved metal from the solution. It is really a large-scale application of the process known as extraction in the chemical laboratory—the separation of a soluble substance from an insoluble one by means of a solvent.

HYPEREUTECTOID STEEL a steel containing more carbon than the eutectoid composition, which is about 0.83 per cent.

HYPOEUTECTOID STEEL a steel containing less carbon than the eutectoid composition, which is about 0.83 per cent.

IDIOMORPHIC CRYSTAL a crystal of definite external shape.

IMPACT TEST a test in which one or more blows are applied suddenly to a specimen. Results usually are expressed in terms of energy absorbed or number of blows (of a given intensity) required to break the specimen. Izod impact or charpy impact are the common methods of studying impact resistance.

INCLUSIONS nonmetallic matter in metals.

INDUCTION HARDENING a surface hardening process based upon the heating effect produced in a piece of steel by placing it in a high-frequency electric field generated in a suitably shaped water-cooled coil of copper wire or tubing.

INGOT IRON an open-hearth product low in carbon, manganese, and impurities.

INGOT MOLD a thick-walled cast-iron mold in which molten metal is cast and allowed to solidify to form an ingot.

INGOTISM a coarse dendritic structure that forms when metals solidify.

INGOTS metal castings of uniform sizes and shapes for subsequent rolling, forging, or processing.

INOCULATION a process of adding some material to molten gray cast iron in the ladle for the purpose of controlling the structure to an extent not possible by control of chemical analysis and other normal variables.

INTERGRANULAR (INTERCRYSTALLINE) FRACTURE metal fractures that follow the crystal boundaries instead of passing through the crystals, as in the usual transcrystalline fracture.

INTERMETALLIC COMPOUND a constituent of alloys that is formed when atoms of two metals combine in certain proportions to form crystals with a different structure from that of either of the metals. The proportions of the two kinds of atoms may be indicated by formulas, as for example (CuZn).

INTERNAL STRESS a residual stress that may exist between different parts of metal products as a result of the differential effects of heating, cooling, working operations, or of constitutional changes in the solid metal. It may be relieved by heating to a low temperature without affecting the mechanical properties.

INTERRUPTED QUENCH a quench that is not carried through to the temperature at which transformation of austenite into martensite commences, but is interrupted at some higher temperature, in order to suppress transformation of austenite into pearlite and at the same time avoid formation of martensite.

INTERSTITIAL SOLID SOLUTION a solid solution in which the solute (stranger) atoms occupy positions in the host lattice and by becoming part of the lattice distort it.

INVESTMENT the refractory material that completely clothes or invests the model in precision casting.

IRON-CARBON CONSTITUTIONAL DIAGRAM a constitutional diagram each point of which represents the composition of steel or cast iron that is in equilibrium and that contains only iron and carbon.

ISOTHERMAL QUENCH an interrupted quench requiring three steps.

ISOTHERMAL TRANSFORMATION DIAGRAMS diagrams that illustrate graphically the changes that take place when steel is cooled at various rates.

ISOTROPIC a material lacking directional properties.

IZOD TEST a notched-bar or impact test in which a notched specimen held in a vise is struck on the end by a striker carried on a pendulum; the energy absorbed in fracture is obtained from the height to which the pendulum rises.

JOLTER (JOLT RAMMING or JAR RAMMING) machine for ramming sand in a flask by repeated jarring or jolting action.

JOMINY TEST a standardized procedure by which the hardenability of a steel is determined.

KILLED STEEL molten steel held in a ladle, furnace, or crucible (and usually treated with aluminum, silicon, or manganese) until no more gas is evolved and the metal is perfectly quiet. When a killed steel solidifies the top surface of the ingot freezes immediately and subsequent shrinkage produces a central pipe. The steel is sound, and free from blowholes and segregation.

KISH graphite thrown out by liquid cast iron in cooling.

LADLE a vessel lined with refractory material; used for conveying molten metal from the furnace to the mold or from one furnace to another.

LAMELLAR PEARLITE pearlite crystals covered by a thin layer of cementite.

LAMINATED STRUCTURE a structure containing alternate layers of different forms of iron, such as pearlite.

LANCE a long slender steel pipe connected to an oxygen line and pushed into the taphole to ignite the carbon and thus melt through the crust of iron which has solidified around the bottom of the hearth.

LAP WELD a term applied to a weld formed by lapping two pieces of metal and then pressing or hammering, particularly to the longitudinal joint produced by a welding process for tubes or pipe in which the edges of the skelp are beveled or scarfed so that when they are overlapped they can be welded together. The product is known as a lap-weld or lap-welded pipe.

LAW OF HORIZONTALS the law used to determine the composition of the liquid of any alloys, in the series of a constitutional diagram, at any stated temperature. It applies to any two-phase heterogeneous field.

LEACHING the extraction of a soluble metal compound from an ore by dissolving in a solvent that does not dissolve the gangue. The metal is subsequently precipitated from the solution.

LEDEBURITE the cementite-austenite eutectic forming at point C on the iron-carbon diagram. During cooling the austenite in ledeburite may transform to ferrite and cementite. It is found in cast iron and high-speed steels.

LEVER ARM PRINCIPLE a principle used to determine the amount of phase or constituent present at any stated temperature on a constitutional diagram.

LIME BOIL that period in the open-hearth process when carbon dioxide gas released from the limestone on the hearth bottom causes the bath to bubble violently.

LIMONITE hydrated ferric oxide ($2 Fe_2O_3 \cdot 3 H_2O$), an important iron ore mineral.

LIQUATION separation of a metal from dross by heating to the melting point of the metal. The metal is permitted to flow out while the dross is left behind.

LIQUIDUS   the upper curve in a constitutional diagram which is the locus of temperatures at which each alloy starts to solidify.

LIXIVIATION   identical with leaching.

LOAM   a coarse, strongly bonded molding sand used for loam and dry sand molding.

LOAM MOLDING   a system of molding, especially for large castings wherein the supporting structure is constructed of brick. Coatings of loam are applied to form the mold face.

LOCAL CELLS   cells set up during corrosion by impurities in the metal, causing a difference in potential and therefore corrosion.

LOOSE PATTERNS   patterns not attached to a frame or rigid body.

LOST WAX PROCESS   a method of casting in which an expendable wax pattern is used for making the molds. See precision casting.

MACHINABILITY   a property of metals that permits them to be cut, turned, broached, or otherwise formed by machine tools.

MACHINE FINISH ALLOWANCE   the allowance made for machining the finished part.

MACHINING STEEL   steel having a carbon content of only 0.1 or 0.2 per cent, which is used widely for carburizing purposes.

MACROETCHING OF IRON AND STEEL   subjecting the metal to the action of a reagent in order to bring out the structure for visual inspection.

MACROSCOPIC   visible either to the naked eye or under low magnifications (up to about 10 diameters).

MACROSTRUCTURE   the structure and internal condition of metals as revealed on a ground or polished (and sometimes etched) specimen, by the naked eye or under low magnifications (up to about 10 diameters).

MAGNAFLUX TEST   a magnetic test used to detect defects on or near the surface of metals.

MAGNESITE   a natural magnesium carbonate used as a basic refractory in open-hearth and other high-temperature furnaces; it is resistant to attack by basic slag.

MAGNETIC PERMEABILITY   the ratio of the magnetic induction of a substance to the magnetizing field to which it is subjected.

MAGNETITE   an iron ore mineral which has the chemical composition Fe_3O_4.

MALLEABILITY   the property of being permanently deformable mechanically by rolling, forging, extrusion, etc., without rupture and without pronounced increase in resistance to deformation (as in case of ductility). Malleability generally increases at elevated temperature.

MALLEABLE IRON   a mixture of iron and carbon including smaller amounts of silicon, manganese, phosphorus, and sulfur, converted structurally by heat treatment into a matrix of ferrite containing nodules of temper carbon.

MALLEABLEIZING   an annealing operation performed on white cast iron partially or wholly to transform the chemically combined carbon to temper carbon, and in some cases wholly to remove the carbon from the iron by decarburization. Temper carbon is free graphitic carbon in the form of rounded nodules composed of an aggregate of minute crystals.

MARTEMPERING   an interrupted quenching process the purpose of which is to produce a fully martensitic structure.

MARTENSITE   a supersaturated solid solution of carbon in ferrite.

MATCH   a form of wood, plaster of Paris, sand, or other material on which an irregular pattern is laid while the drag is rammed.

MATCHPLATE a metal or other plate on which patterns split along the parting line are mounted back to back with the gating system to form an integral piece.

MATRIX the principal substance in which a constituent is embedded.

MATTE a solution of mixed sulfides produced in smelting sulfide ores.

MECHANICAL PROPERTIES properties of a metal determining its behavior under stress.

MECHANICAL WORKING subjecting metal to pressure exerted by rolls, presses, or hammers, to change its form, or to affect the structure and therefore the mechanical and physical properties.

METALLIZING a method used to spray zinc on large articles.

METALLOGRAPHY the branch of metallurgy which deals with the study of the structure and constitution of solid metals and alloys, and the relation of this to properties on the one hand and manufacture and treatment on the other.

METALLOID in metallurgical practice, elements such as carbon, silicon, phosphorus, sulfur, and manganese, which are commonly present in small amounts in iron and steel.

METALLURGIST a technically trained person who has specialized in that science which deals with the separation of metals from the earthy materials with which they are combined.

METALLURGY the art and science of extracting metals from their ores and other metal-bearing products and adapting these metals for human utilization.

MICROSCOPIC visible under a magnification of about 10 diameters or more.

MICROSTRUCTURE the structure and internal condition of metals as revealed in polished, and usually etched, samples when examined under a microscope magnifying 10 diameters or more.

MILL SCALE a pure oxide of iron which forms on the surface of steel as it cools and which is removed when the steel is rolled in the rolling mill.

MILLING reduction in the size of ore particles by crushing and grinding, and classification of these particles by size and specific gravity.

MINERAL an inorganic substance occurring in nature with a characteristic chemical composition, and usually possessing a definite internal atomic structure which may produce a typical external form called a crystal.

MIXER a large vessel used to store molten pig iron coming from the blast furnace until it is required in one of the steel furnaces in the same plant.

MOLD the form containing the cavity into which molten metal is poured to make a casting.

MOLD BOARD the board on which the pattern is placed when ramming the drag.

MOLD WASH usually an aqueous emulsion containing various organic or inorganic compounds or both, which is used to coat the face of a mold cavity. Materials include graphite, silica flour, etc.

MOTTLED white iron structure interspersed with spots or flecks of gray.

MOUNTED PATTERNS patterns mounted in a frame or on a plate.

NITRIDING a hardening operation during which nitrogen is added to iron-base alloys by heating the metal in contact with ammonia gas or other suitable nitrogenous material. Nitriding is conducted at a temperature usually in the range 935°F to 1000°F and produces surface hardening of the metal without quenching.

NORMALIZING heating iron-base alloys to approximately 100°F above the critical temperature range, followed by cooling to below that range in still air at ordinary temperature.

NUCLEI points at which crystals begin to grow during solidification. In general, they are minute crystal fragments formed spontaneously in the melt, but frequently nonmetallic inclusions act as nuclei.

OIL-HARDENING STEELS alloy steels, the S-curves of which lie to the right, and which therefore can be cooled slowly either in oil or air instead of in water.

OPEN-HEARTH FURNACE a refractory-lined, shallow-bath, rectangular furnace in which both hearth and charge are subjected to direct action of the fuel flame. The fuel may be producer gas, natural gas, coke oven gas, powdered coal or oil.

OPTICAL PYROMETER a temperature-measuring device through which the observer sights the heated object and compares its incandescence with that of an electrically heated filament the brightness of which can be regulated.

ORE a natural mineral deposit from which a useful, valuable metal can be extracted profitably.

ORE BOIL that period in the open-hearth process when the iron ore and the iron oxides in the slag act on the carbon of the pig iron to form carbon monoxide gas, causing the bath to froth and rise.

ORE DRESSING a process of mechanically eliminating a portion of the waste material present in the ore, thus increasing the metal content of the product to be sent to the smelter.

OXIDATION in the restricted sense in which the term is used in a discussion of smelting processes, the combination of an element with oxygen to form an oxide, or the combination of an oxide with more oxygen to form a higher oxide.

PACK CARBURIZING a method of carburizing in which the articles to be carburized are clean and packed loosely in a metal box with carbonaceous material or with a commercial carburizing compound.

PANCAKE TEST a rapid method of testing the iron oxide and the basic qualities of a slag by using a small shallow spoon in which the slag forms a pancake.

PARTING LINE the line along which a pattern is divided for molding, or along which the sections of a mold separate.

PARTING SAND sharp or burned sand sprinkled on the joint of the mold to prevent the sand cope and drag from adhering.

PATENTING heating iron-base alloys above the critical temperature range followed by cooling below that range in air, molten lead, or a molten mixture of nitrates and nitrites. It is really normalizing carried out at a high temperature.

PATTERN model of wood, metal, plaster, or other material used in making a mold.

PATTERNMAKER'S SHRINKAGE shrinkage allowance made on all patterns to compensate for the change in dimensions as the solidified casting cools in the mold from freezing temperature of the metal to room temperature. Pattern is made larger by the amount of shrinkage characteristic of the particular metal in the casting and the amount of hindered contraction to be encountered. Rules or scales are available for use.

PEARLITE the lamellar aggregate of ferrite and carbide resulting from the direct transformation of austenite at A_{r_1}. It is recommended that this word be reserved for the microstructures consisting of thin plates or lamellae—that is, those that may have a pearly luster in white light. The lamellae can be very thin and resolvable only with the best microscopic equipment and technique.

PEEN small end of a molder's rammer.

PERITECTIC REACTION a reaction that occurs between a solid and liquid phase resulting in a second solid phase. A typical example is the decomposition of an intermetallic compound before it reaches its melting point.

PERMANENT DEFORMATION deformation which remains after an externally applied load is removed from a body.

PERMANENT MOLD a foundry mold, made of metal or refractory material, that is capable of producing a large number of castings identical in shape.

PERMEABILITY the property in sand molds which permits passage of gases.

PHASE a constituent that is completely homogeneous both physically and chemically, separated from the rest of the alloy by definite bounding surfaces; for example, austenite, ferrite, cementite. Not all constituents are phases, pearlite for example.

PHYSICAL PROPERTIES those properties familiarly discussed in physics, exclusive of those described under mechanical properties; for example, density, electric conductivity, coefficient of thermal expansion.

PICKLING removing oxide scale from metal objects by immersion in a diluted acid bath so as to obtain a chemically clean surface preparatory to cold rolling or wire drawing.

PIG IRON the product obtained by reduction smelting in the blast furnace.

PINHOLES very small gas cavities generally found in alloy castings which

are probably caused by the evolution of occluded gases during the process of solidification. They show up during machining.

PIPE a shrinkage cavity formed in metal (especially ingots) during solidification of the last portion of liquid metal. Contraction of the metal causes this cavity or pipe.

PLAIN CARBON STEEL see CARBON STEEL.

PLASTIC DEFORMATION permanent change in the shape of a piece of metal, or in the constituent crystals, brought about by the application of mechanical force which causes the crystals to slide or glide along slip planes.

PLATES (COMMERCIAL DEFINITION) flat rolled steel.

POLING insertion of timber or green trunks of hardwood trees into the molten metal during fire refining.

POLYCRYSTALLINE an aggregate of crystal grains. Most metals are polycrystalline and therefore stronger than single crystals because of slip interference.

PORPHYRIES large bodies of low-grade ores in which the copper minerals are disseminated relatively uniformly.

POURING BASIN reservoir on top of the mold to receive molten metal.

POWDER METALLURGY the art of converting metals or alloys into powders, plus the compressing or otherwise forming of these powders into desired articles, and the heat treating of these articles to provide desired physical properties.

PRECIPITATION HARDENING (AGE HARDENING, AGING) the phenomenon which results in an increase in hardness with the passage of time at room or elevated temperature. The increase is produced by a change in structure associated with precipitation of a constituent from solid solution along the grain boundaries.

PRECISION CASTING a casting process in which the wax pattern used is destroyed by melting and burning in order to remove it from the mold.

PRIMARY DEPOSITS the first step involved in the accumulation of iron in the earth's crust into iron formations known as *primary* deposits.

PROCESS ANNEALING same as subcritical annealing.

PROPORTIONAL LIMIT the load per unit area beyond which the increases in strain cease to be directly proportional to the increases in stress.

PULLOVER MILL a rolling mill using a single pair of rolls. The metal, after passing through the rolls, is pulled back over the top roll in order to be fed through the mill a second time.

PYROMETALLURGY a metallurgical process which uses fuel as the source of heat.

PYROMETER an instrument for determining elevated temperatures.

QUENCHING rapid cooling of steel from above the critical range by immersion in liquids or gases, or by contact with metal, in order to harden it.

RAISES short vertical shafts extended upward from different depth headings or haulways which are used as chutes for handling the mined material from the upper levels of an iron ore mine.

RAM to pack the sand in a mold.

RAMMER a hand tool for packing the sand of a mold evenly round the pattern. One edge of the rammer is wedge-shaped and is called the peen end; the other edge is flat and is called the butt end.

RECALESCENCE the phenomenon of steel becoming brighter in color, for a limited time, by the steel heating itself spontaneously as it is cooling through the critical range. This causes a retardation of the cooling, which is shown on cooling curves and can be seen by the eye in a darkened room.

RECARBURIZER any carbonaceous material, pig iron, or alloy added to molten steel to increase the carbon content.

RECRYSTALLIZATION formation of new crystals or grains from deformed metal, accomplished by suitable heat treatment.

RED SHORTNESS brittleness in steel when it is red hot.

REDUCING SLAG a slag which aids in the removal of oxygen.

REDUCTION partial or complete removal of oxygen from an oxide.

REFRACTORIES materials capable of resisting high temperatures, changes of temperature, the action of molten metals, and slags, and hot gases carrying solid particles. They are used to line furnaces.

REGENERATIVE SYSTEM a system in which the waste heat of the escaping gases is used to preheat the incoming air. The regenerator is a chamber containing hot firebrick loosely stacked and laid in a checkerboard pattern.

RETAINED AUSTENITE the austenite that remains in tool steels after quenching. It usually can be eliminated by suitable tempering.

REVERBERATORY FURNACE a furnace in which the charge is melted on a shallow hearth by flame passing over the charge and heating a low roof. Firing may be with coal, pulverized coal, oil, or gas. Much of the heating is done by radiation from the roof.

REVERSING MILL a two-high mill in which a bar is passed back and forth between the rolls by reversing the direction of rotation of the rolls.

RIDDLE hand- or power-operated device for removing large particles of sand or foreign material from foundry sand.

RIMMED STEEL an incompletely deoxidized steel normally containing less than 0.25 per cent carbon. A rimmed steel possesses 3 fairly well defined zones in its cross section: (a) an outer wall of clean, solid metal which is lower in carbon than the inner zone; (b) an intermediate zone that contains gas pockets or blowholes that vary in size and extent and that are welded during rolling; and (c) a central zone that contains a considerable concentration of metalloids.

RISER

(a) an opening in the cope into which the metal rises when the mold is filled.

(b) that part of the casting formed in the opening.

(c) reservoir of molten metal attached to the casting to compensate for the internal contraction of the casting as it solidifies. (It also allows dirt to escape and indicates that the mold is full.)

ROASTING the operation of heating sulfide ores in air to convert to oxide.

ROCKWELL HARDNESS TEST a method of determining the hardness of metals by indenting them with a hard steel ball or a diamond cone under a specified load, measuring the depth of penetration, and subtracting the latter from an arbitrary constant. Rockwell hardness numbers are based on the difference between the depths of penetration at major and minor load; the greater this difference, the less the hardness number.

ROD MILL a mill for rolling rods from billets.

RODS wire rods are semifinished hot-rolled rounds of great length, usually coiled, and used principally for drawing to wire.

ROLLING MILLS mills in which a preheated steel ingot is passed between heavy chilled cast steel rolls.

RUNNER that portion of the gate assembly connecting the downgate or sprue with the casting.

s-CURVE same as isothermal transformation diagrams.

SAGGERS or RINGS metal pots in which castings that are to be malleableized are placed for annealing.

SALT BATH a molten bath of special chemical salts used for heating metal, for hardening or tempering. Salt baths give uniform heating and prevent oxidation.

SAND SLINGER molding machine that throws sand into a flask or core box by centrifugal action.

SCARFING (DESEAMING) removal of seams and other surface defects by cutting with the gas torch; also beveling skelp with a cutting tool.

SCLEROSCOPE an instrument for determining hardness which is measured by the drop and rebound of a diamond-tipped hammer.

SCREW-DOWN MECHANISM a device used to adjust the distance between rolls.

SEAM a crack on the surface of metal that has been closed but not welded; usually produced by blowholes which have become oxidized.

SEAMLESS TUBE a tube other than that made by bending over and welding the edges of flat strip.

SEASON CRACKING stress corrosion of brass which has been cold worked. Ammonia aids it. Low temperature annealing relieves the stresses without affecting the mechanical properties.

SECONDARY DEPOSITS iron ore deposits which passed through a process of natural concentration, thus raising their iron content in places.

SECONDARY HARDNESS a further increase in hardness developed by tempering high-alloy steel after quenching.

SEGREGATION nonuniform distribution of impurities, inclusions, and alloying constituents in metals.

SELF-HARDENING STEEL a steel carrying sufficient carbon or alloy content to produce hardening on cooling in air, without the necessity for quenching in oil or water. The alloying elements lower and retard the normal transformation from austenite to pearlite.

SELF-LUBRICATING BEARINGS oil-soaked porous bearings in which the pores serve as reservoirs for the oil which, later, is brought to the surface by capilliary action.

SEMIDEOXIDIZED STEEL a steel which gives off but little gas during solidification in the ingot mold, thus producing steel free from surface blowholes and piping.

SEMIFINISHED STEEL blooms, billets, slabs, sheet bars, rods, and other products, for rerolling or forging.

SEMIKILLED STEEL a type of steel obtained when deoxidation is not complete.

SHEAR STRESSES (TANGENTIAL) stresses effective in a direction along the plane of application.

SHEARED PLATE MILL a mill having horizontal rolls used for rolling ingots and slabs to plates, all margins of which are irregularly formed and require shearing to produce the finished plate.

SHEET MILL a mill that ordinarily rolls sheet bar to sheets.

SHEETS
Cold-rolled: the flat products resulting from cold rolling, after pickling, of sheets previously produced by hot rolling.
Hot-rolled: the flat rolled products resulting from reducing sheet bars on a sheet mill.

SHERARDIZING the process of coating small finished parts of iron and steel, such as nuts, screws, and bolts, with a corrosion-resistant layer of zinc.

SHOCK RESISTANCE the manner in which a metal reacts when subjected to sudden shock.

SHOTTING pouring molten metal through a screen contained in a tall tower before dropping it into water or bins.

SHRINK HOLE a hole or cavity in a casting resulting from contraction and insufficient feed metal, and formed during the time the metal changed from the liquid to the solid state.

SHRINK RULE patternmaker's rule graduated to allow for metal contraction.

SHRINKAGE ALLOWANCE the compensation required in a pattern to take care of the natural shrinkage of a solidified casting as it cools in the mold.

SHRINKAGE CAVITY see SHRINK HOLE.

SHRINKHEAD or HOT TOP the heat-insulated reservoir for excess metal on top of an ingot mold which feeds the shrinkage of the ingot that occurs during solidification.

SHRINKHEAD CASING see HOT COLLAR.

SIEMENS-MARTIN PROCESS another name for the open-hearth process.

SILICA silicon dioxide used to manufacture refractory materials. When the latter contain more than 90 per cent silica they are known as acid refractories (for example, ganister) and are used in open-hearth and other metallurgical furnaces to resist high temperatures and attack by acid slags.

SINTERED CARBIDES see CEMENTED CARBIDES.

SINTERING the fritting together of small particles to form larger particles, cakes, or masses; in case of ores and concentrates it is accomplished by fusion of certain constituents.

SINTERING (POWDER METALLURGY) a heating operation in which a briquette is heated until the particles bond together and until it hardens sufficiently to permit handling and, when necessary, shaping.

SKELP mill steel strip from which tubes are made by drawing through a bell at welding temperature, to produce lap-welded or butt-welded tubes.

SKIM GATE an arrangement which changes the direction of flow of molten metal in the gating system and thereby prevents passage of slag and other extraneous materials beyond that point.

SKIN the surface of a mold or casting.

SKIN-DRIED MOLD a green sand mold, the face of which is sprayed with an additional special bonding material and then dried by rapid application of localized heat to the depth of a fraction of an inch.

SKIPWAY (SKIP HOIST) a steel incline which runs to the top of the blast furnace.

SLAB an ingot reduced, generally by rolling, to a thickness better suited to the operation that follows. A slab, as distinguished from a bloom, has width at least twice its thickness and a minimum thickness of $1\frac{1}{2}$ inches. It is rerolled to plates and to sheet bar. Slabs are classified as semifinished products.

SLAG a nonmetallic covering on molten metal as the result of the combining of impurities contained in the original charge, some ash from the fuel, and any silica and clay eroded from the refactory lining. Except in bottom-pour ladles it is skimmed off prior to pouring the metal.

SLAG HOLE an opening in a furnace for the removal of slag.

SLIP a displacement of a portion of a grain with respect to another portion.

SLIP INTERFERENCE interference of crystals of a metal with each other during slip. Sometimes called work hardening.

SLIP LINES or BANDS the fine dark lines or bands seen on a stressed metal under the microscope, that is, on crystals of metals which have been cold worked.

SLIP PLANES the particular set or sets of crystallographic planes along which slip or sliding takes place in metal and other crystals during the process of plastic deformation.

SMELTING any metallurgical operation in which the metal sought is separated in a state of fusion from the impurities with which it may be chemically combined or physically mixed.

SOAKING PIT an underground furnace in which a stripped ingot is heated or soaked in heat until it is uniformly heated throughout to its rolling temperature.

SOLDER usually an alloy of two or more metals used for joining other metals together by surface adhesion. Most common solder is an alloy of tin and lead; hard solder is composed of copper and zinc.

SOLDERING the lowest temperature method of joining metals without melting.

SOLID PATTERNS one-piece patterns, so constructed that they can be molded with a single joint.

SOLID SOLUTION an alloy in which metals remain dissolved in each other when solid.

SOLIDIFICATION RANGE the temperature range through which metal freezes or solidifies.

SOLIDUS the lower curve in a constitutional diagram which indicates the temperature at which each alloy has completed solidification.

SOLUTION HEAT TREATMENT the operation of heating suitable alloys (for example, duralumin) in order to take the hardening constituent into solution. This is followed by quenching to retain the solid solution, and the alloy is then age-hardened at atmospheric or elevated temperature.

SPACE LATTICE the orderly geometric form into which atoms tend to arrange themselves during the process of crystallization.

SPALLING cracking and flaking of small particles of metal from the surface.

SPARK ARRESTER device over the top of a cupola to prevent emission of sparks.

SPARK TEST classification of steels according to their chemical analysis by visual examination of the sparks thrown off when the steels are held against a high-speed grinding wheel.

SPHEROIDAL or SPHEROIDIZED CEMENTITE a rounded or globular form of carbide resulting from a spheroidizing treatment. The initial structure may be either pearlitic or martensitic.

SPHEROIDIZING any process of heating and cooling steel that produces a rounded or globular form of carbide.

SPIEGELEISEN a pig iron containing 15 to 30 per cent manganese and 4.5 to 5.5 per cent carbon. It is added to steel as a deoxidizing agent and to raise the manganese content of the steel.

SPLIT PATTERN a pattern divided at the parting line or lines to facilitate molding by eliminating coping down by the molder.

SPRUE the channel that conveys the molten metal from the pouring basin to the runner. Also, the metal which solidifies in these channels and is found attached to the casting after the casting has solidified.

STEADITE a hard phosphorus-rich component found in gray iron, containing 10.2 per cent phosphorus and 89.8 per cent iron.

STEEL an alloy of iron and carbon. It contains up to 1.7 per cent carbon plus minor amounts of other elements such as manganese, silicon, phosphorus, sulfur, and oxygen.

STELLITE a series of alloys containing cobalt, chromium, tungsten, and molybdenum in various compositions. Stellites are used for high-speed cutting tools and for protecting surfaces subjected to heavy wear.

STOPE see CONTRACT.

STOVES steel structures that contain firebrick laid loosely in a checkerwork pattern in order to absorb heat and at the same time permit ready flow of gas used to preheat the air entering the blast furnace.

STRAIN HARDENING increase in hardness and yield strength produced by straining metals.

STRAINS (CASTING) strains produced by internal stresses resulting from unequal contraction of the metal as the casting cools.

STRESS internal forces produced by application of external load, tending to displace component parts of the stressed material.

STRESS RELIEF a thermal treatment in which the locked-up stresses in a bar caused by cold working are removed by heating the bar close to but below the lower limit of the critical temperature range, or to approximately 100°F below the tempering temperature.

STRIKE OFF (STRIKE) a straight edge to cut the sand level with the top of the drag or cope flask.

STRIP (COLD-ROLLED) the flat products resulting from cold rolling, after pickling of strip previously produced by hot rolling.

STRIP (HOT-ROLLED) the flat products resulting from reducing sheet bars by hot rolling on a sheet mill; or slabs, blooms, and billets on a continuous strip mill.

STRIP MILL a mill for rolling slabs, blooms, and billets to strip thickness. Commonly a continuous mill with rolls revolving at high speed in order to finish the rolling at sufficiently high temperature.

SUBCRITICAL ANNEALING heating steel to a temperature below its critical temperature, and subsequently cooling at a rate dependent upon the carbon content; also called *process annealing.*

SUBLIMATION vaporization of a solid without intermediate formation of a liquid.

SURFACE TENSION force that causes the surface of a free liquid to assume a spherical shape.

SWAB a piece of hemp or other material to put water in the sand around the pattern before it is rapped and drawn from the mold or to put liquid facing on the surface of a mold or core.

SWABBING dampening the mold surface next to the pattern to strengthen the sand and cause it to be more plastic and consistent.

TAILINGS a waste product from a mill or concentrator.

TAPHOLE a furnace opening through which the refined molten metal flows.

TAPPING the process of removing molten steel from a melting furnace by opening the taphole and allowing the metal to run out into molds or into a ladle.

TEEMING the operation of filling ingot molds from a ladle of molten metal.

TEMPER CARBON carbon in nodular form, characteristic of malleable iron.

TEMPERATURE DIFFERENTIAL the difference in temperature between the center and surface of a metal during the quenching period.

TEMPERING (also termed DRAWING) re-heating hardened steel to some temperature below the lower critical temperature, followed by any desired rate of cooling, in order to decrease the hardness.

TEMPERING MOLDING SAND mixing and moistening molding sand until it sticks together when squeezed in the hand.

TENSILE STRENGTH the maximum normal load, per unit area, which a material is capable of withstanding before rupturing. It is lowest in the annealed state. Also known as *maximum strength* and *ultimate strength.*

TENSILE STRESSES (COMPRESSIVE STRESSES) stresses effective in a direction perpendicular to the direction of application.

TENSION TEST application of a pulling force to a specimen of material and measurement of the reactions that occur. In steel these reactions occur in two distinct phases: the elastic phase wherein the material is not permanently deformed by the pulling force, and the plastic or yield phase wherein the material becomes either permanently deformed or ruptured.

THERMOCOUPLE a combination of dissimilar metallic conductors so joined that they produce an electromotive force when the junctions are at different temperatures. The junction can be maintained at the temperature which it is desired to measure in terms of the thermoelectric current produced.

THREE-HIGH MILL a mill having three horizontal rolls, one above another. The piece being rolled goes in one direction through the bottom and middle rolls and returns through passes in the middle and top rolls. A three-high mill performs the same service as a two-high reversing mill.

THREE-HIGH PLATE MILL a three-high mill in which the upper and lower rolls are of the same diameter but the middle roll is smaller. The latter can be raised or lowered to work with either of the rolls above.

TIME-TEMPERATURE TRANSFORMATION CURVES same as isothermal transformation diagrams.

TIN PLATE thin sheet steel covered with an adherent layer of tin formed by passing the steel through a bath of molten tin. It resists atmospheric oxidation and attack by many organic acids.

TOOL STEEL a steel used for cutting tools. Tool steels are those which contain more than 0.6 per cent carbon and nominally 0.25 per cent silicon and 0.25 per cent manganese.

TOUGHNESS the ability to withstand load without breaking.

TRANSCRYSTALLINE FRACTURE (FAILURE) the normal type of failure observed in metals. The line of fracture passes through the crystals, and not around the boundaries as in intercrystalline fracture.

TRANSFORMATION a constitutional change in a solid metal; for example, the change from gamma to alpha iron or the formation of pearlite from austenite.

TRANSFORMATION RANGE the range of temperatures at which changes in phase of iron-carbon alloys occur.

TRANSFORMATION TEMPERATURES the temperatures at which changes in phase of iron-carbon alloys occur.

TRANSITION POINT the temperature at which one crystalline form of a substance is converted into another solid modification.

TRANSVERSE FISSURE a physical rupture through or practically through the horizontal section of an ingot.

TRANSVERSE WEAKNESS a weakness through the horizontal section of an ingot.

TREE the pattern formed when several wax patterns are joined together by means of wax in precision casting.

T.T.T. CURVE same as isothermal transformation diagram.

TUYERES openings through which the air blast enters any metallurgical furnace.

TWO-HIGH MILL a mill having two horizontal rolls generally used for rolling rails, structural shapes, bars, etc.

TWO-HIGH REVERSIBLE MILL a two-high mill in which it is possible to reverse the direction in which the rolls are rotating so that the cross section of the ingot is reduced each time the metal passes back and forth.

UNIVERSAL PLATE MILL a mill having horizontal and vertical rolls. The horizontal rolls control the thickness, and the vertical rolls control the width during the rolling of ingots and slabs into plates.

VENT (VENT HOLE) an opening in a mold or core to permit the escape of steam and gases.

VENT ROD (WIRE) a piece of wire or bar used to form the vents in sand.

VENT WAX wax in rod shape placed in the core during manufacture. In the oven the wax is melted out, leaving a vent or passage.

VIBRATOR a device that jars or vibrates the pattern or matchplate as it is withdrawn from the sand.

WASH ORES ores that require concentration by removal of the sand remaining after crushing by washing with water in agitators.

WATER-HARDENING STEELS low-carbon and low-alloy steels that must be quenched in water for hardening.

WELDABILITY the ease with which simple metal parts made of similar and dissimilar metals may be joined together in order to form complicated structures.

WELDING the process of joining two metals by a similar metal at a temperature above the melting point of the metal; suitable only for low carbon steels.

WELDING ROD filler metal in the form of a wire or rod, used in electric welding when the electrode itself does not furnish the filler metal.

WELL or CRUCIBLE that section of the cupola furnace which lies below the tuyeres.

WHITE or HARD IRON iron of suitable composition in which the castings, later to be malleableized, are originally cast. Carbon is in the combined form; hence its white fracture and name.

WHITE METAL BEARING ALLOYS alloys in which lead, tin, and cadmium are the major elements.

WIDMÄNSTATEN STRUCTURE a pattern formed when one or more ferrite grains alternate with areas of pearlite of the same general shape when a low carbon steel composed of large austenite grains is cooled at a moderately fast rate.

WIND BOX chamber surrounding the cupola furnace at the tuyeres, to equalize the volume and pressure of the blast and deliver it to the tuyeres.

WORK HARDENING hardening that takes place in a metal when work of any sort, such as bending, rolling, hammering, drawing, punching, and the like, is done at a temperature below that at which recrystallization

takes place. Lead, tin, and zinc are not appreciably hardened by cold working, because they can recrystallize at room temperature.

WORKING FACES sidewise extensions in an iron ore mine.

WORKING PERIOD (REFINING PERIOD) the interval between the end of the lime boil and the time of tapping in the open-hearth process or electric steel process. During this period the molten metal is refined or worked by the action of the slag, additions of ore, limestone, and the like in order to obtain the steel desired.

WROUGHT IRON (ASTM DEFINITION) a ferrous material, aggregated from a solidifying mass of pasty particles of highly refined metallic iron with which, without subsequent fusion, is incorporated a minutely and uniformly distributed quantity of slag.

YIELD POINT the load per unit area at which a marked increase in deformation of the specimen occurs without increase of load; the stress at which there occurs a marked increase in strain without an increase in stress.

YIELD STRENGTH the load per unit area at which a material exhibits a specified permanent deformation or a specified elongation under load.

ZYGLO a highly fluorescent nondestructive penetrant inspection test applied to nonmagnetic materials to detect flaws.

Index

A CATALOGUE OF SELECTED DOVER BOOKS
IN ALL FIELDS OF INTEREST

A CATALOGUE OF SELECTED DOVER BOOKS
IN ALL FIELDS OF INTEREST

LEATHER TOOLING AND CARVING, Chris H. Groneman. One of few books concentrating on tooling and carving, with complete instructions and grid designs for 39 projects ranging from bookmarks to bags. 148 illustrations. 111pp. 7⅞ x 10.
23061-9 Pa. $2.50

THE CODEX NUTTALL, A PICTURE MANUSCRIPT FROM ANCIENT MEXICO, as first edited by Zelia Nuttall. Only inexpensive edition, in full color, of a pre-Columbian Mexican (Mixtec) book. 88 color plates show kings, gods, heroes, temples, sacrifices. New explanatory, historical introduction by Arthur G. Miller. 96pp. 11⅜ x 8½. 23168-2 Pa. $7.50

AMERICAN PRIMITIVE PAINTING, Jean Lipman. Classic collection of an enduring American tradition. 109 plates, 8 in full color—portraits, landscapes, Biblical and historical scenes, etc., showing family groups, farm life, and so on. 80pp. of lucid text. 8⅜ x 11¼. 22815-0 Pa. $4.00

WILL BRADLEY: HIS GRAPHIC ART, edited by Clarence P. Hornung. Striking collection of work by foremost practitioner of Art Nouveau in America: posters, cover designs, sample pages, advertisements, other illustrations. 97 plates, including 8 in full color and 19 in two colors. 97pp. 9⅜ x 12¼. 20701-3 Pa. $4.00
22120-2 Clothbd. $10.00

THE UNDERGROUND SKETCHBOOK OF JAN FAUST, Jan Faust. 101 bitter, horrifying, black-humorous, penetrating sketches on sex, war, greed, various liberations, etc. Sometimes sexual, but not pornographic. Not for prudish. 101pp. 6½ x 9¼. 22740-5 Pa. $1.50

THE GIBSON GIRL AND HER AMERICA, Charles Dana Gibson. 155 finest drawings of effervescent world of 1900-1910: the Gibson Girl and her loves, amusements, adventures, Mr. Pipp, etc. Selected by E. Gillon; introduction by Henry Pitz. 144pp. 8¼ x 11⅜. 21986-0 Pa. $3.50

STAINED GLASS CRAFT, J.A.F. Divine, G. Blachford. One of the very few books that tell the beginner exactly what he needs to know: planning cuts, making shapes, avoiding design weaknesses, fitting glass, etc. 93 illustrations. 115pp. 22812-6 Pa. $1.50

CREATIVE LITHOGRAPHY AND HOW TO DO IT, Grant Arnold. Lithography as art form: working directly on stone, transfer of drawings, lithotint, mezzotint, color printing; also metal plates. Detailed, thorough. 27 illustrations. 214pp.

21208-4 Pa. $3.00

DESIGN MOTIFS OF ANCIENT MEXICO, Jorge Enciso. Vigorous, powerful ceramic stamp impressions — Maya, Aztec, Toltec, Olmec. Serpents, gods, priests, dancers, etc. 153pp. 6⅛ x 9¼.

20084-1 Pa. $2.50

AMERICAN INDIAN DESIGN AND DECORATION, Leroy Appleton. Full text, plus more than 700 precise drawings of Inca, Maya, Aztec, Pueblo, Plains, NW Coast basketry, sculpture, painting, pottery, sand paintings, metal, etc. 4 plates in color. 279pp. 8⅜ x 11¼.

22704-9 Pa. $4.50

CHINESE LATTICE DESIGNS, Daniel S. Dye. Incredibly beautiful geometric designs: circles, voluted, simple dissections, etc. Inexhaustible source of ideas, motifs. 1239 illustrations. 469pp. 6⅛ x 9¼.

23096-1 Pa. $5.00

JAPANESE DESIGN MOTIFS, Matsuya Co. Mon, or heraldic designs. Over 4000 typical, beautiful designs: birds, animals, flowers, swords, fans, geometric; all beautifully stylized. 213pp. 11⅜ x 8¼.

22874-6 Pa. $5.00

PERSPECTIVE, Jan Vredeman de Vries. 73 perspective plates from 1604 edition; buildings, townscapes, stairways, fantastic scenes. Remarkable for beauty, surrealistic atmosphere; real eye-catchers. Introduction by Adolf Placzek. 74pp. 11⅜ x 8¼.

20186-4 Pa. $2.75

EARLY AMERICAN DESIGN MOTIFS, Suzanne E. Chapman. 497 motifs, designs, from painting on wood, ceramics, appliqué, glassware, samplers, metal work, etc. Florals, landscapes, birds and animals, geometrics, letters, etc. Inexhaustible. Enlarged edition. 138pp. 8⅜ x 11¼.

22985-8 Pa. $3.50
23084-8 Clothbd. $7.95

VICTORIAN STENCILS FOR DESIGN AND DECORATION, edited by E.V. Gillon, Jr. 113 wonderful ornate Victorian pieces from German sources; florals, geometrics; borders, corner pieces; bird motifs, etc. 64pp. 9⅜ x 12¼.

21995-X Pa. $2.75

ART NOUVEAU: AN ANTHOLOGY OF DESIGN AND ILLUSTRATION FROM THE STUDIO, edited by E.V. Gillon, Jr. Graphic arts: book jackets, posters, engravings, illustrations, decorations; Crane, Beardsley, Bradley and many others. Inexhaustible. 92pp. 8⅛ x 11.

22388-4 Pa. $2.50

ORIGINAL ART DECO DESIGNS, William Rowe. First-rate, highly imaginative modern Art Deco frames, borders, compositions, alphabets, florals, insectals, Wurlitzer-types, etc. Much finest modern Art Deco. 80 plates, 8 in color. 8⅜ x 11¼.

22567-4 Pa. $3.00

HANDBOOK OF DESIGNS AND DEVICES, Clarence P. Hornung. Over 1800 basic geometric designs based on circle, triangle, square, scroll, cross, etc. Largest such collection in existence. 261pp.

20125-2 Pa. $2.50

150 MASTERPIECES OF DRAWING, edited by Anthony Toney. 150 plates, early 15th century to end of 18th century; Rembrandt, Michelangelo, Dürer, Fragonard, Watteau, Wouwerman, many others. 150pp. 8⅜ x 11¼. 21032-4 Pa. $3.50

THE GOLDEN AGE OF THE POSTER, Hayward and Blanche Cirker. 70 extraordinary posters in full colors, from Maîtres de l'Affiche, Mucha, Lautrec, Bradley, Cheret, Beardsley, many others. 9⅜ x 12¼. 22753-7 Pa. $4.95
21718-3 Clothbd. $7.95

SIMPLICISSIMUS, selection, translations and text by Stanley Appelbaum. 180 satirical drawings, 16 in full color, from the famous German weekly magazine in the years 1896 to 1926. 24 artists included: Grosz, Kley, Pascin, Kubin, Kollwitz, plus Heine, Thöny, Bruno Paul, others. 172pp. 8½ x 12¼. 23098-8 Pa. $5.00
23099-6 Clothbd. $10.00

THE EARLY WORK OF AUBREY BEARDSLEY, Aubrey Beardsley. 157 plates, 2 in color: Manon Lescaut, Madame Bovary, Morte d'Arthur, Salome, other. Introduction by H. Marillier. 175pp. 8½ x 11. 21816-3 Pa. $3.50

THE LATER WORK OF AUBREY BEARDSLEY, Aubrey Beardsley. Exotic masterpieces of full maturity: Venus and Tannhäuser, Lysistrata, Rape of the Lock, Volpone, Savoy material, etc. 174 plates, 2 in color. 176pp. 8½ x 11. 21817-1 Pa. $4.00

DRAWINGS OF WILLIAM BLAKE, William Blake. 92 plates from Book of Job, Divine Comedy, Paradise Lost, visionary heads, mythological figures, Laocoön, etc. Selection, introduction, commentary by Sir Geoffrey Keynes. 178pp. 8½ x 11. 22303-5 Pa. $3.50

LONDON: A PILGRIMAGE, Gustave Doré, Blanchard Jerrold. Squalor, riches, misery, beauty of mid-Victorian metropolis; 55 wonderful plates, 125 other illustrations, full social, cultural text by Jerrold. 191pp. of text. 8⅛ x 11. 22306-X Pa. $5.00

THE COMPLETE WOODCUTS OF ALBRECHT DÜRER, edited by Dr. W. Kurth. 346 in all: Old Testament, St. Jerome, Passion, Life of Virgin, Apocalypse, many others. Introduction by Campbell Dodgson. 285pp. 8½ x 12¼. 21097-9 Pa. $6.00

THE DISASTERS OF WAR, Francisco Goya. 83 etchings record horrors of Napoleonic wars in Spain and war in general. Reprint of 1st edition, plus 3 additional plates. Introduction by Philip Hofer. 97pp. 9⅜ x 8¼. 21872-4 Pa. $3.00

ENGRAVINGS OF HOGARTH, William Hogarth. 101 of Hogarth's greatest works: Rake's Progress, Harlot's Progress, Illustrations for Hudibras, Midnight Modern Conversation, Before and After, Beer Street and Gin Lane, many more. Full commentary. 256pp. 11 x 14. 22479-1 Pa. $7.00
23023-6 Clothbd. $13.50

PRIMITIVE ART, Franz Boas. Great anthropologist on ceramics, textiles, wood, stone, metal, etc.; patterns, technology, symbols, styles. All areas, but fullest on Northwest Coast Indians. 350 illustrations. 378pp. 20025-6 Pa. $3.50

MOTHER GOOSE'S MELODIES. Facsimile of fabulously rare Munroe and Francis "copyright 1833" Boston edition. Familiar and unusual rhymes, wonderful old woodcut illustrations. Edited by E.F. Bleiler. 128pp. 4½ x 6⅜. 22577-1 Pa. $1.00

MOTHER GOOSE IN HIEROGLYPHICS. Favorite nursery rhymes presented in rebus form for children. Fascinating 1849 edition reproduced in toto, with key. Introduction by E.F. Bleiler. About 400 woodcuts. 64pp. 6⅞ x 5¼. 20745-5 Pa. $1.00

PETER PIPER'S PRACTICAL PRINCIPLES OF PLAIN & PERFECT PRONUNCIATION. Alliterative jingles and tongue-twisters. Reproduction in full of 1830 first American edition. 25 spirited woodcuts. 32pp. 4½ x 6⅜. 22560-7 Pa. $1.00

MARMADUKE MULTIPLY'S MERRY METHOD OF MAKING MINOR MATHEMATICIANS. Fellow to Peter Piper, it teaches multiplication table by catchy rhymes and woodcuts. 1841 Munroe & Francis edition. Edited by E.F. Bleiler. 103pp. 4⅝ x 6.
22773-1 Pa. $1.25
20171-6 Clothbd. $3.00

THE NIGHT BEFORE CHRISTMAS, Clement Moore. Full text, and woodcuts from original 1848 book. Also critical, historical material. 19 illustrations. 40pp. 4⅝ x 6. 22797-9 Pa. $1.00

THE KING OF THE GOLDEN RIVER, John Ruskin. Victorian children's classic of three brothers, their attempts to reach the Golden River, what becomes of them. Facsimile of original 1889 edition. 22 illustrations. 56pp. 4⅝ x 6⅜.
20066-3 Pa. $1.25

DREAMS OF THE RAREBIT FIEND, Winsor McCay. Pioneer cartoon strip, unexcelled for beauty, imagination, in 60 full sequences. Incredible technical virtuosity, wonderful visual wit. Historical introduction. 62pp. 8⅜ x 11¼. 21347-1 Pa. $2.50

THE KATZENJAMMER KIDS, Rudolf Dirks. In full color, 14 strips from 1906-7; full of imagination, characteristic humor. Classic of great historical importance. Introduction by August Derleth. 32pp. 9¼ x 12¼. 23005-8 Pa. $2.00

LITTLE ORPHAN ANNIE AND LITTLE ORPHAN ANNIE IN COSMIC CITY, Harold Gray. Two great sequences from the early strips: our curly-haired heroine defends the Warbucks' financial empire and, then, takes on meanie Phineas P. Pinchpenny. Leapin' lizards! 178pp. 6⅛ x 8⅜. 23107-0 Pa. $2.00

WHEN A FELLER NEEDS A FRIEND, Clare Briggs. 122 cartoons by one of the greatest newspaper cartoonists of the early 20th century — about growing up, making a living, family life, daily frustrations and occasional triumphs. 121pp. 8½ x 9½.
23148-8 Pa. $2.50

THE BEST OF GLUYAS WILLIAMS. 100 drawings by one of America's finest cartoonists: The Day a Cake of Ivory Soap Sank at Proctor & Gamble's, At the Life Insurance Agents' Banquet, and many other gems from the 20's and 30's. 118pp. 8⅜ x 11¼. 22737-5 Pa. $2.50

CATALOGUE OF DOVER BOOKS

THE BEST DR. THORNDYKE DETECTIVE STORIES, R. Austin Freeman. The Case of Oscar Brodski, The Moabite Cipher, and 5 other favorites featuring the great scientific detective, plus his long-believed-lost first adventure — 31 New Inn — reprinted here for the first time. Edited by E.F. Bleiler. USO 20388-3 Pa. $3.00

BEST "THINKING MACHINE" DETECTIVE STORIES, Jacques Futrelle. The Problem of Cell 13 and 11 other stories about Prof. Augustus S.F.X. Van Dusen, including two "lost" stories. First reprinting of several. Edited by E.F. Bleiler. 241pp.
20537-1 Pa. $3.00

UNCLE SILAS, J. Sheridan LeFanu. Victorian Gothic mystery novel, considered by many best of period, even better than Collins or Dickens. Wonderful psychological terror. Introduction by Frederick Shroyer. 436pp. 21715-9 Pa. $4.00

BEST DR. POGGIOLI DETECTIVE STORIES, T.S. Stribling. 15 best stories from EQMM and The Saint offer new adventures in Mexico, Florida, Tennessee hills as Poggioli unravels mysteries and combats Count Jalacki. 217pp. 23227-1 Pa. $3.00

EIGHT DIME NOVELS, selected with an introduction by E.F. Bleiler. Adventures of Old King Brady, Frank James, Nick Carter, Deadwood Dick, Buffalo Bill, The Steam Man, Frank Merriwell, and Horatio Alger — 1877 to 1905. Important, entertaining popular literature in facsimile reprint, with original covers. 190pp. 9 x 12.
22975-0 Pa. $3.50

ALICE'S ADVENTURES UNDER GROUND, Lewis Carroll. Facsimile of ms. Carroll gave Alice Liddell in 1864. Different in many ways from final Alice. Handlettered, illustrated by Carroll. Introduction by Martin Gardner. 128pp. 21482-6 Pa. $1.50

ALICE IN WONDERLAND COLORING BOOK, Lewis Carroll. Pictures by John Tenniel. Large-size versions of the famous illustrations of Alice, Cheshire Cat, Mad Hatter and all the others, waiting for your crayons. Abridged text. 36 illustrations. 64pp. 8¼ x 11.
22853-3 Pa. $1.50

AVENTURES D'ALICE AU PAYS DES MERVEILLES, Lewis Carroll. Bué's translation of "Alice" into French, supervised by Carroll himself. Novel way to learn language. (No English text.) 42 Tenniel illustrations. 196pp. 22836-3 Pa. $2.50

MYTHS AND FOLK TALES OF IRELAND, Jeremiah Curtin. 11 stories that are Irish versions of European fairy tales and 9 stories from the Fenian cycle — 20 tales of legend and magic that comprise an essential work in the history of folklore. 256pp.
22430-9 Pa. $3.00

EAST O' THE SUN AND WEST O' THE MOON, George W. Dasent. Only full edition of favorite, wonderful Norwegian fairytales — Why the Sea is Salt, Boots and the Troll, etc. — with 77 illustrations by Kittelsen & Werenskiöld. 418pp.
22521-6 Pa. $4.00

PERRAULT'S FAIRY TALES, Charles Perrault and Gustave Doré. Original versions of Cinderella, Sleeping Beauty, Little Red Riding Hood, etc. in best translation, with 34 wonderful illustrations by Gustave Doré. 117pp. 8⅛ x 11. 22311-6 Pa. $2.50

EARLY NEW ENGLAND GRAVESTONE RUBBINGS, Edmund V. Gillon, Jr. 43 photographs, 226 rubbings show heavily symbolic, macabre, sometimes humorous primitive American art. Up to early 19th century. 207pp. 8⅜ x 11¼.
21380-3 Pa. $4.00

L.J.M. DAGUERRE: THE HISTORY OF THE DIORAMA AND THE DAGUERREOTYPE, Helmut and Alison Gernsheim. Definitive account. Early history, life and work of Daguerre; discovery of daguerreotype process; diffusion abroad; other early photography. 124 illustrations. 226pp. 6⅙ x 9¼.
22290-X Pa. $4.00

PHOTOGRAPHY AND THE AMERICAN SCENE, Robert Taft. The basic book on American photography as art, recording form, 1839-1889. Development, influence on society, great photographers, types (portraits, war, frontier, etc.), whatever else needed. Inexhaustible. Illustrated with 322 early photos, daguerreotypes, tintypes, stereo slides, etc. 546pp. 6⅛ x 9¼.
21201-7 Pa. $5.95

PHOTOGRAPHIC SKETCHBOOK OF THE CIVIL WAR, Alexander Gardner. Reproduction of 1866 volume with 100 on-the-field photographs: Manassas, Lincoln on battlefield, slave pens, etc. Introduction by E.F. Bleiler. 224pp. 10¾ x 9.
22731-6 Pa. $5.00

THE MOVIES: A PICTURE QUIZ BOOK, Stanley Appelbaum & Hayward Cirker. Match stars with their movies, name actors and actresses, test your movie skill with 241 stills from 236 great movies, 1902-1959. Indexes of performers and films. 128pp. 8⅜ x 9¼.
20222-4 Pa. $2.50

THE TALKIES, Richard Griffith. Anthology of features, articles from Photoplay, 1928-1940, reproduced complete. Stars, famous movies, technical features, fabulous ads, etc.; Garbo, Chaplin, King Kong, Lubitsch, etc. 4 color plates, scores of illustrations. 327pp. 8⅜ x 11¼.
22762-6 Pa. $6.95

THE MOVIE MUSICAL FROM VITAPHONE TO "42ND STREET," edited by Miles Kreuger. Relive the rise of the movie musical as reported in the pages of Photoplay magazine (1926-1933): every movie review, cast list, ad, and record review; every significant feature article, production still, biography, forecast, and gossip story. Profusely illustrated. 367pp. 8⅜ x 11¼.
23154-2 Pa. $6.95

JOHANN SEBASTIAN BACH, Philipp Spitta. Great classic of biography, musical commentary, with hundreds of pieces analyzed. Also good for Bach's contemporaries. 450 musical examples. Total of 1799pp.
EUK 22278-0, 22279-9 Clothbd., Two vol. set $25.00

BEETHOVEN AND HIS NINE SYMPHONIES, Sir George Grove. Thorough history, analysis, commentary on symphonies and some related pieces. For either beginner or advanced student. 436 musical passages. 407pp.
20334-4 Pa. $4.00

MOZART AND HIS PIANO CONCERTOS, Cuthbert Girdlestone. The only full-length study. Detailed analyses of all 21 concertos, sources; 417 musical examples. 509pp.
21271-8 Pa. $4.50

THE FITZWILLIAM VIRGINAL BOOK, edited by J. Fuller Maitland, W.B. Squire. Famous early 17th century collection of keyboard music, 300 works by Morley, Byrd, Bull, Gibbons, etc. Modern notation. Total of 938pp. 8⅜ x 11.
ECE 21068-5, 21069-3 Pa., Two vol. set $14.00

COMPLETE STRING QUARTETS, Wolfgang A. Mozart. Breitkopf and Härtel edition. All 23 string quartets plus alternate slow movement to K156. Study score. 277pp. 9⅜ x 12¼.
22372-8 Pa. $6.00

COMPLETE SONG CYCLES, Franz Schubert. Complete piano, vocal music of Die Schöne Müllerin, Die Winterreise, Schwanengesang. Also Drinker English singing translations. Breitkopf and Härtel edition. 217pp. 9⅜ x 12¼.
22649-2 Pa. $4.50

THE COMPLETE PRELUDES AND ETUDES FOR PIANOFORTE SOLO, Alexander Scriabin. All the preludes and etudes including many perfectly spun miniatures. Edited by K.N. Igumnov and Y.I. Mil'shteyn. 250pp. 9 x 12.
22919-X Pa. $5.00

TRISTAN UND ISOLDE, Richard Wagner. Full orchestral score with complete instrumentation. Do not confuse with piano reduction. Commentary by Felix Mottl, great Wagnerian conductor and scholar. Study score. 655pp. 8⅛ x 11.
22915-7 Pa. $10.00

FAVORITE SONGS OF THE NINETIES, ed. Robert Fremont. Full reproduction, including covers, of 88 favorites: Ta-Ra-Ra-Boom-De-Aye, The Band Played On, Bird in a Gilded Cage, Under the Bamboo Tree, After the Ball, etc. 401pp. 9 x 12.
EBE 21536-9 Pa. $6.95

SOUSA'S GREAT MARCHES IN PIANO TRANSCRIPTION: ORIGINAL SHEET MUSIC OF 23 WORKS, John Philip Sousa. Selected by Lester S. Levy. Playing edition includes: The Stars and Stripes Forever, The Thunderer, The Gladiator, King Cotton, Washington Post, much more. 24 illustrations. 111pp. 9 x 12.
USO 23132-1 Pa. $3.50

CLASSIC PIANO RAGS, selected with an introduction by Rudi Blesh. Best ragtime music (1897-1922) by Scott Joplin, James Scott, Joseph F. Lamb, Tom Turpin, 9 others. Printed from best original sheet music, plus covers. 364pp. 9 x 12.
EBE 20469-3 Pa. $6.95

ANALYSIS OF CHINESE CHARACTERS, C.D. Wilder, J.H. Ingram. 1000 most important characters analyzed according to primitives, phonetics, historical development. Traditional method offers mnemonic aid to beginner, intermediate student of Chinese, Japanese. 365pp.
23045-7 Pa. $4.00

MODERN CHINESE: A BASIC COURSE, Faculty of Peking University. Self study, classroom course in modern Mandarin. Records contain phonetics, vocabulary, sentences, lessons. 249 page book contains all recorded text, translations, grammar, vocabulary, exercises. Best course on market. 3 12" 33⅓ monaural records, book, album.
98832-5 Set $12.50

MANUAL OF THE TREES OF NORTH AMERICA, Charles S. Sargent. The basic survey of every native tree and tree-like shrub, 717 species in all. Extremely full descriptions, information on habitat, growth, locales, economics, etc. Necessary to every serious tree lover. Over 100 finding keys. 783 illustrations. Total of 986pp.
20277-1, 20278-X Pa., Two vol. set $8.00

BIRDS OF THE NEW YORK AREA, John Bull. Indispensable guide to more than 400 species within a hundred-mile radius of Manhattan. Information on range, status, breeding, migration, distribution trends, etc. Foreword by Roger Tory Peterson. 17 drawings; maps. 540pp. 23222-0 Pa. $6.00

THE SEA-BEACH AT EBB-TIDE, Augusta Foote Arnold. Identify hundreds of marine plants and animals: algae, seaweeds, squids, crabs, corals, etc. Descriptions cover food, life cycle, size, shape, habitat. Over 600 drawings. 490pp.
21949-6 Pa.$5.00

THE MOTH BOOK, William J. Holland. Identify more than 2,000 moths of North America. General information, precise species descriptions. 623 illustrations plus 48 color plates show almost all species, full size. 1968 edition. Still the basic book. Total of 551pp. 6½ x 9¼. 21948-8 Pa. $6.00

AN INTRODUCTION TO THE REPTILES AND AMPHIBIANS OF THE UNITED STATES, Percy A. Morris. All lizards, crocodiles, turtles, snakes, toads, frogs; life history, identification, habits, suitability as pets, etc. Non-technical, but sound and broad. 130 photos. 253pp. 22982-3 Pa. $3.00

OLD NEW YORK IN EARLY PHOTOGRAPHS, edited by Mary Black. Your only chance to see New York City as it was 1853-1906, through 196 wonderful photographs from N.Y. Historical Society. Great Blizzard, Lincoln's funeral procession, great buildings. 228pp. 9 x 12. 22907-6 Pa. $6.00

THE AMERICAN REVOLUTION, A PICTURE SOURCEBOOK, John Grafton. Wonderful Bicentennial picture source, with 411 illustrations (contemporary and 19th century) showing battles, personalities, maps, events, flags, posters, soldier's life, ships, etc. all captioned and explained. A wonderful browsing book, supplement to other historical reading. 160pp. 9 x 12. 23226-3 Pa. $4.00

PERSONAL NARRATIVE OF A PILGRIMAGE TO AL-MADINAH AND MECCAH, Richard Burton. Great travel classic by remarkably colorful personality. Burton, disguised as a Moroccan, visited sacred shrines of Islam, narrowly escaping death. Wonderful observations of Islamic life, customs, personalities. 47 illustrations. Total of 959pp. 21217-3, 21218-1 Pa., Two vol. set$10.00

INCIDENTS OF TRAVEL IN CENTRAL AMERICA, CHIAPAS, AND YUCATAN, John L. Stephens. Almost single-handed discovery of Maya culture; exploration of ruined cities, monuments, temples; customs of Indians. 115 drawings. 892pp.
22404-X, 22405-8 Pa., Two vol. set $8.00

CONSTRUCTION OF AMERICAN FURNITURE TREASURES, Lester Margon. 344 detail drawings, complete text on constructing exact reproductions of 38 early American masterpieces: Hepplewhite sideboard, Duncan Phyfe drop-leaf table, mantel clock, gate-leg dining table, Pa. German cupboard, more. 38 plates. 54 photographs. 168pp. 8⅜ x 11¼. 23056-2 Pa. $4.00

JEWELRY MAKING AND DESIGN, Augustus F. Rose, Antonio Cirino. Professional secrets revealed in thorough, practical guide: tools, materials, processes; rings, brooches, chains, cast pieces, enamelling, setting stones, etc. Do not confuse with skimpy introductions: beginner can use, professional can learn from it. Over 200 illustrations. 306pp. 21750-7 Pa. $3.00

METALWORK AND ENAMELLING, Herbert Maryon. Generally conceeded best all-around book. Countless trade secrets: materials, tools, soldering, filigree, setting, inlay, niello, repoussé, casting, polishing, etc. For beginner or expert. Author was foremost British expert. 330 illustrations. 335pp. 22702-2 Pa. $3.50

WEAVING WITH FOOT-POWER LOOMS, Edward F. Worst. Setting up a loom, beginning to weave, constructing equipment, using dyes, more, plus over 285 drafts of traditional patterns including Colonial and Swedish weaves. More than 200 other figures. For beginning and advanced. 275pp. 8¾ x 6⅜. 23064-3 Pa. $4.00

WEAVING A NAVAJO BLANKET, Gladys A. Reichard. Foremost anthropologist studied under Navajo women, reveals every step in process from wool, dyeing, spinning, setting up loom, designing, weaving. Much history, symbolism. With this book you could make one yourself. 97 illustrations. 222pp. 22992-0 Pa. $3.00

NATURAL DYES AND HOME DYEING, Rita J. Adrosko. Use natural ingredients: bark, flowers, leaves, lichens, insects etc. Over 135 specific recipes from historical sources for cotton, wool, other fabrics. Genuine premodern handicrafts. 12 illustrations. 160pp. 22688-3 Pa. $2.00

THE HAND DECORATION OF FABRICS, Francis J. Kafka. Outstanding, profusely illustrated guide to stenciling, batik, block printing, tie dyeing, freehand painting, silk screen printing, and novelty decoration. 356 illustrations. 198pp. 6 x 9. 21401-X Pa. $3.00

THOMAS NAST: CARTOONS AND ILLUSTRATIONS, with text by Thomas Nast St. Hill. Father of American political cartooning. Cartoons that destroyed Tweed Ring; inflation, free love, church and state; original Republican elephant and Democratic donkey; Santa Claus; more. 117 illustrations. 146pp. 9 x 12.
22983-1 Pa. $4.00
23067-8 Clothbd. $8.50

FREDERIC REMINGTON: 173 DRAWINGS AND ILLUSTRATIONS. Most famous of the Western artists, most responsible for our myths about the American West in its untamed days. Complete reprinting of Drawings of Frederic Remington (1897), plus other selections. 4 additional drawings in color on covers. 140pp. 9 x 12.
20714-5 Pa. $3.95

How to Solve Chess Problems, Kenneth S. Howard. Practical suggestions on problem solving for very beginners. 58 two-move problems, 46 3-movers, 8 4-movers for practice, plus hints. 171pp. 20748-X Pa. $2.00

A Guide to Fairy Chess, Anthony Dickins. 3-D chess, 4-D chess, chess on a cylindrical board, reflecting pieces that bounce off edges, cooperative chess, retrograde chess, maximummers, much more. Most based on work of great Dawson. Full handbook, 100 problems. 66pp. 7⅞ x 10¾. 22687-5 Pa. $2.00

Win at Backgammon, Millard Hopper. Best opening moves, running game, blocking game, back game, tables of odds, etc. Hopper makes the game clear enough for anyone to play, and win. 43 diagrams. 111pp. 22894-0 Pa. $1.50

Bidding a Bridge Hand, Terence Reese. Master player "thinks out loud" the binding of 75 hands that defy point count systems. Organized by bidding problem—no-fit situations, overbidding, underbidding, cueing your defense, etc. 254pp. EBE 22830-4 Pa. $2.50

The Precision Bidding System in Bridge, C.C. Wei, edited by Alan Truscott. Inventor of precision bidding presents average hands and hands from actual play, including games from 1969 Bermuda Bowl where system emerged. 114 exercises. 116pp. 21171-1 Pa. $1.75

Learn Magic, Henry Hay. 20 simple, easy-to-follow lessons on magic for the new magician: illusions, card tricks, silks, sleights of hand, coin manipulations, escapes, and more —all with a minimum amount of equipment. Final chapter explains the great stage illusions. 92 illustrations. 285pp. 21238-6 Pa. $2.95

The New Magician's Manual, Walter B. Gibson. Step-by-step instructions and clear illustrations guide the novice in mastering 36 tricks; much equipment supplied on 16 pages of cut-out materials. 36 additional tricks. 64 illustrations. 159pp. 6⅝ x 10. 23113-5 Pa. $3.00

Professional Magic for Amateurs, Walter B. Gibson. 50 easy, effective tricks used by professionals —cards, string, tumblers, handkerchiefs, mental magic, etc. 63 illustrations. 223pp. 23012-0 Pa. $2.50

Card Manipulations, Jean Hugard. Very rich collection of manipulations; has taught thousands of fine magicians tricks that are really workable, eye-catching. Easily followed, serious work. Over 200 illustrations. 163pp. 20539-8 Pa. $2.00

Abbott's Encyclopedia of Rope Tricks for Magicians, Stewart James. Complete reference book for amateur and professional magicians containing more than 150 tricks involving knots, penetrations, cut and restored rope, etc. 510 illustrations. Reprint of 3rd edition. 400pp. 23206-9 Pa. $3.50

The Secrets of Houdini, J.C. Cannell. Classic study of Houdini's incredible magic, exposing closely-kept professional secrets and revealing, in general terms, the whole art of stage magic. 67 illustrations. 279pp. 22913-0 Pa. $2.50

THE MAGIC MOVING PICTURE BOOK, Bliss, Sands & Co. The pictures in this book move! Volcanoes erupt, a house burns, a serpentine dancer wiggles her way through a number. By using a specially ruled acetate screen provided, you can obtain these and 15 other startling effects. Originally "The Motograph Moving Picture Book." 32pp. 8¼ x 11. 23224-7 Pa. $1.75

STRING FIGURES AND HOW TO MAKE THEM, Caroline F. Jayne. Fullest, clearest instructions on string figures from around world: Eskimo, Navajo, Lapp, Europe, more. Cats cradle, moving spear, lightning, stars. Introduction by A.C. Haddon. 950 illustrations. 407pp. 20152-X Pa. $3.00

PAPER FOLDING FOR BEGINNERS, William D. Murray and Francis J. Rigney. Clearest book on market for making origami sail boats, roosters, frogs that move legs, cups, bonbon boxes. 40 projects. More than 275 illustrations. Photographs. 94pp. 20713-7 Pa. $1.25

INDIAN SIGN LANGUAGE, William Tomkins. Over 525 signs developed by Sioux, Blackfoot, Cheyenne, Arapahoe and other tribes. Written instructions and diagrams: how to make words, construct sentences. Also 290 pictographs of Sioux and Ojibway tribes. 111pp. 6⅛ x 9¼. 22029-X Pa. $1.50

BOOMERANGS: HOW TO MAKE AND THROW THEM, Bernard S. Mason. Easy to make and throw, dozens of designs: cross-stick, pinwheel, boomabird, tumblestick, Australian curved stick boomerang. Complete throwing instructions. All safe. 99pp. 23028-7 Pa. $1.50

25 KITES THAT FLY, Leslie Hunt. Full, easy to follow instructions for kites made from inexpensive materials. Many novelties. Reeling, raising, designing your own. 70 illustrations. 110pp. 22550-X Pa. $1.25

TRICKS AND GAMES ON THE POOL TABLE, Fred Herrmann. 79 tricks and games, some solitaires, some for 2 or more players, some competitive; mystifying shots and throws, unusual carom, tricks involving cork, coins, a hat, more. 77 figures. 95pp. 21814-7 Pa. $1.25

WOODCRAFT AND CAMPING, Bernard S. Mason. How to make a quick emergency shelter, select woods that will burn immediately, make do with limited supplies, etc. Also making many things out of wood, rawhide, bark, at camp. Formerly titled Woodcraft. 295 illustrations. 580pp. 21951-8 Pa. $4.00

AN INTRODUCTION TO CHESS MOVES AND TACTICS SIMPLY EXPLAINED, Leonard Barden. Informal intermediate introduction: reasons for moves, tactics, openings, traps, positional play, endgame. Isolates patterns. 102pp. USO 21210-6 Pa. $1.35

LASKER'S MANUAL OF CHESS, Dr. Emanuel Lasker. Great world champion offers very thorough coverage of all aspects of chess. Combinations, position play, openings, endgame, aesthetics of chess, philosophy of struggle, much more. Filled with analyzed games. 390pp. 20640-8 Pa. $3.50

SLEEPING BEAUTY, illustrated by Arthur Rackham. Perhaps the fullest, most delightful version ever, told by C.S. Evans. Rackham's best work. 49 illustrations. 110pp. 7⅞ x 10¾. 22756-1 Pa. $2.00

THE WONDERFUL WIZARD OF OZ, L. Frank Baum. Facsimile in full color of America's finest children's classic. Introduction by Martin Gardner. 143 illustrations by W.W. Denslow. 267pp. 20691-2 Pa. $2.50

GOOPS AND HOW TO BE THEM, Gelett Burgess. Classic tongue-in-cheek masquerading as etiquette book. 87 verses, 170 cartoons as Goops demonstrate virtues of table manners, neatness, courtesy, more. 88pp. 6½ x 9¼. 22233-0 Pa. $1.50

THE BROWNIES, THEIR BOOK, Palmer Cox. Small as mice, cunning as foxes, exuberant, mischievous, Brownies go to zoo, toy shop, seashore, circus, more. 24 verse adventures. 266 illustrations. 144pp. 6⅝ x 9¼. 21265-3 Pa. $1.75

BILLY WHISKERS: THE AUTOBIOGRAPHY OF A GOAT, Frances Trego Montgomery. Escapades of that rambunctious goat. Favorite from turn of the century America. 24 illustrations. 259pp. 22345-0 Pa. $2.75

THE ROCKET BOOK, Peter Newell. Fritz, janitor's kid, sets off rocket in basement of apartment house; an ingenious hole punched through every page traces course of rocket. 22 duotone drawings, verses. 48pp. 6⅞ x 8⅜. 22044-3 Pa. $1.50

PECK'S BAD BOY AND HIS PA, George W. Peck. Complete double-volume of great American childhood classic. Hennery's ingenious pranks against outraged pomposity of pa and the grocery man. 97 illustrations. Introduction by E.F. Bleiler. 347pp. 20497-9 Pa. $2.50

THE TALE OF PETER RABBIT, Beatrix Potter. The inimitable Peter's terrifying adventure in Mr. McGregor's garden, with all 27 wonderful, full-color Potter illustrations. 55pp. 4¼ x 5½. USO 22827-4 Pa. $1.00

THE TALE OF MRS. TIGGY-WINKLE, Beatrix Potter. Your child will love this story about a very special hedgehog and all 27 wonderful, full-color Potter illustrations. 57pp. 4¼ x 5½. USO 20546-0 Pa. $1.00

THE TALE OF BENJAMIN BUNNY, Beatrix Potter. Peter Rabbit's cousin coaxes him back into Mr. McGregor's garden for a whole new set of adventures. A favorite with children. All 27 full-color illustrations. 59pp. 4¼ x 5½. USO 21102-9 Pa. $1.00

THE MERRY ADVENTURES OF ROBIN HOOD, Howard Pyle. Facsimile of original (1883) edition, finest modern version of English outlaw's adventures. 23 illustrations by Pyle. 296pp. 6½ x 9¼. 22043-5 Pa. $2.75

TWO LITTLE SAVAGES, Ernest Thompson Seton. Adventures of two boys who lived as Indians; explaining Indian ways, woodlore, pioneer methods. 293 illustrations. 286pp. 20985-7 Pa. $3.00

HOUDINI ON MAGIC, Harold Houdini. Edited by Walter Gibson, Morris N. Young. How he escaped; exposés of fake spiritualists; instructions for eye-catching tricks; other fascinating material by and about greatest magician. 155 illustrations. 280pp. 20384-0 Pa. $2.50

HANDBOOK OF THE NUTRITIONAL CONTENTS OF FOOD, U.S. Dept. of Agriculture. Largest, most detailed source of food nutrition information ever prepared. Two mammoth tables: one measuring nutrients in 100 grams of edible portion; the other, in edible portion of 1 pound as purchased. Originally titled Composition of Foods. 190pp. 9 x 12. 21342-0 Pa. $4.00

COMPLETE GUIDE TO HOME CANNING, PRESERVING AND FREEZING, U.S. Dept. of Agriculture. Seven basic manuals with full instructions for jams and jellies; pickles and relishes; canning fruits, vegetables, meat; freezing anything. Really good recipes, exact instructions for optimal results. Save a fortune in food. 156 illustrations. 214pp. 6⅛ x 9¼. 22911-4 Pa. $2.50

THE BREAD TRAY, Louis P. De Gouy. Nearly every bread the cook could buy or make: bread sticks of Italy, fruit breads of Greece, glazed rolls of Vienna, everything from corn pone to croissants. Over 500 recipes altogether. including buns, rolls, muffins, scones, and more. 463pp. 23000-7 Pa. $3.50

CREATIVE HAMBURGER COOKERY, Louis P. De Gouy. 182 unusual recipes for casseroles, meat loaves and hamburgers that turn inexpensive ground meat into memorable main dishes: Arizona chili burgers, burger tamale pie, burger stew, burger corn loaf, burger wine loaf, and more. 120pp. 23001-5 Pa. $1.75

LONG ISLAND SEAFOOD COOKBOOK, J. George Frederick and Jean Joyce. Probably the best American seafood cookbook. Hundreds of recipes. 40 gourmet sauces, 123 recipes using oysters alone! All varieties of fish and seafood amply represented. 324pp. 22677-8 Pa. $3.00

THE EPICUREAN: A COMPLETE TREATISE OF ANALYTICAL AND PRACTICAL STUDIES IN THE CULINARY ART, Charles Ranhofer. Great modern classic. 3,500 recipes from master chef of Delmonico's, turn-of-the-century America's best restaurant. Also explained, many techniques known only to professional chefs. 775 illustrations. 1183pp. 6⅝ x 10. 22680-8 Clothbd. $17.50

THE AMERICAN WINE COOK BOOK, Ted Hatch. Over 700 recipes: old favorites livened up with wine plus many more: Czech fish soup, quince soup, sauce Perigueux, shrimp shortcake, filets Stroganoff, cordon bleu goulash, jambonneau, wine fruit cake, more. 314pp. 22796-0 Pa. $2.50

DELICIOUS VEGETARIAN COOKING, Ivan Baker. Close to 500 delicious and varied recipes: soups, main course dishes (pea, bean, lentil, cheese, vegetable, pasta, and egg dishes), savories, stews, whole-wheat breads and cakes, more. 168pp. USO 22834-7 Pa. $1.75

COOKIES FROM MANY LANDS, Josephine Perry. Crullers, oatmeal cookies, chaux au chocolate, English tea cakes, mandel kuchen, Sacher torte, Danish puff pastry, Swedish cookies — a mouth-watering collection of 223 recipes. 157pp.

22832-0 Pa. $2.00

ROSE RECIPES, Eleanour S. Rohde. How to make sauces, jellies, tarts, salads, potpourris, sweet bags, pomanders, perfumes from garden roses; all exact recipes. Century old favorites. 95pp.

22957-2 Pa. $1.25

"OSCAR" OF THE WALDORF'S COOKBOOK, Oscar Tschirky. Famous American chef reveals 3455 recipes that made Waldorf great; cream of French, German, American cooking, in all categories. Full instructions, easy home use. 1896 edition. 907pp. 6⅝ x 9⅜.

20790-0 Clothbd. $15.00

JAMS AND JELLIES, May Byron. Over 500 old-time recipes for delicious jams, jellies, marmalades, preserves, and many other items. Probably the largest jam and jelly book in print. Originally titled May Byron's Jam Book. 276pp.

USO 23130-5 Pa. $3.00

MUSHROOM RECIPES, André L. Simon. 110 recipes for everyday and special cooking. Champignons à la grecque, sole bonne femme, chicken liver croustades, more; 9 basic sauces, 13 ways of cooking mushrooms. 54pp.

USO 20913-X Pa. $1.25

FAVORITE SWEDISH RECIPES, edited by Sam Widenfelt. Prepared in Sweden, offers wonderful, clearly explained Swedish dishes: appetizers, meats, pastry and cookies, other categories. Suitable for American kitchen. 90 photos. 157pp.

23156-9 Pa. $2.00

THE BUCKEYE COOKBOOK, Buckeye Publishing Company. Over 1,000 easy-to-follow, traditional recipes from the American Midwest: bread (100 recipes alone), meat, game, jam, candy, cake, ice cream, and many other categories of cooking. 64 illustrations. From 1883 enlarged edition. 416pp.

23218-2 Pa. $4.00

TWENTY-TWO AUTHENTIC BANQUETS FROM INDIA, Robert H. Christie. Complete, easy-to-do recipes for almost 200 authentic Indian dishes assembled in 22 banquets. Arranged by region. Selected from Banquets of the Nations. 192pp.

23200-X Pa. $2.50